PROGRAMMABLE INTEGRATED PHOTONICS

Programmable Integrated Photonics

José Capmany and Daniel Pérez

OXFORD
UNIVERSITY PRESS

OXFORD
UNIVERSITY PRESS

Great Clarendon Street, Oxford, OX2 6DP,
United Kingdom

Oxford University Press is a department of the University of Oxford.
It furthers the University's objective of excellence in research, scholarship,
and education by publishing worldwide. Oxford is a registered trade mark of
Oxford University Press in the UK and in certain other countries

First Edition published in 2020

Impression: 1

Published in the United States of America by Oxford University Press
198 Madison Avenue, New York, NY 10016, United States of America

British Library Cataloguing in Publication Data
Data available

Library of Congress Control Number: 2020931719

ISBN 978–0–19–884440–2

Printed and bound by
CPI Group (UK) Ltd, Croydon, CR0 4YY

Contents

List of Abbreviations

2D	two-dimensional
3D	three-dimensional
ACF	amplitude comparison function
ADC	analogue-to-digital converter
ANN	artificial neural network
APD	avalanche photodiode
ASIC	application-specific integrated circuit
ASPIC	application-specific photonic integrated circuit
AWG	arbitrary waveform generation
BBB	basic building block
BS	beamsplitter
BUD	basic unit delay
BUL	basic unit length
CAB	configurable analogue block
CBB	compound building block
CCS	closed-loop control system
CLA	configurable logic array
CLE	configurable logic element
CLIPP	contactless integrated photonic probe
CMOS	complementary metal oxide semiconductor
CNM	Centro Nacional de Microelectrónica
CPR	cross-port ratio
CROW	coupled resonator optical waveguide

CT	Cooley–Tukey
CT	crosstalk coefficient
DAC	digital-to-analogue converter
DBR	distributed Bragg reflector
DC	directional coupler
DDDC	dual-drive directional coupler
DDL	dispersive delay line
DFB	distributed feedback
DFT	discrete Fourier transform
DPMZM	dual parallel Mach–Zehnder modulator
EO	electro-optic
ER	extinction ratio
FCA	free-carrier absorption
FFT	fast Fourier transform
FGL	first-grade leakage
FHD	flame hydrolysis deposition
FIR	finite impulse response
FPAA	field programmable analog array
FPGA	field programmable gate array
FPPGA	field programable photonic gate array
FSR	free spectral range
FWM	four wave mixing
GFM	generic foundry model
GIM	generic integration model
GPIB	general purpose interface bus
HIC	high-index contrast
HPB	high-performance block

I/O	input/output
IIR	infinite impulse response
IL	insertion loss
IMO	InP membrane on silicon
InP	indium phosphide
IoT	Internet of Things
IQP	integrated quantum photonics
ITU	International Telecommunication Union
LIC	low-index contrast
LOC	local oscillator
LOC	linear optical component
LPCVD	low-pressure chemical vapour deposition
LUC	linear universal component
MEDA	multi-channel electronic driver array
MEM	micro-electromechanical
MEMA	multi-channel electronic monitor array
MFD	mode field diameter
MIMO	multiple input/multiple output
MMI	multimode interference
mm-wave	millimetre wave
MRR	micro-ring resonator
MWP	microwave photonics
MZI	Mach–Zehnder interferometer
NF	noise figure
NRE	non-recurring engineering
ODSB-CS	optical double side band-carrier suppressed
OE	opto-electronic

OEO	opto-electronic oscillator
OIU	optical interference unit
ONU	optical non-linear unit
ORR	optical ring resonator
ORSE	optical routing and switching element
OSA	optical spectrum analyser
OSSBM	optical single-side band modulation
PCB	printed circuit board
PIC	photonic integrated circuit
PILOSS	path independent loss
PIP	programmable integrated photonics
PLB	programmable logic block
PLC	planar lightwave circuit
PM	phase modulator
PNN	processing network node
PPAB	programmable photonic analogue block
QT	quantum transport
RAMZI	ring-assisted MZI
RBS	reconfigurable beamsplitter
RF	radiofrequency
RPI	reconfigurable photonic interconnect
RR	ring resonator
SCISSORs	side-coupled integrated spaced sequence of optical resonators
SCT	separate carrier tuning
SDL	switched delay line
SDN	software defined network
SFDR	spurious free dynamic range

SFWM	spontaneous four wave mixing
SGL	second-grade leakage
SISO	single input/single output
SNSPD	superconducting nanowire single-photon detector
SOA	semiconductor optical amplifier
SOI	silicon on insulator
SPDC	spontaneous parametric down-conversion
SR	software radio
SVD	singular value decomposition
TBU	tuneable basic unit
TE	transverse electric
TIA	trans-impedance amplifier
TM	transvere magnetic
TPA	two-photon absorption
TPR	through-port ratio
TTD	true-time delay
TTODL	true-time optical delay line
TDC	tuneable directional coupler
UMZI	unbalanced MZI
UWB	ultra-wideband
VLDL	variable-length delay line
VPVL	variable-propagation-velocity line
WDM	wavelength-division multiplexing
WGMR	whispering gallery mode resonator
ZGL	zero-grade leakage

1

Introduction to Programmable Integrated Photonics

Programmable integrated photonics (PIP) aims at designing common integrated optical hardware configurations, which—by suitable programming—can implement a variety of functionalities that can be elaborated for basic or more complex operations in many application fields. It follows a different approach to that of application specific photonic integrated circuits (ASPICs), which have dominated during the last few decades. The interest in PIP is driven by the surge of a considerable number of emerging applications in the fields of telecommunications, quantum information processing, sensing and neurophotonics that will require flexible, reconfigurable, low-cost, compact and low-power-consuming devices, much in the same way as field programmable gate array (FPGA) devices operate in electronics. This chapter serves as a general introduction to the book and reviews the main basic principles and recent advances in PIP, including fabrication platforms, design principles, architecture choices, challenges and limitations. Finally, it provides a brief introduction to the applications of this new field.

1.1 An Introduction to Reconfigurable Systems

1.1.1 Introductory Remarks

Reconfigurable systems are systems with soft-definable features i.e. soft hardware that can be tuned, reshaped or otherwise altered at will to suit the purposes of their users [1]. Reconfigurability is now commonplace in electronics components, circuits and systems and the FPGA device—the paradigm of electronic reconfigurable systems—is one of the best-known examples. Indeed, Freeman's FPGA patent application filed in 1989 [2] highlights the nature of a reconfigurable system as being an ensemble of exchangeable elements.

Initially conceived to compete with application specific integrated circuits (ASICs), after almost thirty years since their inception FPGAs have practically replaced ASICs in most applications [3]. FPGA devices have gained this wide coverage because of the benefits of Moore's law, cost reductions due to non-recurring engineering costs and the

Programmable Integrated Photonics. José Capmany and Daniel Pérez, Oxford University Press (2020). © José Capmany and Daniel Pérez.
DOI: 10.1093/oso/9780198844402.001.0001

trend in fusing practical engineering and design principles that can effectively manage the flexibility required in the so-called *software-x* approaches; these include, among others, software-defined radios [4], software-defined networks [5], cloud computing [6] and data centre concepts [7]. All in all, the FPGA concept has shown the potential of reconfigurability to become a disruptive technology in information and communication systems [8].

Furthermore, electronic programmable systems under development also encompass new subsystems [3], including small sets of resident reconfigurable analogue processing blocks, digital-to-analogue and analogue-to-digital conversion stages, etc.

But is there sense in considering the extension of programmable systems to other application scenarios, e.g. reconfigurable circuitry at radio frequencies, antenna structures and photonics? Currently, none of these fields have reached the degree of integration maturity of electronics in general, and FPGAs in particular.

Broadly speaking, the design of reconfigurable systems needs to consider two important aspects. The first, *designing for reconfigurability*, refers to the set of ideas for embedding flexibility within systems through the methodical incorporation of exploitable switch and configuration mechanisms. The second, *designing with reconfigurability*, refers to the processes by which user designs are embedded, often through design automation, in reconfigurable fabrics.

1.1.2 Advantages of Reconfigurable Systems

Reconfigurable systems are sometimes criticized because they require overhead, which may lower performance and add complexity, possibly reducing reliability. These considerations must be traded off against their advantages when considering their use for a particular application or field.

Reconfigurable systems bring a considerable number of benefits, which can be summarized as follows:

a) *Flexible reshaping of finite resources*: A reconfigurable system can be considered as a finite collection of resources that can be reshaped at will, leading to advantages in mass extreme customisation, reduction of non-recurring engineering expenses, economic savings through inventory collapse, design rectification and functional update and iterative refinement to accommodate evolution.
b) *Robustness and resilience*: Reconfigurability results in systems that are tolerant to faults and manufacturing defects by exploiting unused configurable resources. Redundancy approaches can be implemented provided that enough spatial resources are available in the programmable system. Furthermore, by software programming it is possible to create self-healing and/or cooperative multi-tasking systems.
c) *Achieving 'x on demand'*: Reconfigurability brings the advantage of creating systems quickly. Prebuilt parts can be personalized rapidly, overcoming the long fabrication cycles of dedicated systems or chips. This is especially important in electronic microfabrication.

d) *Infinite resources through timesharing*: Reconfigurable systems can be thought of as a set of infinitely repurposable components leading to temporal reuse.

The interested reader is directed to the chapter references for more information on these and other concepts pertaining to the general properties of reconfigurable systems.

The rest of the book explores and considers the potential and advantages that reconfigurability brings to photonics and specifically to integrated photonic circuits.

1.2 Principles of Programmable Integrated Photonics

PIP [9]–[24] is a new paradigm that aims at designing common integrated optical hardware configurations, which by suitable programming can implement a variety of functionalities that can be widely exploited. By means of external control signals Programmability enables both chip reconfiguration for multi-function operation as well as chip optimization against non-ideal operation due to fluctuations in environmental conditions and fabrication errors. Programming also allows activation of parts of the chip that are not essential for the implementation of a given functionality but that can be of help in reducing noise levels through the diversion of undesired reflections.

PIP is therefore a transversal concept inspired by similar approaches, which are already employed in other technology fields. For instance, in electronics, FPGA devices enable a much more flexible universal operation as compared to ASICs [3]. In communications, software defined networks (SDN) enable the exploitation and reconfiguration of a common set of resources provided by a network hardware infrastructure to ensure an optimum configuration in demand of time-varying requirements set upon several quality and bandwidth performance indicators. Finally, in wireless transmission, software radio (SR) allows the emulation of different specific radiofrequency (RF) receivers with a single hardware platform.

In the area of photonics, the PIP approach aims to provide a complementary approach to that based on ASPICs, which has been dominant during the last years. The objective is to leverage the universal properties of this approach and seek similar advantages to those FPGAs bring over ASICs in electronics (Table 1.1).

PIP has recently raised the interest of many research groups worldwide, justified by the surge of a number of emerging applications that are and will be calling for true flexibility and reconfigurability as well as low-cost, compact and low-power-consuming devices. One area in which considerable seminal work has been produced is in quantum information technologies, where PIP can open avenues to large-scale quantum gates and boson sampling circuits based on unitary matrix transformations [9], [10], [13], [15].

In the field of telecommunications, PIP can be instrumental in a series of signal processing functionalities, such as the implementation of arbitrary mode converters [25], [26], fibre-wireless interfacing devices [27] and broadband switches [28], which can also form the basis for computer interconnections [29]. In the field of sensing, PIP can lead to a generic class of programmable measuring devices [30], which might be successfully integrated as a building block in the future Internet of Things (IoT).

Table 1.1 *Basic features of ASIC and FPGA approaches in electronics.*

	ASIC	FPGA
Time to Market	Slow	Fast
Non-recurring Engineering	Very High	Low
Unit Cost	Low	Medium
Design Flow	Complex	Simple
Performance	High	Medium
Application Flexibility/Versatility	Very Low	High
Power Consumption	Low	High
Size	Low	Medium

The success of PIP relies on designing suitable basic building blocks (BBBs) able to carry elementary signal processing operations and interconnection hardware architectures that can offer a very high spatial regularity and the possibility of independently setting (with a very low power consumption) the interconnection state of each connecting element. The most popular BBBs proposed so far are based on elementary 2×2 tunable photonic coupling components capable of providing independent setting of the power coupling ratio and the phase shift experienced by the incoming signals from two input waveguides in their transition to the two output waveguide ports of the devices. The matrix tools to analyse such elementary BBBs and the main proposed layouts (beamsplitters, Mach–Zehnder interferometers (MZIs), directional couplers, etc.) are covered in Chapter 2.

Reconfigurable ASPICs are the first evolutionary step in programmable photonics. These are designed to carry a given functionality and feature some degree of flexibility that is enabled by the possibility of tuning some of its internal parameters by means of appropriate control signals. Another distinctive feature of these circuits resides in the fixed nature of their physical topology and designs. Chapter 3 covers some of the most significant issues pertaining to this class of circuits without aiming to be exhaustive.

The interconnection architectures for the BBBs fall broadly into two categories. The first includes the so-called multi-port interferometers and has been historically the first to be developed, in particular, to enable unitary linear optical transformations that are at the heart of quantum circuits, reconfigurable neurophotonic systems and Fourier-based optical signal processors [24]. The most popular multi-port interferometer designs are the triangular multi-port interferometer concept proposed by Reck and colleagues [21] and subsequently developed for integrated optics by Miller [11], [12] and the more recent rectangular multi-port interferometer proposed by Clements and colleagues [22], [23]. Chapter 4 provides a full description of these designs and also covers more advanced topics such as their applications to general mode converters, universal coupling subsystems and 3D fast interferometers. Multi-port interferometers can enable any kind of linear optics transformations as long as feedforward-only propagation is assumed. While this covers a wide range of applications it does not enable the programming of resonant structures where simultaneous feedforward and feedbackward propagation are required.

Chapter 5 coves integrated waveguide meshes [17]–[20], the second category of interconnection architectures and which overcome this limitation. They provide regular and periodic geometries, formed by replicating a unit cell, which can take the form of a square, hexagon or triangle among other configurations. Each side of the cell is formed by two integrated waveguides connected by means of a beamsplitter/tunable coupler that can be operated by means of an output control signal as a crossbar switch or as a variable coupler with independent power division ratio and phase shift. A mesh formed by a suitable amount of unit cells can be programmed to implement a wide variety of functionalities much in the same way as an FPGA operates in electronics [17], [20]. This chapter defines the main figures of merit to benchmark the performance of different waveguide mesh designs and provide a comparative analysis.

The design of efficient and practical waveguide meshes entails addressing a considerable number of challenges and limitations. Some of these reside in the pure physical characteristics of the material platforms employed in their construction and mainly affect their insertion losses through signal coupling efficiency and propagation and scattering losses. Others are connected to the inherent fabrication errors and fluctuations in the operating values of the BBB parameters and call for active control and monitoring of the waveguide meshes. A third class of challenge is connected to the design and programming of scalable structures containing ever-increasing unit cells. These challenges are addressed in detail in Chapter 6, which also covers some of the most interesting experimental results reported so far.

Present and future envisaged applications of PIP are covered in the two following chapters of the book: Chapter 7 reviews the applications of PIP to emerging quantum systems, including quantum computing, boson sampling and quantum simulation, and Chapter 8 focuses on PIP applications in classical systems. It covers salient results in different areas, including telecommunications and switching, RF and analogue photonics, neurophotonics and artificial intelligence, computer and processor interconnections, optical signal processing and sensors.

Chapter 9 delves into the future of PIP by addressing the topic of field programable photonics gate arrays (FPPGAs), which play a similar role to that of FPGAs in electronics. It discusses the basic hardware design principles as well as design flow and technology mapping concepts.

The rest of this first chapter provides a short review of the main topics covered in the book. Section 1.2 introduces the integrated multi-port interferometer and the waveguide mesh concepts including the description of its BBB or tunable basic unit (TBU), which can be implemented by means of a beamsplitter/tunable coupler. It shows how TBUs can be programmed to operate in a cross/bar state or as a tunable coupling device providing independent amplitude and phase values.

Section 1.3 discusses the main material platforms that can be employed in the fabrication of PIP circuits, provides some experimental results reported both in silicon and silicon nitride material platforms and outlines some of the practical challenges to be overcome in future designs.

Section 1.4 addresses the operation of PIP circuits, including the inclusion of PIP circuits in an extended environment, where it has to be interfaced to RF and electronic parts. It considers how waveguide meshes can be programmed to implement both traditional signal processing structures, such as finite and infinite impulse response filters, delay lines, beamforming networks, as well as more advanced multi-port interferometric devices enabling linear matrix optics functionalities.

Section 1.5 reviews some of the salient classical and quantum applications of PIP systems, and Section 1.6 provides a very brief introduction to the concept of field programmable photonic arrays.

1.3 Integrated Multi-port Interferometers and Waveguide Meshes

1.3.1 Basic Concepts

Integrated multi-port interferometers and waveguide meshes are 2D structures where a unitary cell composed of a single TBU (in the first case) or a subset of TBUs (in the second) is spatially replicated [12][17]–[22], [31]. Several examples are shown in Figure 1.1.

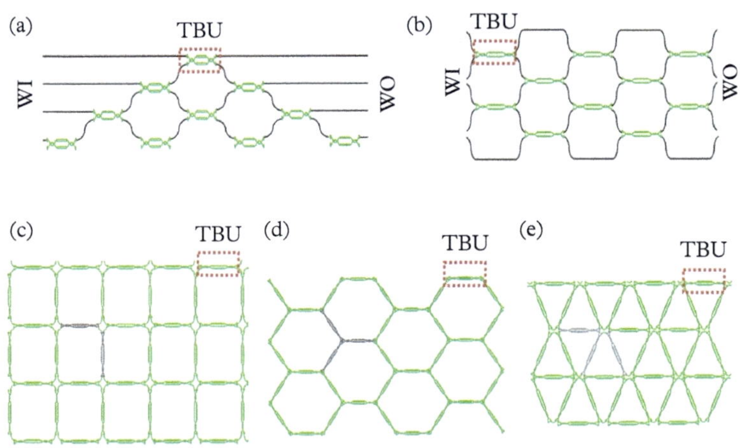

Figure 1.1 *Different integrated multiport interferometer and waveguide mesh arrangements: (a) Triangular Feedforward proposed by Reck at al [21] and Reformulated by Miller [12], (b) Rectangular Feedforward proposed by Clements at al [22], (c) Squared Feedforward/backward [17], (d) Hexagonal Feedforward/backward [18], [31] and (d) Triangular Feedforward/backward [18], [31]. WI/WO: Waveguide inputs/outputs.*

Each unitary cell is implemented by one or more sets of integrated waveguide pairs coupled by means of a TBU. Chapter 2 shows how its core can be a balanced MZI or a directional coupler among other possible options. The application of external electrical signals to the TBU allows the independent amplitude and phase control of the photonic signals coupled between the two waveguides. In particular, each TBU in the mesh can be configured to operate either as an optical crossbar switch or as an intermediate power divider. In this way, the combination of different TBUs in the 2D grid—each individually configured as desired—enables the synthesis of any kind of optical core circuit topology, including finite (FIR) and infinite impulse response (IIR) multi-port interferometers and filters. Figure 1.1(a) and Figure 1.1(b) correspond to meshes representing multi-port interferometers that will be addressed in Chapter 4. Figure 1.1(a) shows the triangular interferometer proposed by Reck and colleagues [21] and Figure 1.1(b) the rectangular interferometer proposed by Clements and colleagues [22]. These have been widely employed in the implementation of quantum circuits [9]–[15]. Note that they allow only feedforward propagation of light so they are limited to the implementation of FIR multi-port interferometer filters. Figures 1.1(c)–1.1(e) show the main reported designs for waveguide meshes allowing for both feedforward and feedbackward propagation. Here the unit cell is composed by several TBUs following a geometrical configuration: square, hexagonal and triangular, respectively. These are the most flexible waveguide mesh configurations allowing the implementation of both FIR and IIR multi-port interferometers and filters. Although each mesh topology has inherent advantages, Chapter 4 and 5 shows how the hexagonal mesh is potentially the most flexible approach for implementing the PIP concept.

1.3.2 Tunable Basic Units

The basic building block in the waveguide mesh is the TBU [12], [17], [18], [21], [22]. Figure 1.2(a) shows an example of a hexagonal cell implemented by nine TBUs arranged in three trilattice structures. Each TBU can be programmed to operate in one of the three modes shown in Figure 1.2(b). The central element of the TBU shown in Figure 1.2(c) is a 2×2 waveguide coupling element or tunable coupler, which can be implemented using different photonic components, including a balanced MZI (Figure 1.2(d1)) or a dual drive directional coupler (Figure 1.2(d2) [32]), among other alternatives.

1.3.3 TBU Programming

TBUs based on 2×2 coupling elements in general and on MZI and dual drive directional couplers in particular can be programmed to achieve independent amplitude and phase shift settings. Here we illustrate the process for the MZI approach, but the reader will find detailed information regarding the dual drive directional coupler approach in Chapter 2. Referring to Figure 1.2(d1), the TBU can be programmed to implement three different states: cross state switch (light path connects in_1 to out_2 and in_2 to out_1), bar state switch (light path connects in_1 to out_1 and in_2 to out_2) and tunable splitter. For example, a balanced MZI loaded with heaters on both arms has a splitting

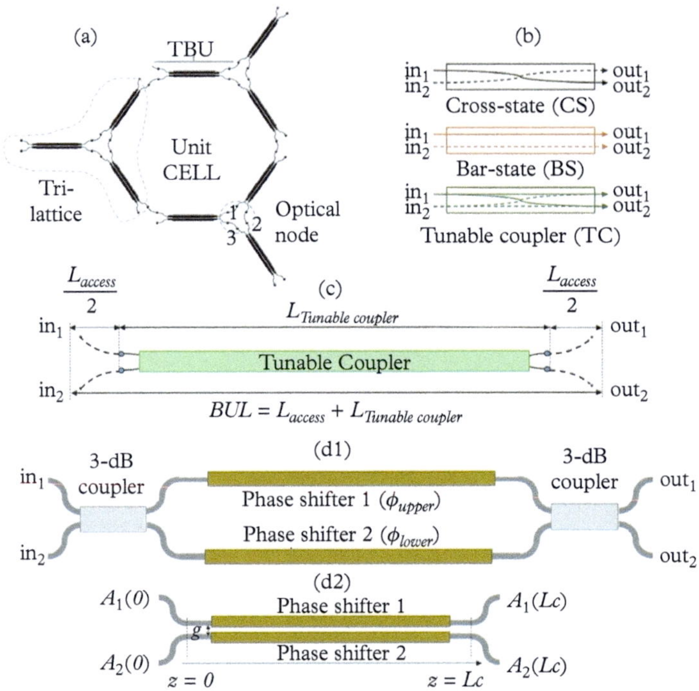

Figure 1.2 *(a) Hexagonal Unit Cell. (b) Signal flow for the different TBU configurations (c) Schematic of a general tunable coupler acting as the basic building block of the mesh: Tunable Basic Unit (TBU). The Basic Unit Length is illustrated as the sum of the tunable coupler length and the arc length of the access waveguides. A particular case of a tunable coupler implemented by (d1) a MZI and (d2) Integrated dual-drive tunable directional coupler.*

ratio obtained by increasing the effective index due to the Joule effect in the upper or lower arm, which produce a φ_{upper} and φ_{lower} phase shift, respectively. Once set, a common drive in both heaters will provide a common phase shift, leading to independent control of the amplitude ratio and the common phase. The device matrix is defined in Chapter 2, by [12], [21], [23]:

$$h_{TBU} = je^{j\Delta}\begin{pmatrix}\sin\theta & \cos\theta \\ \cos\theta & -\sin\theta\end{pmatrix}\gamma, \tag{1.1}$$

where, θ is $\left(\varphi_{upper} - \varphi_{lower}\right)/2$ and Δ is $\left(\varphi_{upper} + \varphi_{lower}\right)/2$. The optical power coupling factor K is then defined as $cos^2(\theta)$. Finally, γ is a general loss term that includes the propagation and insertion losses of the access and tunable coupler waveguides and the 3-dB couplers, respectively. For practical applications and in the case of MZI-based TBUs a more detailed description of the TBU operation is desired—one that takes into account the departure from the ideal 3-dB splitting ratio of the input and output couplers in the MZI, and the uneven insertion losses in each of its arms. This matrix representation has been derived by Mower and colleagues [33] and then applied in conjunction with random phase statistical representations of θ and Δ to analyse the impact of imperfect TBUs over the fidelity of several quantum gate circuits and the impact that their reconfigurability brings in improving this figure of merit, and Chapter 6 provides more detail. It also considers how TBU reconfigurability brings the additional potential for correcting fabrication errors and reports demonstrations over classical circuit configurations [14], [20].

1.4 Fabrication Technologies

PIP circuits rely principally on optical waveguides and therefore the same material platforms and approaches that are employed in the implementation of ASPICs are applicable here. In addition to the low cost implicit in the economies of scale resulting from replicating the same architecture, further cost savings can be made by aligning the fabrication processes with current state-of-the-art generic integration (GIM) and generic foundry (GFM) models [34]. In the GIM a small set of standardised BBBs are provided to control the basic properties of light, after which more complex circuits or compound building blocks (CBBs) are built by connecting several BBBs. In the GFM the foundry provides a shared open access to its generic integration process through multi-project wafers, where different designs or ASPICs from different users are combined on the same wafer; this allows cost sharing and other additional advantages, e.g. on-wafer testing. Since GIM and GFM allow a broad range of functionalities to be synthesised from a small set of BBBs they have been identified as the only path leading to mass production and subsequent cost reduction for PIC manufacturing, where figures in the order of 10 €/mm^2 can already be reached at volumes of 1,000 chips for InP technology [34].

To date, three material platforms: indium phosphide (InP), silicon on insulator (SOI) and silicon nitride ($Si_3N_4 - SiO_2$) have reached the required degree of maturity to be considered as viable options for the implementation of PIPs, either monolithic or hybrid. These operate mainly in the 1.3 and 1.55µm wavelength regions. The next few sections briefly review the salient features of these and we refer the interested reader to the abundant literature in the field.

1.4.1 Indium Phosphide PIC Technology

InP is a compound III–V semiconductor material and the only technology capable of the monolithic integration of active (i.e. featuring optical amplification) and passive photonic components [34]–[36]. A variety of techniques, including butt-joint regrowth, selective area growth, offset and dual quantum well placing and quantum well intermixing, can be employed to integrate regions with different absorption/gain properties along a single waveguide. Waveguide types include shallow (low loss and long bending radius) and deeply etched (higher losses and lower bending radius) designs (see upper row in Figure 1.3).

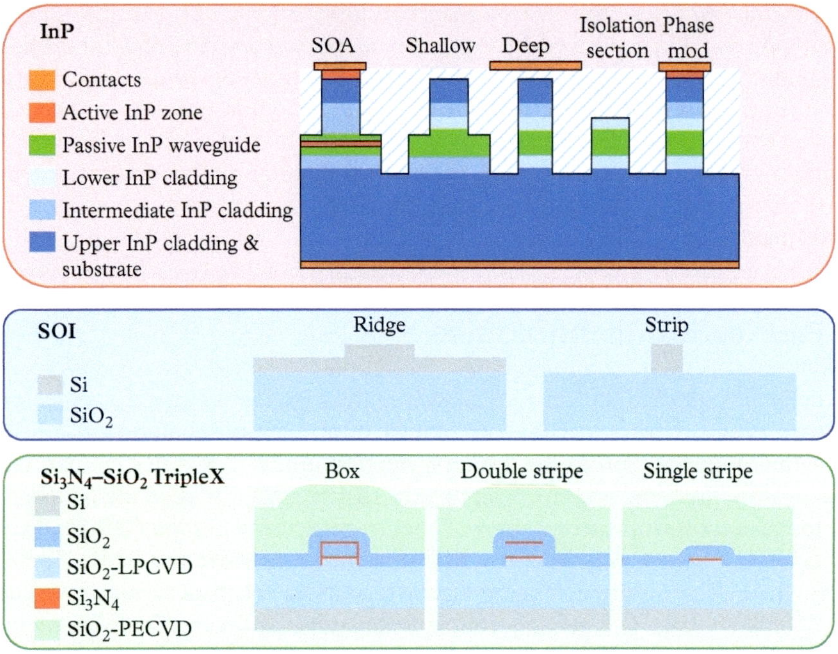

Figure 1.3 *Waveguide structures for the InP (upper row) SOI (intermediate row) and* Si_3N_4*-*SiO_2 *(lower row) platforms.*

Shallow waveguides have typical widths of 2 μm and this figure is approximately 1.5 μm for deeply etched designs. Typical losses are around 1.5 and 3 dB/cm for shallow and deeply etched waveguides, respectively, and this is usually quoted as one of the disadvantages of this technology. However, these losses can be reduced to figures below 1 dB/cm by removing the p-type top layer in the non-active parts of the chip. Minimum bending radii are in between 10 and 500 μm for deeply etched and shallow etched waveguides, respectively. Another quoted limitation of the InP platform is related to the difficulty of its integration with electronics into the same chip. However, highly promising work is under way towards the integration of the full photonic functionality in a single InP membrane on silicon (IMOs) that could be implemented both in a CMOS or an InP fab, merging InP photonics with silicon electronics.

InP provides the most complete list of available components for integration, including [34] passives, optical filters, arrayed waveguide gratings, semiconductor optical amplifiers, saturable absorbers, optical sources (Fabry–Perot, distributed Bragg reflector (DBR), distributed feedback (DFB) and mode-locked lasers), detectors (pin single and balanced configurations), amplitude and phase modulators and polarization-handling devices. Moreover, it has shown record performance in integrated component count on the same chip, with figures exceeding 450, and now targeting 1000 [36]. GIM and GFM have been developed in Europe organized by the JePPIX platform [37] (www.jeppix.eu) that, apart from bringing together Europe's key players in the field, offers semi-commercial access.

1.4.2 Silicon Photonics PIC Technology

Also known as SOI [38]–[41], silicon (Si) photonics is a semiconductor technology where components are etched/patterned/fabricated in a 180–220 nm Si layer placed on top of a 1–3 μm insulator layer. Si passives are formed by an initial few mask layers through partial and/or full Si etching steps, after which multiple ion implantations are conducted for 'active' devices such as Ge photodetectors and Si modulators. Coupling into and outside of the chip can be performed via edge couplers (with typical losses of 1dB/facet) or vertically, via Si surface gratings (3.5–6 dB/coupler with 40–70 nm 3-dB bandwidth). The main advantage of SOI technology resides in its potential compatibility with CMOS fabrication processes and the infrastructure used in micro-electronics, and thus in the potential for monolithic integration of the electronic and photonic parts of the chip. Refractive index contrast is over 100 per cent ($n = 3.4$ for Si and $n = 1.45$ for SiO_2), leading to small footprint circuits.

Two main types of waveguides are available (see intermediate row in Figure 1.3). Ridge or shallow (1–8 μm width) waveguides exhibit relatively low losses down to 0.1–0.5 dB/cm, but are limited in bending radius to around 100 μm. Strip waveguides (500 nm width) exhibit much higher losses (1–3 dB/cm) but support lower values for minimum bending radius (5–20 μm). Integration density on a chip is currently below 100 components [42], but the component count integration trend is exceeding the rate given by Moore's law. Several basic photonic components are available in monolithic SOI, including passives, such as arrayed waveguide gratings and optical filters, Ge photodetectors, ring and travelling-wave electro-refractive modulators (up to 50 GHz). GFM and GIM are being developed in Europe where access is granted through the Europractice platform (www.europractice-ic.com), which also provides several BBBs. The main disadvantage of monolithic SOI technology is that it does not support optical sources and other active components like optical amplifiers. To overcome this limitation, III–V functionalities have to be integrated into the SOI platform by means of either molecular [42] or adhesive [43] wafer bonding. This approach, known as hybrid silicon or hybrid integration technology, has succeeded in incorporating DFB and mode-locked lasers, optical amplifiers, electro-absorption (>70 GHz), and Mach–Zehnder (>25 GHz) modulators, photodetectors (>35 GHz) and polarization handling components into SOI chips. However, GFM and GIM have not yet been developed for hybrid integration technology.

1.4.3 Silicon Nitride Si_3N_4-SiO_2 PIC Technology

Silicon nitride waveguide technology is based on a combination of stoichiometric silicon nitride (Si_3N_4) as waveguide layers, filled by and encapsulated with silica (SiO_2) as cladding layers grown on a silicon wafer [44], [45]. SiO_2 and Si_3N_4 layers are fabricated with CMOS-compatible industrial standard low-pressure chemical vapour deposition

(LPCVD) equipment that enables cost-effective volume production. A special technology known as TriPleX^{TM}[36], developed by the Dutch company LioniX International (www.lionix-international.com), allows the fabrication of waveguides with minimised surface roughness that allows high refractive index contrast (20–30 per cent) and low scattering losses.

The TriPleX^{TM} platform offers seven waveguide cross-sectional geometries [36]. Their salient characteristics @1550 are: low-index contrast (LIC) box-shaped (1 μm × 1 μm) that features reduced birefringence, ≤ 0.2 dB/cm losses and minimum bending radius (R_{min}) around 500 μm and mode field diameter (MFD) of 3.6 × 3.6 μm^2; high-index contrast (HIC) featuring R_{min} of 150 μm and MFD 1.4 × 1.4 m^2; single-stripe (R_{min} of 2000 μm, <0.03 dB/cm and MFD 4.7 × 2.9 μm^2); double-stripe (1 μm × 1 μm, R_{min} of 100 μm, MFD 1.6 × 1.7 μm^2) obtained by removing the sidewalls of the box configuration, which leads to significant reduction of the waveguide losses that are now in the <0.1 dB/cm range. In/out-coupling is achieved by means of adiabatically tapered spot-size converters to LIC double-stripe cross-sections with < 1 dB coupling loss. Several fundamental building blocks are available, including the optical waveguide, tuning elements and directional and MMI couplers. From these, more complex subsystems have been demonstrated. For the double-stripe geometry, a library of standard optical components with predictable characteristics is available, and is currently being offered as a multi-project wafer service run by the JePPIX platform. The main disadvantage of this technology is that no optical sources, detectors, amplifiers and modulators are available. The integration of these components requires a hybrid approach with separately fabricated InP platform chips [45]. The recent development of InP device stamping techniques [46] and InP membranes on silicon [47] opens the path for the compact and versatile implementation of this required hybrid integration approach. In fact, coupling these two platforms together with low insertion loss is targeted in the JePPIX roadmap [37].

1.4.4 Waveguide Meshes Fabricated in Silicon Photonics

Silicon photonics is one of the most attractive integration platforms for programmable waveguide meshes because it allows high-volume fabrication as well as high-integration densities due to its low refractive index contrast and moderate propagation losses of between 0.5–2.5 dB/cm.

Silicon feedforward waveguide meshes have been reported for the implementation of mode conversion operations [26], [27]. Ribeiro and colleagues [26] reported the implementation and demonstration of a 4 × 4-port universal linear optical circuit based on a triangular multi-port interferometer arrangement [21], where the silicon photonic circuit, combined with electronic control and software feedback, can perform any linear operation between its input and output ports. The circuit (Figure 1.4(a) and 1.4(b)) consisted of a network of thermally tunable symmetric MZIs with phase and amplitude control, in-circuit optical power monitors and local software-controlled feedback loops. More recently, Annoni and colleagues [27] reported a 4 × 4 mode unscrambler based also on a triangular Reck multi-port interferometer arrangement (Figure 1.4(c)). The

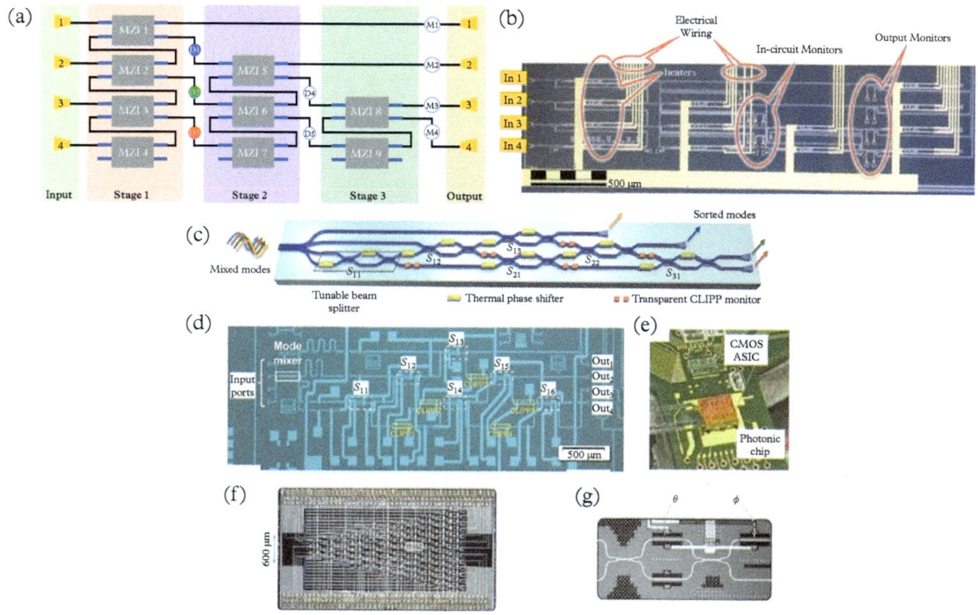

Figure 1.4 *Silicon Feedforward waveguide mesh circuits. Schematic (a) and microscope image (b) of the integrated implementation of the mode converter circuit using MZIs after [26]. Guided-wave implementation (c), microscope image (d) and assembled carrier (e) of the mode unscramber using MZIs reported in [27]. Processor composed of 88 MZIs, 26 input modes, 26 output modes and 176 phase shifters (f) reported in [36] and detail of the integrated beamplitters implemented using a 3dB MZI (g).*

chip was able to automatically unscramble optical beams that had been arbitrarily mixed in a multi-mode waveguide, undoing the scattering and mixing between the spatial modes. The structure incorporated a set of transparent light detectors integrated in a photonic chip (Figure 1.4(d)); these light detectors were used to directly monitor the evolution of each mode along the mesh and allowed sequential tuning and adaptive individual feedback control of each beamsplitter. Electronic reconfiguration and set-up was enabled by an assorted complementary metal oxide semiconductor (CMOS) ASIC (Figure 1.4(e)). More recently, Harris and colleagues [48] reported a reconfigurable chip built upon a waveguide mesh of integrated beamsplitters (Figure 1.4(f)) implemented using integrated MZIs (Figure 1.4(g)). The chip includes eighty-eight MZIs and 176 phase shifters and, by suitable programming of the latter, it has been employed to emulate over 64,000 different quantum particle and phase transport experiments.

Regarding feedforward/backward meshes, a recent contribution [20] reported the operation of a waveguide mesh composed of seven hexagonal cells (thirty thermally tuned TBUs) fabricated in SOI. The chip photograph is shown in Figure 1.5(a). The device was fabricated at the Southampton Nanofabrication Centre at the University of Southampton. SOI wafers with a 220-nm-thick silicon overlayer and a 3-μm-thick buried oxide layer were used (for more details on fabrication and testing see [20]).

Despite the simplicity of the layout (Figure 1.5(a)), the seven-cell structure is capable of implementing over 100 different circuits for general purpose optical processing (basic MZI, FIR transversal filters, basic tunable ring cavities and IIR filters, as well as compound structures such as coupled resonator optical waves (CROWs) and side-coupled integrated spaced sequence of resonators (SCISSORs), true time delay lines and optical coherent interferometry. The basic delay was 13.5 ps, given by a BUL of 975 μm and a group index of 4.18.

1.4.5 Waveguide Meshes Fabricated in Silicon Nitride

Silicon nitride (Si_3N_4) waveguide meshes benefit from the integration of low propagation loss waveguides between 0.00045–1.5 dB/cm with moderate integration density values. Figure 1.5(c) shows the basic layout and photograph of the programmable optical chip architecture connecting thermally tuned MZI devices in a square-shaped mesh network grid proposed by Zhuang and colleagues [17]. The structure, fabricated in Si_3N_4, comprised two square cells, was fully programmable and was employed to demonstrate simple FIR and IIR impulse response filters with single and/or double input/output ports of synthetized ORRs (see Section 1.4.6). A very recent work reported a mesh based on forty thermally tuned TBUs (Figure 1.5(b)). Here, the shape of the TBU was redesigned to achieve a more compact layout and increase the component integration density. This chip has been fabricated in a multi-project wafer run at a fabrication platform developed

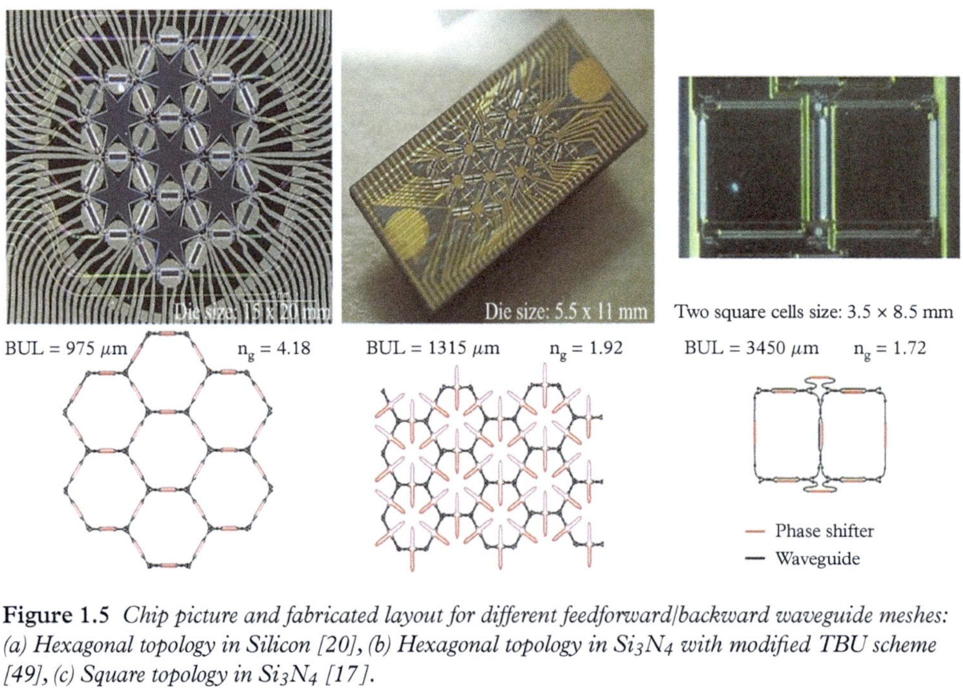

Figure 1.5 *Chip picture and fabricated layout for different feedforward/backward waveguide meshes: (a) Hexagonal topology in Silicon [20], (b) Hexagonal topology in Si_3N_4 with modified TBU scheme [49], (c) Square topology in Si_3N_4 [17].*

by VLC Photonics and the Centro Nacional de Microelectrónica (CNM) [49]. Based on previous and current fabrication runs, a maximum free spectral range (FSR) of 60 GHz is expected, given by a basic time delay of 8.42 ps (group index of 1.92 and a BUL of 1315 μm).

1.4.6 Salient Experimental Results for Feedforward/backward Meshes: A Review

Figure 1.6(a) shows the basic layout and photograph of the programmable optical chip architecture connecting MZI devices in the square-shaped mesh network grid proposed by Zhuang and colleagues [17], which was fabricated in Si_3N_4, comprised two square cells and featured an FSR of 14 GHz. By appropriate programming of this processor, Zhuang and colleagues demonstrated bandpass filters with a tunable centre frequency that spans two octaves (1.6–6 GHz) and a reconfigurable band shape (including flat-top resonance with up to passband–stopband 25 dB extinction). They also demonstrated notch filters with up to 55 dB rejection ratio, Hilbert transformers and tunable delay lines (Figure 1.6(b)). The basic delay was greater than 19.7 ps, given by a BUL of 3450 μm and a group index of 1.72.

Regarding the hexagonal waveguide mesh circuit in silicon, by suitably tuning the TBUs in the seven-cell hexagonal waveguide mesh, the authors were able to program a wide variety of PIC topologies and design parameters [20], [23]. For example, Figure 1.7 illustrates a single cavity optical ring resonator with a cavity length given by 6 BULs. The figure shows (a) the waveguide mesh configurations (with the TBU device status according to the colour code previously described), (b) the circuit layout and (c) the modulus response for the OUT_1 port. The measured results correspond to different values of K_1 and K_2, which settle the positions of the zero and the pole. The IIR filter tunability (Figure 1.7 (d)) is achieved by exploiting the fact that the coupling constant and the phase shift in any TBU of the mesh can be adjusted independently. Hence, any TBU inside the cavity can be operated as a constant-amplitude phase shifter. Finally, Figure 1.7(e) shows the time response of the ring resonator when the critical coupling is achieved.

More complex filter programming by incorporating more cavities and delay line paths could also be achieved over the same structure and are reported in [20]. For example, Figure 1.8 shows the results obtained when the mesh is programmed to implement double- and triple-cavity filters. Figure 1.8(a) illustrates a CROW structure where two cascaded ring resonators (input: IN, output: OUTPUT 1) implement a series of reconfigurable filters by arbitrarily moving its zeros and poles. One of the TBUs identified by an asterisk (*) was kept unbiased. Nevertheless, this is an example of how TBUs can be configured in order to extract non-ideal leaking due to optical crosstalk from the circuit. Figure 1.8(b) shows the reflection response of two coupled ring resonators in a SCISSOR configuration (input: IN, output: OUTPUT 1). This circuit structure is widely used for dispersion compensation and reconfigurable filtering. Figure 1.8(c) shows the transmission response (input: IN, output: OUTPUT 2) of a CROW comprising three ring resonators.

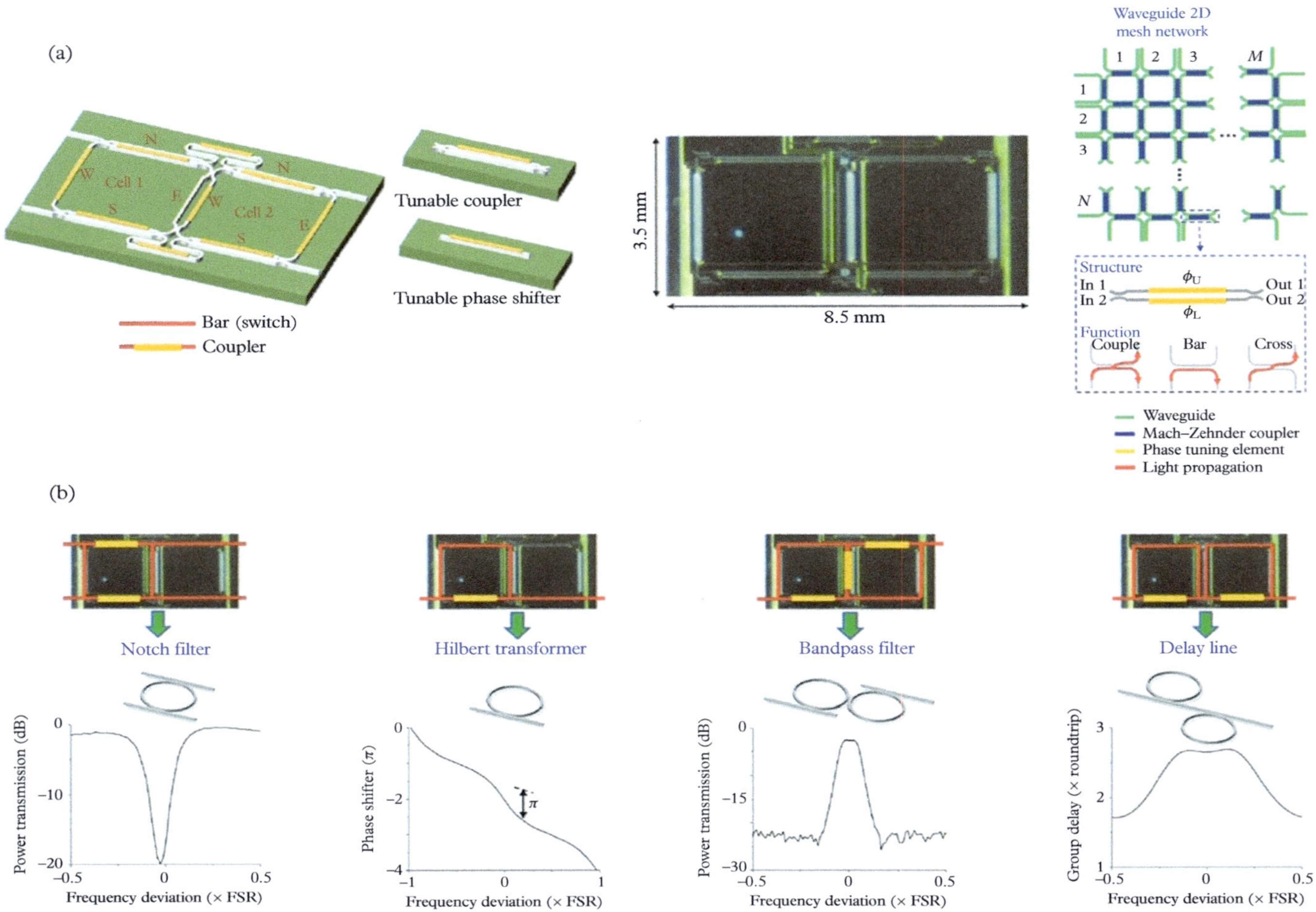

Figure 1.6 *(a) Schematic and a photo of the Si3N4 waveguide technology (TriPleX) chip implementing a 2 square cell waveguide mesh reported in [17]. (b) Experimental results for different programmed circuit configurations obtained by varying phase-tuning elements in the chip and the measurements of their corresponding frequency responses.*

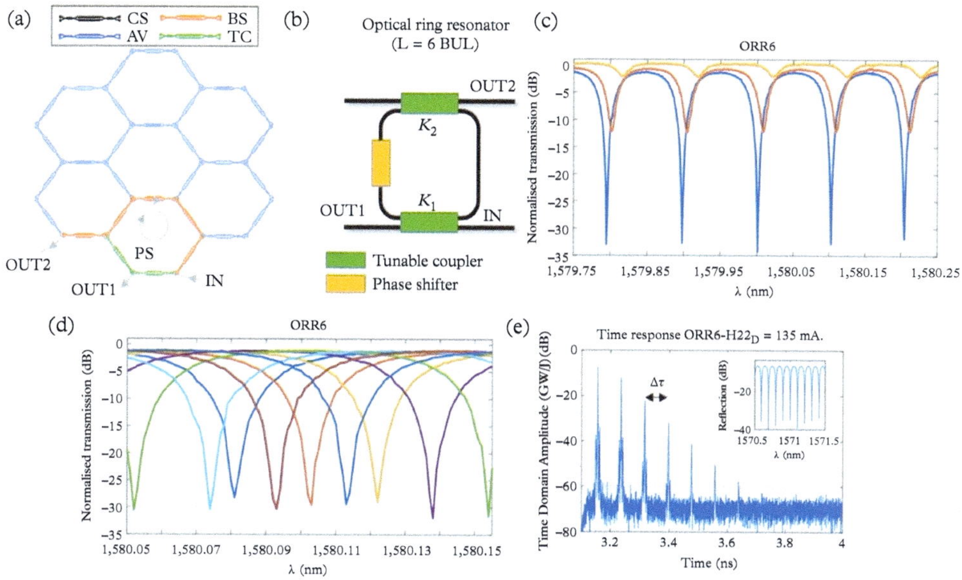

Figure 1.7 *Experimental results for 6-BUL ring resonator IIR and FIR+IIR filters (after [20]). (a) Waveguide mesh connection diagram and color code CS=Cross Switch, BS=Bar Switch, TC= Tunable coupler, AV=Available (unused), (b) circuit layout and (c) measured modulus transfer function for a IIR filter for different values of the coupling constants K_1 and K_2, (d) IIR filter along a full spectral period for different values of the optical ring resonator round-trip phase shift, (e) time response for critical coupling condition. TBU colout code is the following.*

Recently, the same configuration has been employed for the implementation of time-domain true time delay lines and assorted circuits [32]. The upper part of Figure 1.9 illustrates the programmed configuration of two discrete delay lines enabled by the hexagonal mesh featuring paths of 5 and 11 BULs, respectively.

Note that the amplitudes of the different delayed pulse replicas are different, which is due to the different losses experienced by the signal as it propagates through different paths. In particular, the logarithmic power response vs length (time) was measured to be approximately linear with a decaying rate of 0.59 dB per BUL (per 13.5 ps). Errors in the fitting are mainly related to variations in grating coupler losses (±0.5 dB) and to a lesser extent in TBU losses (<0.1 dB), [20]. Note that different input/output ports combinations were employed. Nevertheless, these amplitudes can be equalized (though not shown in the figure) using the ability of TBUs to independently tune their coupling constant values if a splitting tree configuration is required.

Linear matrix transformations can be implemented either by feedforward-only waveguide meshes or by feedforward/backward designs. For instance, Figure 1.10 shows the measured 4 × 4 mode mixing matrix transformations implemented by the circuit reported by Annoni and colleagues [27]. After the mode mixing operation evenly distributing the powers of all the four modes among the four chip inputs by means of

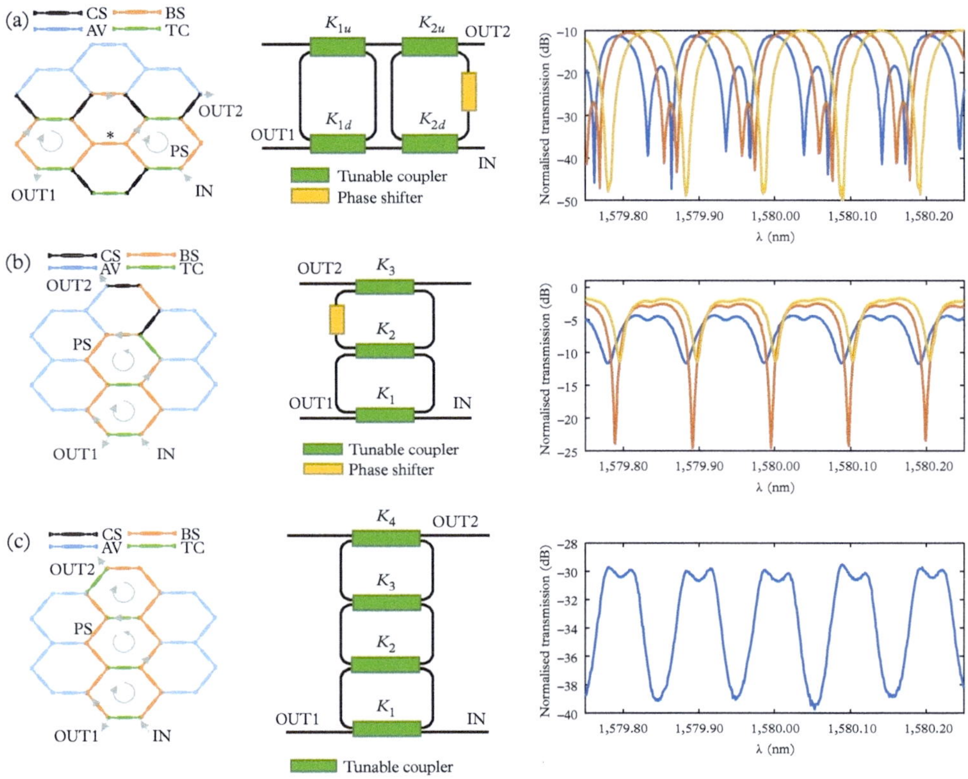

Figure 1.8 *Experimental results for tunable multicavity ring filters. Waveguide mesh connection diagram (left), circuit layout (middle) and measured modulus (right) for different values of the coupling constants K_1 and K_2 in the case of a), a 6-BUL side-coupled integrated spaced sequences of optical resonator (SCISSOR) filter; b), a 6-BUL Coupled resonator waveguide (CROW) filter; c), a 6-BUL CROW. For each case, the first column shows the 7-cell hexagonal waveguide mesh configuration, where each Mach Zehnder Interferometer (MZI) device is represented by a given colour following the code given in Figure 1.7. The second column shows the layout of the implemented structure, while the third column shows the measured modulus for the synthesized configuration where the input is in the IN port and the outputs are the OUT1 and OUT2 ports. BUL: basic unit length, PS: phase shifter, CS: cross state, BS: bar state, AV: available, TC: tuneable coupler.*

matrix H, the circuit was reconfigured to select different full mode sorting to different outputs according to different values of its transformation H_{mesh}.

The hexagonal waveguide mesh chip could also be programmed for implementing several 3×3 and 4×4 linear unitary transformations [25]. These are relevant examples of signal processing tasks that are needed in different applications. Figure 1.11 illustrates, for example, an experimental demonstration of a 3×3 linear unitary transformation corresponding to a three-way beamsplitter and a discrete Fourier transform (DFT) configuration. In all cases, the measured results show an excellent agreement with the

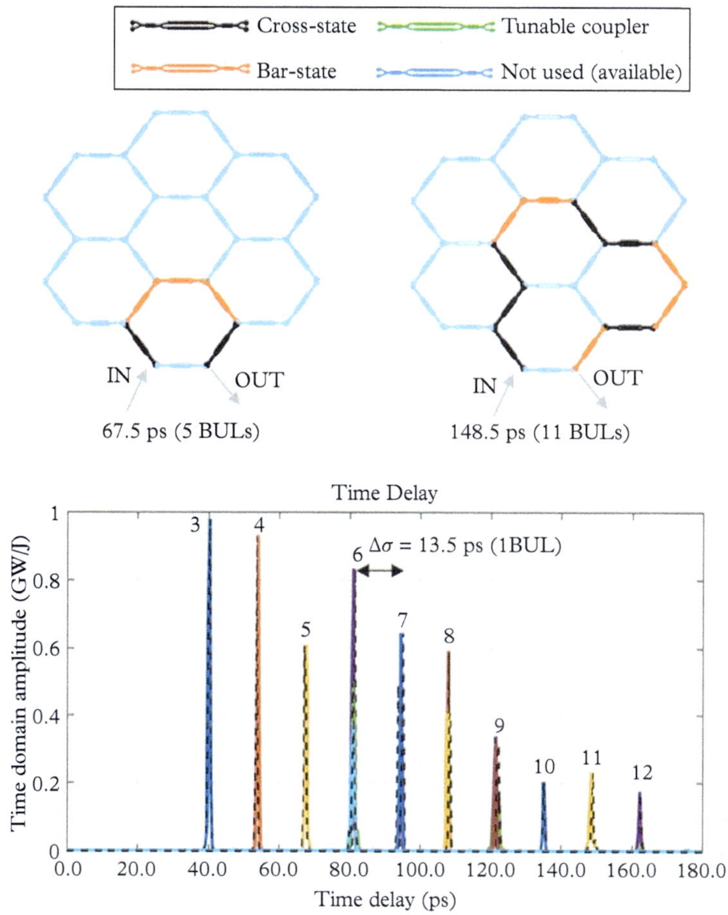

Figure 1.9 *Discrete optical delay lines implemented in a 7-cell layout: (up) settings for two different time delays corresponding to 5 and 11 BULs respectively. (Bottom) Measured delays up to 12 BULs. After [32].*

targeted matrices for the operation wavelength of $\lambda = 1580$ nm with an extinction ratio >25 dB between the 1 and 0 coefficients. The required values for the coupling constants and phases of the TBUs used in these implementations were obtained by the synthesis algorithm adaptation explained in Section 1.4.7. The resulting coefficients are translated into the required injected currents to the phase shifters according to the calibration curves obtained for each TBU during the chip characterisation.

More complex 4×4 transformations were demonstrated including that corresponding to the C-NOT universal quantum gate as shown in Figure 1.12. Note again that the same waveguide mesh allows for a fast reconfiguration of these transformers to perform different operations according to the synthesized values for $N \times N$ their TBU parameters.

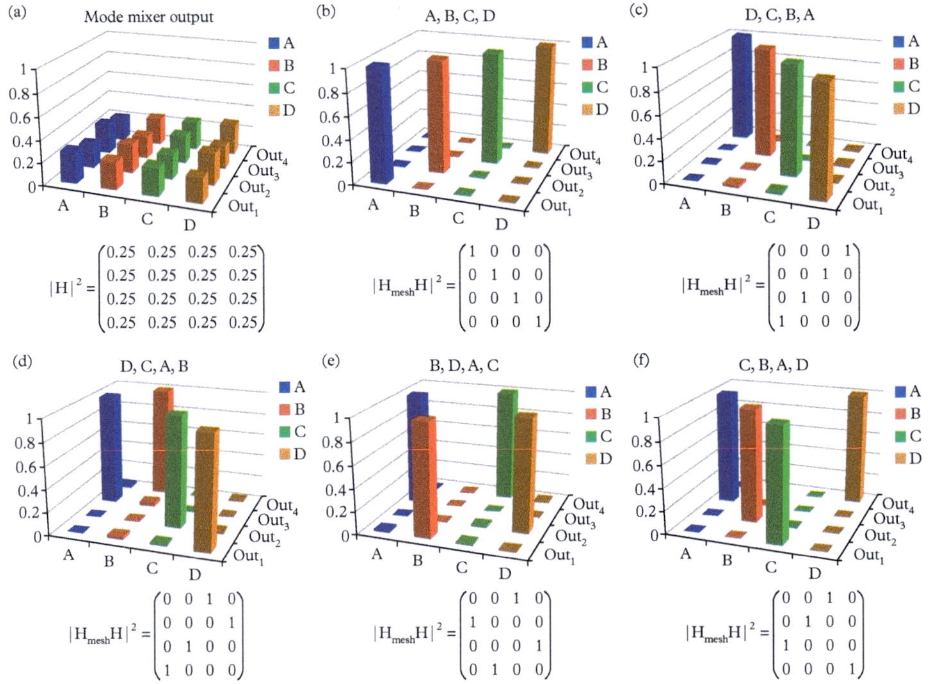

Figure 1.10 *On-chip mode sorting operation of a 4x4 feedforward-only 4x4 chip reported in [27]. The mesh transmission matrix Hmesh can be configured in order to sort the reconstructed modes {A,B,C,D} arbitrarily at the output ports {Out1,Out2,Out3,Out4} of the mesh after mode scrambling by mode mixer H, evenly spreading the power in the input waveguides of the mesh (a), (b-f) different sorting configurations (b) A,B,C,D; (c) D,C,B,A; (d) D, C, B, A; (e) C,A, D,B; (f) C,B,A,D.*

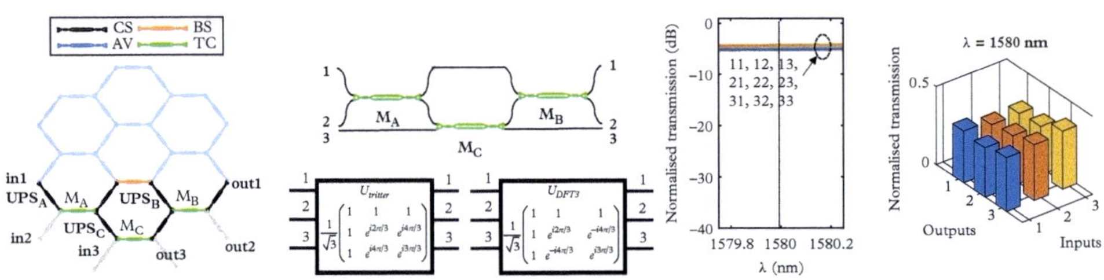

Figure 1.11 *Unitary transformation configuration of a 3 x 3 interferometer based on a rectangular arrangement. Left Column: 7-cell configuration (CS = TBU in cross state, BS = MZI in bar state, TC = TBU in tunable coupler state, AV = TBU not employed), Central-Left Column: circuit layout of the implemented interferometer, Central-Right Column: spectral measurement of all input/output port connections, Right Column: normalized bar diagram of the resulting unitary matrix for λ = 1580 nm (after [23]).*

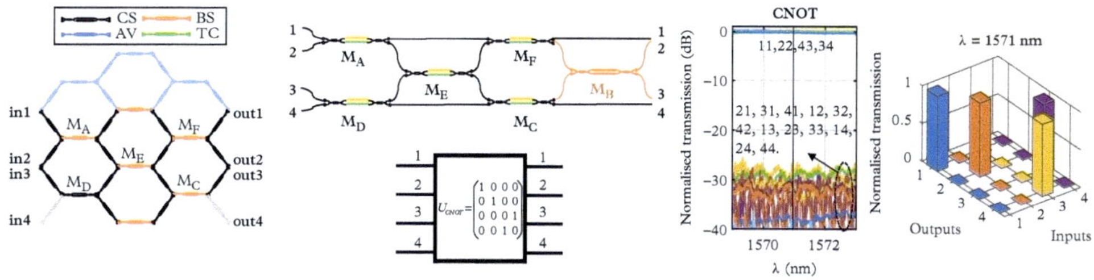

Figure 1.12 *Configuration of a 4 x 4 interferometer based on a rectangular arrangement to implement a C-NOT transformation, Left Column 7-cell configuration (CS = MZI in cross state, BS = MZI in bar state, TC = MZI in tunable coupler state, AV = MZI not employed), Central-Left Column : circuit layout of the implemented interferometer, Central-Right Column : Spectral measurement of all input/output port connections. Righ Column: Normalized Bar diagram of the resulting unitary matrix for λ = 1571nm. (After [23]).*

1.4.7 Introduction to Practical Limitations

Ideally, a higher number of TBUs in a PIP processor results in a more versatile waveguide mesh circuit. However, in practice, there exist footprint limitations together with several sources of degradation that must be considered: accumulated losses, imperfect coupling splitting ratios, phase control, parasitic back-reflections, loss imbalances, fabrication errors (gradients through the circuit in thickness or temperature) and drift in time [19], [23]. These factors will be thoroughly addressed and discussed in Chapter 6, but here we provide a brief introduction.

Several works reporting the integration of a high-density MZI arrangement for matrix-switching operations have succeeded at integrating more than 450 structures in a single die in an SOI platform, exceeding the limits set by Moore's law [50].

When designing programmable waveguide meshes, the designer faces an important miniaturisation trade-off: minimum delay and accumulated losses. The BUL and the group index will determine the minimum delay. For low refractive index difference platforms, the BUL is mainly limited by the tuning mechanism length and the 3-dB coupler lengths. 3-dB couplers in silicon can be reduced to less than 50 μm [51], [52], including the bend sections, while the heaters can be reduced to 62 μm [53]. With the inclusion of bends and straight waveguide sections to increase the distance between both arms of the TBU to decrease thermal crosstalk, a total BUL of 240 μm seems potentially achievable. Assuming a typical SOI group index of 4.18, this is translated to maximum FSRs of around 150 and 50 GHz for the synthesis of MZIs and ORRs, respectively, in the hexagonal waveguide mesh topology. However, a reduction of the BUL implies that the signal must go through a greater number of TBUs to obtain a desired delay. If the 3-dB couplers limit the overall IL of the TBU, this miniaturisation trade-off must be highly considered. In fact, the main limitation of these structures resides in the number of accumulated losses. Assuming 0.1-dB loss 3-dB couplers, a path equivalent to 50 BULs will experience additional 10-dB losses than a classic waveguide of the same length.

The tuning mechanism impact on the final TBU length, tuning crosstalk and power consumption will limit the maximum number of active TBUs at the operational mesh and its performance. The use of alternative tuning mechanisms micro-electromechanical systems (MEMS), piezoelectrics, or electromechanics are promising solutions to reduce the power consumption while enabling a reduction of the distance between the two TBU arms.

Other sources of limitation are connected to fabrication and biasing errors in the TBUs implementing the PIP circuit. Even if these are tightly controlled unavoidable drifts in the circuit parameter setting will arise due to environmental factors. Thus, active monitoring and corrective schemes will be necessary in PIPs (see Chapter 6).

1.5 Operation Modes and Programming

1.5.1 Waveguide Mesh Allocation in a PIP Processor

While the 2D waveguide mesh is the key central element required in PIP, there are other components that are needed to configure a programmable processor [27], [31]. The left part of Figure 1.13 represents a simplified version of the general photonic processor architecture for a wide range of applications (see Chapter 5 for the complete system version).

All the elements are connected to the 2D reconfigurable photonic integrated waveguide mesh in such a way that they produce the desired processing engines as well as dynamically connect the internal and the external elements required for different functionalities. Figure 1.13 shows the processor includes both passive and active photonic components, interface ports with electronic control signals and RF driving input and output ports. Also, pure input/output optical ports can be directly accessed.

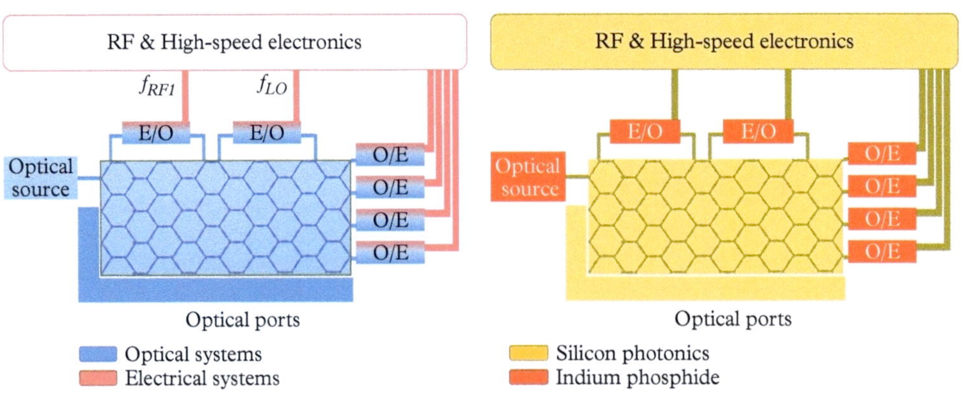

Figure 1.13 *(Left) General photonic integrated processor architecture and (right) candidate fabrication platforms for each subsystem (after [31]).*

The right side of Figure 1.13 illustrates a possible configuration based on a hybrid design approach. In this case, the low loss silicon passive platform (yellow colour) is chosen to implement the passive devices, while InP (red colour) is used for the active devices. Note that an array of optical amplifiers in this platform might be required to overcome the large conversion losses when moving from the RF to the optical domain. These losses are mainly related to the conversion efficiency of modulators and photodetectors as well as the propagation losses. Optical amplifiers should also be implemented in InP and their optimum allocation is under current research. The recent development of InP device stamping techniques [46] and InP membranes on silicon [47] opens the path for the compact and versatile implementation of this required hybrid integration approach.

1.5.2 Operation Modes for PIP Processors

PIP architectures support four different modes of operation as far as the input/output signals are concerned: electrical/electrical, electrical/optical, optical/electrical and optical/optical operations (Figure 1.14).

Electrical/electrical operations are typically employed in microwave photonic (MWP) functionalities, e.g. reconfigurable RF filtering, instantaneous frequency measurement, frequency mixing, RF and millimetre-wave arbitrary signal generation. It requires the processor to enable an optical source, electro-optic (EO) and opto-electronic (OE) converters as well as the reconfigurable optical core. Figure 1.14(a) illustrates the signal

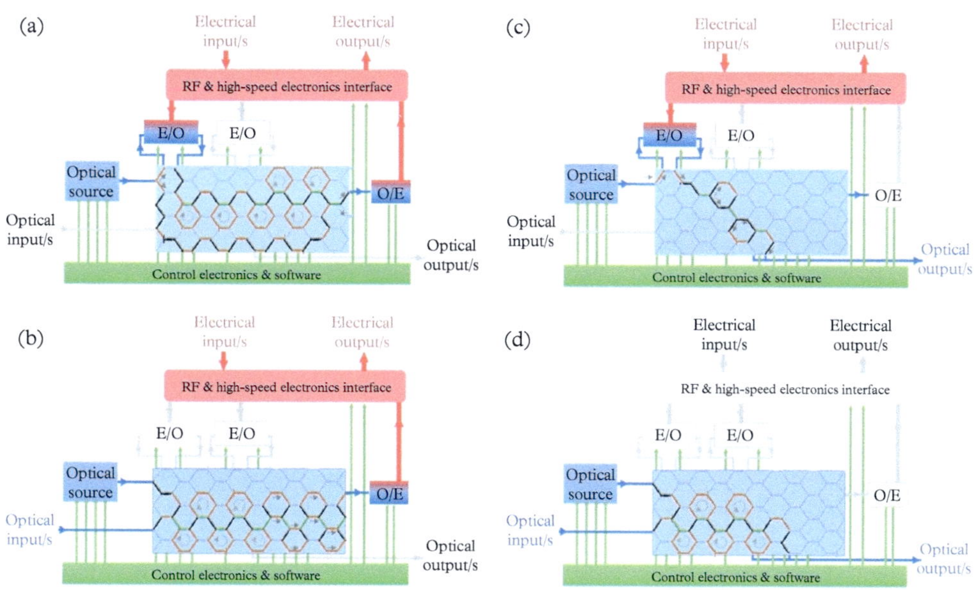

Figure 1.14 *Different modes of Operation for the Programmable Multifunctional Nanophotonics Processor: (a) Electrical/Electrical, (b) Electrical/Optical, (c) Optical/Electrical and (d) Optical/optical.*

flow for these operations. Note that if a second modulator is integrated, it can be enabled to perform frequency-mixing operations based on the cascade of two EO modulators. Sometimes, the processed signal has to be distributed over a particular fibre link length after generation and/or processing. The processor can leverage the inherent properties of optical fibres for distribution purposes. The electrical/optical mode is widely employed in radio-over-fibre MWP links. At the receiver point of the link, another multi-purpose MWP processor can be employed. In this case, the receiver would be working in optical/electrical mode, processing the signal before the photo-detection. Optionally, the receiver also can enable its own optical source to act as a local oscillator. Both modes of operation are displayed in Figures 1.14(b) and 1.14(c), respectively. The last mode of operation, illustrated in Figure 1.14(d), is the optical/optical. In this case, the input signal can be processed directly in the optical domain. Optical channel management can then perform common optical processing operations such as add/drop, switching and broadcasting. Note that all the previous modes of operation may coexist for a certain multi-task functionality. For example, a modulated signal could be divided after being processed and both distributed through the optical ports and down-converted by the photodetectors.

1.5.3 Circuit Synthesis and Emulation

PIP Processors can directly implement and/or emulate a variety of circuit classes. Table 1.2 summarizes the main features for both the feedforward-only and feedforward/backward configurations.

Table 1.2 *Circuit emulation capability of feedforward-only and feedforward/backward waveguide meshes.*

	Feedforward-only Waveguide Meshes	Feedforward/Backward Waveguide Meshes
Traditional FIR, IIR and FIR+IIR SISO and 2 × 2 Filters	Only FIR	Yes
Multiport Non-recirculating N × N Unitary Interferometers	Yes, but either triangular or rectangular, depending on the configuration	Yes, both triangular and rectangular configurations
Multiport non-unitary M × N matrix transformers	Yes, but either triangular or rectangular, depending on the configuration	Yes, both triangular and rectangular configurations
Multiport feedbackward subsystems	No	Yes

From Table 1.2, it follows that feedforward/backward configurations provide the highest operation flexibility and thus, the following briefly introduces the synthesis procedures for this option. More detailed information about synthesis procedures for multiport interferometers and waveguide meshes is provided in Chapter 4 and Chapter 5 respectively.

1.5.4 Synthesis and Emulation of SISO and 2 × 2 input/output circuits

2D integrated waveguide meshes can be programmed to operate as a standard single input/single output (SISO) or a 2 × 2 signal processor [17], [18], [19]. In principle, it has been demonstrated that internal connections set by proper biasing of the intermediate TBUs can enable the programming of a wide variety of functionalities including: FIR filtering implemented either by 2 × 2 unbalanced MZI lattice structures and transversal filters, IIR filters (including single and multi-cavity resonators) and more complex hybrid structures such as CROWs and SCISSORs. Adequate programming also allows the implementation of tunable true time delay lines.

The available synthesis methods for the specific hardware SISO and 2 × 2 configurations that can be emulated using the 2D integrated waveguide mesh can be applied by developing a suitable procedure, which translates the results provided by the synthesis equations into specific parameter values of the TBUs that are needed to implement the waveguide coupling points in the emulated layout. This is possible for all the main FIR, IIR and combined FIR+IIR discrete-time signal processing hardware configurations employed in practice. For example, FIR filters are based either on cascades/lattices of 3-dB tunable MZIs or in transversal filter configurations. For both FIR filter alternatives, synthesis and recursive scaling algorithms have been developed in the literature [54] directly applicable since the hexagonal waveguide mesh can directly implement both 3 dB-tuneable MZI cascade lattices and transversal filter configurations. For IIR filters, either simple/compound optical ring cavities of ring-loaded 3-dB tuneable MZI cascades are employed. Again, synthesis algorithms have been reported in the literature [54] that are directly applicable, since the hexagonal waveguide mesh can directly implement either simple or multiple cavity ring filters or ring-loaded 3-dB tunable MZI cascades.

As an example, Figure 1.15 shows the settings and programming of a hexagonal 2D integrated waveguide mesh to provide different CROW and SCISSOR structures [20]. Here, each cavity is defined by 6 TBUs, setting the free spectral range. Chapter 5 provides a detailed discussion of the synthesis procedures for these kinds of circuits.

1.5.5 Synthesis and Emulation of Multi-port and Multiple Input-Multiple Output (MIMO) Non-recirculating N × N Unitary Interferometers

Multiport waveguide interferometers can be straightforwardly implemented through the triangular and rectangular arrangements of TBUs following the layouts proposed in [21] and [22], respectively. The procedure for implementation is discussed in detail in

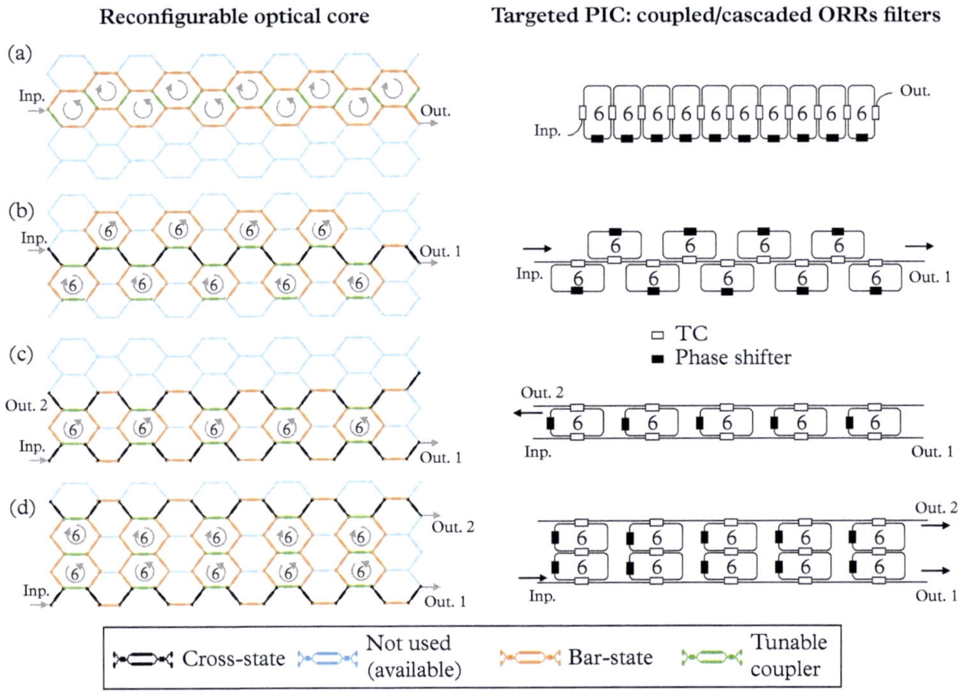

Figure 1.15 *(Left) Settings for CROW & SCISSOR filter implementations in the hexagonal mesh core for (Right) (a) 10th-order CROW, (b) 9th-order Single channel SCISSOR, (c) 5th-order Double channel SCISSOR, and (e) twisted double channel SCISSOR. 6-BUL cavities are specified.*

Chapter 4. Furthermore, they can also be emulated using a 2D integrated hexagonal waveguide meshes. This means that the detailed synthesis and recursive scaling algorithms developed by Reck/Miller for triangular configurations [12], [21] and by Clements and colleagues for rectangular [22] configurations can be applied, provided that they are adapted to the hexagonal waveguide mesh configuration. This section discusses these adaptations briefly and the interested reader is referred to Chapter 5 for a thorough discussion on the methods reported in the literature [20], [23], [31].

Multiple input/multiple output (MIMO) processors implement an arbitrary unitary transformation. In essence, this task is equivalent to that of a linear optics device, which transforms a series of N orthogonal modes ($|\varphi_I\rangle$) into the corresponding N orthogonal modes at the output ($|\varphi_O\rangle$)[11], [12], [23]. This transformation is defined by means of a unitary matrix $U(|\varphi_O\rangle = \mathrm{U}|\varphi_I\rangle)$. Linear transformations are the fundamental building block of many applications in quantum information and communication systems, switching and routing, microwave photonics and optical channel management and supervision. The 2D hexagonal integrated waveguide mesh enables the implementation of the two layout versions of the universal linear interferometer proposed in the literature. The first case corresponds to a Reck–Miller triangular arrangement interferometer [21].

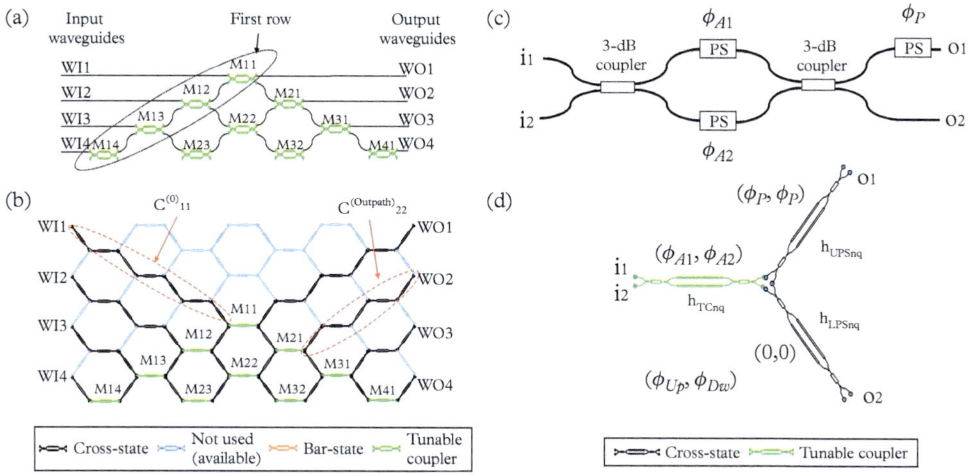

Figure 1.16 *Universal interferometers: (a) Classical triangular arrangement and (b) hexagonal mesh based implementation of a 4x4 interferometer. (c) Beamsplitter for the classical approach and (d) corresponding beamspliter implementation with 3 TBUs for the hexagonal waveguide Mesh.*

Figure 1.16(a) displays an example of a 4×4 interferometer implemented by means of a triangular arrangement of beamsplitters and Figure 1.16(b) shows the equivalent structure implemented on a hexagonal waveguide mesh. Each beamsplitter can set a certain splitting ratio and a relative phase to the upper output. Reck and colleagues and Miller have developed algorithms to program and configure the triangular arrangement so it can implement any desired linear unitary transformation [21]. For example, in order to adapt the synthesis algorithm developed by Miller to the hexagonal waveguide mesh we first must consider the possible different phase contributions due to the different access paths established between the interferometer inputs and the internal processing elements that form the triangular arrangement of beam splitters and, from these, to the different outputs.

These different phase contributions must be compensated. Then, we need to establish an equivalent configuration—using the available elements in the hexagonal waveguide mesh—to the MZI with a phase shifter in the upper output port employed by Miller (Figure 1.16(c)). In our case (Figure 1.16(d)), the equivalent 'beamsplitter' is implemented using a TBU for the tunable coupler (with a transfer matrix defined by h_{TC} as in Eqn. (1.1)), followed by two TBUs, which are biased in cross state and employed as output connections. In the latter, the upper TBU also implements a phase shifter and is defined by the transfer matrix h_{UPS}. The lower TBU is defined by the transfer matrix h_{LPS}. Miller's synthesis algorithm is based on writing any of the input basis functions as a linear combination of each input port or rectangular functions ($|\varphi_{1n}\rangle$), and configuring sequentially each row of beam couplers for each input mode. These input modes are obtained from the columns of the Hermitian adjoint of the matrix U.

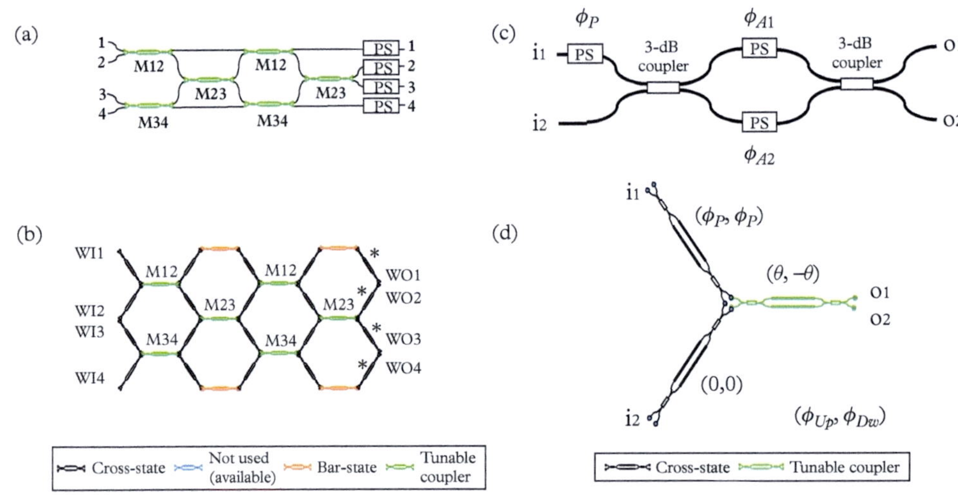

Figure 1.17 *Universal interferometer: a) Rectangular arrangement of a 4×4 interferometer, (b) Equivalent implementation using the hexagonal waveguide mesh. c) MZI and phase shifter arrangement proposed in [22] for the implementation of the beam splitters in the rectangul arrangement. d) Equivalent configuration for the MZI and phase shifter arrangement using the elements available in the hexagonal cell. (φ_{Up}, φ_{Dw}) are the upper and lower phase shift applied to each TBU arm.*

The novel multi-port interferometer configuration based on a rectangular arrangement proposed by Clements and colleagues [22] can also be emulated using the hexagonal 2D integrated waveguide mesh. Figure 1.17(a) displays an example illustrating the implementation of a 4 × 4 multi-port interferometer. Figure 1.17(b) shows the equivalent structure implemented on a hexagonal waveguide mesh. In the algorithm developed by Clements and colleagues [22], each beamsplitter (Figure 1.17(c)) sets a certain splitting ratio and a relative phase sequentially to program and configure the whole rectangular arrangement so it can implement any desired linear unitary transformation efficiently.

1.5.6 Synthesis and Emulation of M × N Non-unitary Interferometers

In a more general scope, 2D integrated waveguide meshes can implement any singular value decomposition (SVD) of an M × N matrix *D*. The SVD, described in detail in Chapter 2, states that an M × N matrix D can be decomposed as [12]:

$$D = VD_{diag}U^{\dagger} \tag{1.2}$$

Where U and V are N × N and M × M unitary matrices and D_{diag} is a M × N diagonal matrix. Figure 1.18 illustrates a 4 × 4 SVD transformation using the 2D hexagonal integrated waveguide mesh and the triangular interferometer configuration. More details

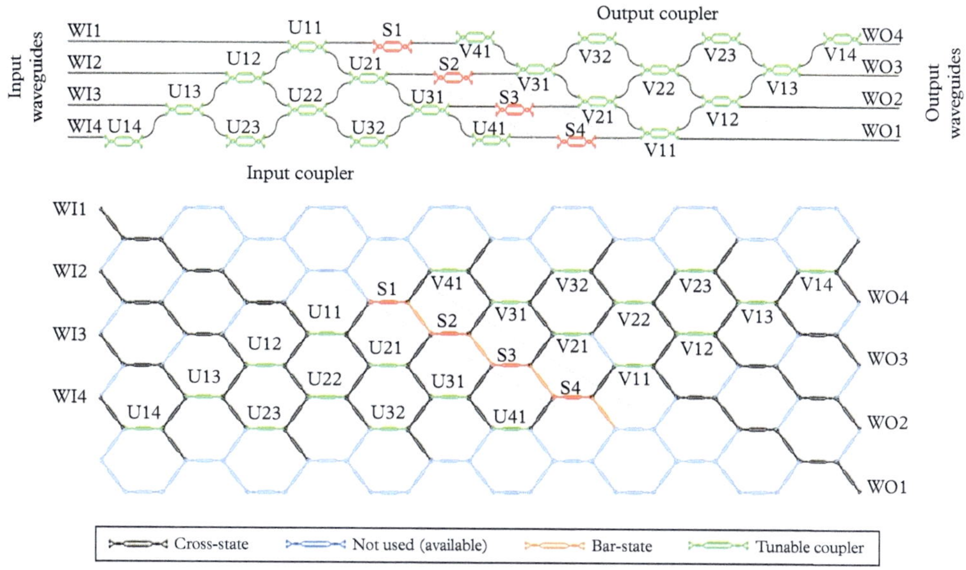

Figure 1.18 *Triangular interferometer layout (Upper) and 2D hexagonal waveguide mesh implementation of a SVD for a 4x4 matrix transformation.*

and thorough discussion on the application of the SVD method for the implementation of multi-port interferometers and the emulation MIMO transformations are provided in Chapter 4 and Chapter 5, respectively.

1.6 Applications of PIP

As previously mentioned, PIP circuits based on 2D integrated feedforward/backward waveguide meshes can find applications in myriad new emerging fields. Chapter 8 provides a detailed discussion regarding their use in telecommunications and networking in processor cores implementing the main required functionalities in MWP systems and radio-over-fibre transmission. In particular, PIP provides a potential alternative to a number of ASPICs that have been reported for delay lines [55], [56], tunable bandpass and notch filtering [56]–[67], reconfigurable signal processing [64]–[66], beamsteering [67]–[70], microwave beamsplitters [71] and arbitrary RF waveform generators [72], to cite some examples. In all these applications, based on ultrafast implementations mainly using ring resonators, speed, bandwidth, tunability and footprint need to be optimal [73]–[75]. The waveguide mesh concept fully addresses all of them. Furthermore, programmability allows not only for the structure reconfiguration but also for a tuning of circuit parameters in response to fabrication errors and time drifts.

This configuration can also emulate a triangular or rectangular multi-port interferometer, which could be employed for mode unscrambling in a similar way to the configuration reported in [26]. Another interesting field of application is switching and interconnection [76], [77]. Hexagonal waveguide meshes emulating rectangular multi-port interferometers can be programmed to manage different optical channels enabling broadcasting, add/drop configurations, multiplexing and demultiplexing functions, to cite but a few. A different device can be obtained if we add optical ports in not only the left and right side of the arrangement, but also in the top and bottom sides of the rectangular arrangement—a feature enabled by the hexagonal mesh topology. The main difference is that functionalities that are more compact can be achieved with this configuration. Consider that, for the standard rectangular arrangement, add/drop functionality would require *N* input ports equal to the number of channel inputs (I) and add channels (A). In the same way, the number of outputs ports will be equal to the number of output channels (O) and drop channels (D).

Figure 1.19 (top) illustrates this new configuration that places add and drop channels in the upper and bottom part, respectively, enabling a more efficient device. The TBUs from H1 to H5 are set in bar state performing the add/drop operation for

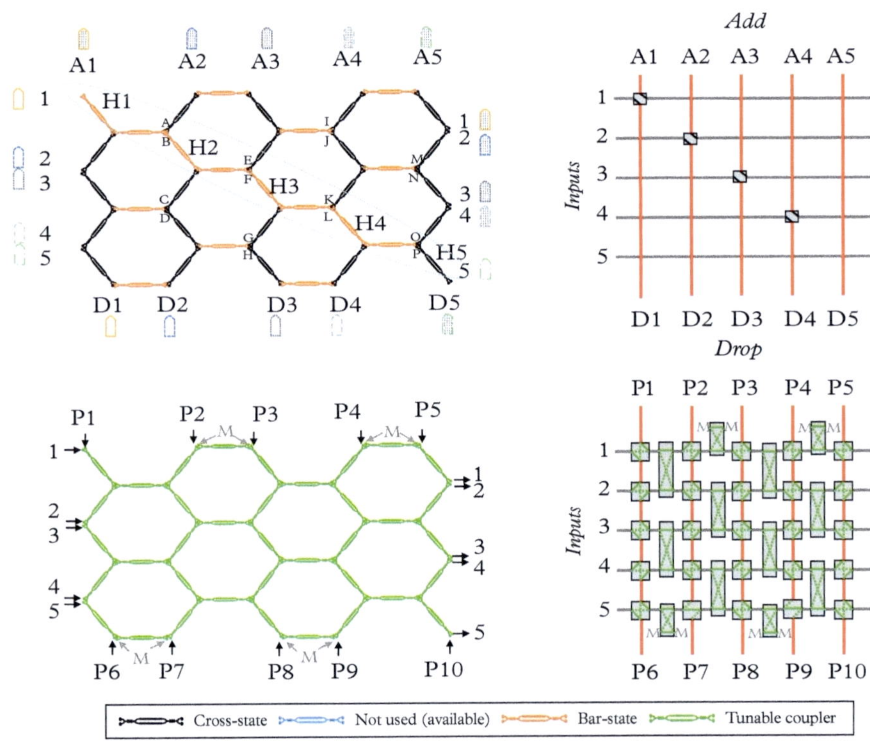

Figure 1.19 *Waveguide mesh settings for channel management application: (Upper) Add-drop configurations for channels 1–4. Channel 5 bypasses the device. (Lower) A fully reconfigurable channel management station allowing broadcasting, add/droppping, channel combinations and demultiplexing. M: Signal monitoring points.*

the matrix illustrated in Figure 1.19 (top/right). In particular, it corresponds to an add (drop) operation for A1(D1) to A4(D4), while channel 5 bypasses the device. Figure 1.19 (lower) illustrates the fully reconfigurable interconnection matrix that can be programmed. Each block is a tunable coupler and can be configured as a switch or define a desired splitting ratio in order to enable broadcasting, multiplexing, demultiplexing or switching operations.

The implementation of arbitrary unitary matrices enables the emulation of any linear transformation, which is a fundamental operation in many other fields of application. For instance, in the area of quantum information, N × N unitary transformations support the implementation of simple and complex logic gates [78]–[84], the emulation of boson sampling [85]–[87] circuits and quantum lab on a chip [88], to cite a few applications. As Chapter 7 shows, waveguide meshes open the path for reconfigurable large-scale integrated quantum information systems with a potential to supersede current approaches based on static configurations [89]. Furthermore, since meshes can be implemented in completely symmetric configurations they naturally enable reversible operation, which is essential for quantum information processing. In computer processor interconnections, reconfigurable broadband inter-processor and computer interconnections are fundamental in high-performance computing and data centres [90]. Photonic linear transformations provide a clean, interference-free and high-speed option for core processor resource management [91]. In the field of optical signal processing, linear transformations that can be supported by PIP processors based on 2D waveguide meshes include several operations that are central to optical signal processing—for example, the optical FFT [92], Hilbert transformation [93], integrators and differentiators [94], [95]. In neurophotonics, unitary (N × M) and non-unitary (N × M) matrix transformations are fundamental building blocks preceding non-linear threshold operations in neural networks, spike and reservoir computing [96], [97]. Some of these applications are reviewed in Chapter 8. The availability of PIP processors opens an interesting and exciting research avenue in this emerging field. In the area of biophotonic sensors, PIP support of simple and MIMO interferometric structures for lab-on-a-chip and multi-parameter sensing applications enable the future implementation of multi-parameter integrated photonic sensing [98], [99]. Finally, in advanced physics, waveguide mesh PIP provides a programable 2D platform to implement different topological systems, e.g. multi-ring cavity structures, to support research in synthetic dimensions and devices based on topological insulator principles [100]–[102].

1.7 Field Programmable Photonic Gate Arrays

As discussed, PIP seeks the design of common integrated optical hardware configurations, which can implement a wide variety of functionalities by suitable programming. While configurations and design principles for programmable circuits based on the cascade of either beamsplitters or integrated MZIs offer versatile hardware solutions to the implementation of programmable circuits, none defines a complete architectural

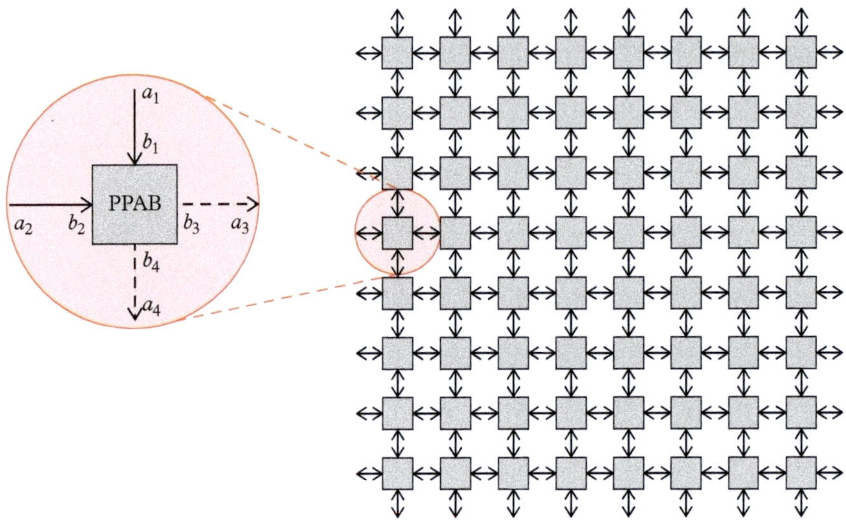

Figure 1.20 *Schematic diagram example of the proposed FPPGA device. The zoom shows a detail of the Programmable Photonic Analog Block as it pertains to the left-up to right-bottom direction of propagation.*

solution of a photonics device that could be programmed for the implementation of arbitrary simple, complex or even simultaneous circuits.

In electronics, this concept is sustained by field programmable gate arrays (FPGAs) and field programmable analogue arrays (FPAAs) and following a similar rationale behind the principles of these devices we propose the implementation of a similar concept that can be realized by combining a set of programmable photonics analogue blocks (PPABs) and a set of reconfigurable photonic interconnects (RPIs) implemented over a photonic chip. This element, which we call field programmable photonics gate array (FPPGA; see Chapter 9), can implement one or various simultaneous photonics circuits and/or linear multi-port transformations by the appropriate programming of its resources (i.e. PPABs and RPIs) and the selection of its input and output ports.

Figure 1.20 shows the high-level concept of the proposed FPPGA.

It consists of a set of PPABs and RPIs implemented through an array of photonic waveguide elements grown on a photonic chip substrate. The waveguide elements that compose the RPIs have programmable features as well and can propagate light in both directions. The layout in Figure 1.20 does not presuppose any particular waveguide array geometry and the square layout depicted is for illustration purposes only.

Although several configurations can be considered for the PPABs, Chapter 9 illustrates the concept with very basic 2 × 2 (two input ports/two output ports PPABs) units. The scheme of such PPAB is shown in the inset of Figure 1.20 for a particular axis orientation and with no internal coupling paths. We consider four options (A, B, C and D), all of which can be obtained from A by convenient rotations (B is A rotated 90°, C is A rotated

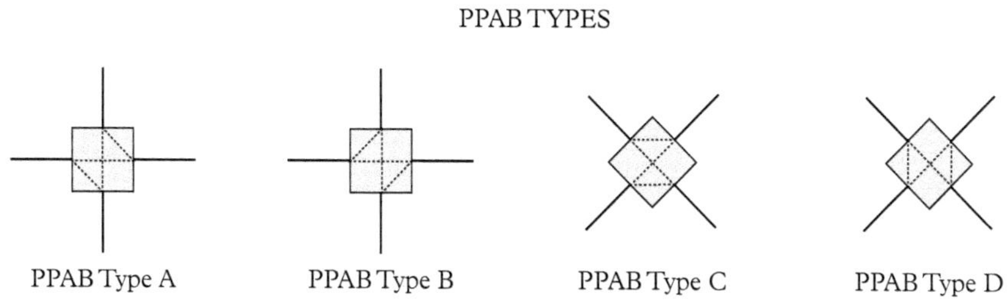

Figure 1.21 *Four types of 2x2 PPAB units considered for the implementation of FPPAs and their internal signal coupling layouts shown in broken lines.*

45° and D is A rotated −45°). Figure 1.21 shows the possible options. The role of the PPAB is to provide tunable independent power coupling ratios and phase shifts, and this is discussed in further detail in Chapter 9, which also covers the important issue of design flow and technology mapping that allows the implementation of advanced PIP concepts.

1.8 Chapter Summary

This opening chapter has introduced and reviewed the main principles and recent advances in the field of PIP, including fabrication platforms, design principles, architecture choices, challenges and limitations and applications of this new field. The objective was to provide to the reader a brief but comprehensive insight about the field and to provide the necessary links between the main topics in this field and the remaining chapters in the book.

REFERENCES

[1] J. Lyke, C.G. Christodolou, G. Alonzo Vera and A.H. Edwards, An Introduction to Reconfigurable Systems. (2015) *Proc IEEE* 103, 291–317.
[2] R.H. Freeman, Configurable Electrical Circuit Having Configurable Logic Elements and Configurable Interconnects, US Patent US4870302, (1989).
[3] S.M. Trimberger, Three ages of FPGAs: A Retrospective on the First Thirty Years of FPGA Technology. (2015) *Proc IEEE* 103, 318–31.
[4] J. Mitola, The Software Radio Architecture. (1995) *IEEE Commun Mag* 33, 26–38.
[5] B. Nunes, et al., A Survey of Software-defined Networking: Past, Present, Future of Programmable Networks. (2014) *IEEE Commun Surveys & Tutorials* 16, 1617–34.
[6] C. Papagianni, et al., On the Optimal Allocation of Virtual Resources in Cloud Computing Networks. (2013) *IEEE Trans Comput* 62, 1060–71.
[7] K. Kerpez and G. Ginis, Software-defined Access Network (SDAN). *Proc 48th Ann Conf Inform Sci Sys (CISS)*, 1–16. Princeton, New Jersey, 2014.
[8] C.M. Cristensen, *The Innovator's Dilemma: When New Technologies Cause Great Firms to Fail*, Boston, MA: Harvard Business School Press, 1997.

[9] A. Peruzzo, et al., Multimode Quantum Interference of Photons in Multiport Integrated Devices. (2011) *Nat Commun* 2, 224.

[10] B.J. Metcalf, et al., Multiphoton Quantum Interference in a Multiport Integrated Photonic Device. (2013) *Nat Commun* 4, 1356.

[11] D.A.B. Miller, Self-aligning Universal Beam Coupler. (2013) *Opt Express* 21, 6360–70.

[12] D.A.B. Miller, Self-configuring Universal Linear Optical Component. (2013) *Photonics Res* 1, 1–15.

[13] J. Carolan, et al., Universal Linear Optics. (2015) *Science* 349, 711.

[14] D.A.B. Miller, Perfect Optics with Imperfect Components. (2015) *Optica* 2, 747–50.

[15] N.C. Harris, et al., Quantum Transport Simulations in a Programmable Nanophotonic Processor. (2017) *Nat Photonics* 11, 447–52.

[16] O. Graydon, Birth of the Programmable Optical Chip. (2016) *Nat Photonics* 10.

[17] L. Zhuang, et al., Programmable Photonic Signal Processor Chip for Radiofrequency Applications. (2015) *Optica* 2, 854–9.

[18] D. Pérez, I. Gasulla, J. Capmany and R.A. Soref, Reconfigurable Lattice Mesh Designs for Programmable Photonic Processors. (2016) *Opt Express* 24, 12093–106.

[19] J. Capmany, I. Gasulla and D. Pérez, Microwave Photonics: The Programmable Processor. (2016) *Nat Photonics* 10, 6–8.

[20] D. Pérez, et al., Multipurpose Silicon Photonics Signal Processor Core. (2017) *Nat Communications* 8, 636.

[21] M. Reck, A. Zeilinger, H.J. Bernstein and P. Bertani, Experimental Realization of Any Discrete Unitary Operator. (1994) *Phys Rev Lett* 73, 58–61.

[22] W. R. Clements, et al., Optimal Design for Universal Multiport Interferometers. (2016) *Optica* 3, 1460–5.

[23] D. Pérez, et al., Silicon Photonics Rectangular Universal Interferometer. (2017) *Lasers and Photon Rev* 11, 1700219.

[24] Y. Shen, et al., Deep Learning with Coherent Nanophotonic Circuits. (2017) *Nat Photonics* 11, 441–6.

[25] A. Ribeiro, A. Ruocco, L. Vanacker and W. Bogaerts, Demonstration of a 4×4-port universal linear circuit. (2016) *Optica* 3, 1348–57.

[26] A. Annoni, et al., Unscrambling Light—Automatically Undoing Strong Mixing Between Modes. (2017) *Light Sci Appl* 6, e17110.

[27] J. Capmany, I. Gasulla and D. Pérez, Towards Programmable Microwave Photonics Processors. (2018) *IEEE J Lightwave Tech* 26, 519–32.

[28] L.N. Chen, E. Hall, L. Theogarajan and J. Bowers, Photonic Switching for Data Center Applications. (2011) *IEEE Photon J* 3, 834–44.

[29] D.A.B. Miller, Silicon Photonics: Meshing Optics with Applications. (2017) *Nat Photonics* 11, 403–4.

[30] N. Thomas-Peter, et al., Integrated Photonic Sensing. (2011) *New J Phys* 13, 055024.

[31] D. Pérez, *Integrated Microwave Photonic Processors Using Waveguide Mesh Cores*. PhD Thesis, Universitat Politècnica de València, 2017.

[32] D. Pérez, E. Sánchez and J. Capmany, Programmable True-Time Delay Lines Using Integrated Waveguide Meshes. (2018) *IEEE J Lightwave Tech* 36, 4591–4601.

[33] J. Mower, et al., High-fidelity Quantum State Evolution in Imperfect Photonic Integrated Circuits. (2015) *Phys Rev A* 92, 032322.

[34] M.K. Smit, et al., An Introduction to InP-based Generic Integration Technology. (2014) *Semicond Sci Technol* 29(8), 083001.

[35] L.A. Coldren, et al., High Performance InP-Based Photonic ICs—A Tutorial. (2011) *IEEE J Lightwave Tech* 29, 554–70.
[36] F. Kish, et al., From Visible Light-Emitting Diodes to Large-Scale III–V Photonic Integrated Circuits. (2013) *Proceedings of the IEEE* 101, 2255–70.
[37] JePPIX Roadmap 2015. Available from http://www.jeppix.eu.
[38] M. Hochberg and T. Baehr-Jones, Towards Fabless Silicon Photonics. (2010) *Nat Photonics* 4, 492–4.
[39] W. Bogaerts, M. Fiers and P. Dumon, Design Challenges in Silicon Photonics. (2014) *IEEE J Sel Top Quantum Electron* 20, 1–8.
[40] R. Soref, 'The Past, Present, and Future of Silicon Photonics. (2006) *IEEE J Sel Top Quantum Electron* 12, 1678–87.
[41] L. Chrostowski and M. Hochberg, *Silicon Photonics Design: From Devices to Systems*, Cambridge University Press, 2015.
[42] M.J.R. Heck, et al., Hybrid Silicon Photonic Integrated Circuit Technology. (2013) *IEEE J Sel Top Quantum Electron* 19, 6100117–6100117.
[43] S. Keyvaninia, et al., Ultra-thin DVS-BCB Adhesive Bonding of III-V Wafers, Dies and Multiple Dies to a Patterned Silicon-On-Insulator Substrate. (2013) *Opt Mater Express* 3, 35–46.
[44] R.G. Heideman, M. Hoekman, and F. Schreuder, TriPleX™-based Integrated Optical Ring Resonators for Lab-on-a-Chip- and Environmental Detection. (2012) *IEEE J Sel Top Quantum Electron* 18, 1583–96.
[45] C.G.H. Roeloffzen, et al., 'Silicon Nitride Microwave Photonic Circuits,' (2013) *Opt Express* 21, 22937–61.
[46] B. Corbett, et al., Transfer Print Techniques for Heterogeneous Integration of Photonic Components. *Prog Quant Electron* 2017, 52, 1–17.
[47] J.G.M. Van der Tol, et al., Indium Phosphide Integrated Photonics in Membranes. *IEEE J Sel T Quantum Electron* 2018, 24, 6100809.
[48] N. Harris, et al., Quantum Transport Simulations in a Programmable Nanophotonic Processor. *Nat Photonics* 2017, 11, 447–53.
[49] G. Micó, et al., C-band Linear Propagation Properties for a 300 nm Film Height Silicon Nitride Photonics Platform. *European Conference on Integrated Optics*. 3–5 April, Eindhoven, Netherlands, 2017.
[50] D. Celo, et al., 32×32 Silicon Photonic Switch. *OptoElectronics and Communications Conference & IEEE Photonics in Switching*. 3–7 July, Niigata, Japan, 2016.
[51] Z. Sheng, et al., A Compact and Low-loss MMI Coupler Fabricated with CMOS Technology. (2012) *IEEE Photonics J* 4, 22272–7.
[52] G.W. Cong, et al., Demonstration of a 3-dB Directional Coupler with Enhanced Robustness to Gap Variations for Silicon Wire Waveguide. (2014) *Opt Express* 22, 2051–9.
[53] N.C. Harris, et al., Efficient, Compact and Low Loss Thermo-optic Phase Shifter in Silicon. (2014) *Opt Express* 22, 10487–93.
[54] C.K. Madsen and J.H. Zhao, *Optical Filter Design and Analysis: A Signal Processing Approach*, John Wiley & Sons Inc., 1999.
[55] L. Zhuang, et al., Low-loss, High-index-contrast Si3N4/SiO2 Optical Waveguides for Optical Delay Lines in Microwave Photonics Signal Processing. (2011) *Opt Express* 19, 23162–70.
[56] M. Burla, et al., On-chip CMOS Compatible Reconfigurable Optical Delay Line with Separate Carrier Tuning for Microwave Photonic Signal Processing. (2011) *Opt Express* 19, 21475–84.

[57] L. Zhuang, et al., Novel Low-loss Waveguide Delay Lines Using Vernier Ring Resonators for On-chip Multi-λ Microwave Photonic Signal Processors. (2013) *Laser Photon Rev* 7, 994–1002.
[58] A. Choudhary, et al., Tailoring of the Brillouin Gain for On-chip Widely Tunable and Reconfigurable Broadband Microwave Photonic Filters. (2016) *Opt Lett* 41, 436–9.
[59] K.Y. Tu, et al., Silicon RF-photonic Filter and Down-converter. (2010) *J Lightwave Technol* 28, 3019–28.
[60] J.S. Fandiño, P. Muñoz, D. Doménech and J. Capmany. Monolithic Integrated Photonic Microwave Filter. (2017) *Nat Photonics* 11, 124–9.
[61] M.S. Rasras, et al., Demonstration of a Tunable Microwave-photonic Notch Filter Using Low-loss Silicon Ring Resonators. (2009) *J Lightwave Technol* 27, 2105–10.
[62] J. Sancho, et al., Integrable Microwave Filter Based on a Photonic Crystal Delay Line. (2012) *Nat Commun* 3, 1075.
[63] P. Orlandi, et al., Photonic Integrated Filter with Widely Tunable Bandwidth. (2014) *J Lightwave Technol* 32, 897–907.
[64] E.J. Norberg, et al., Programmable Photonic Lattice Filters in InGaAsP-InP. (2010) *IEEE Photon Technol Lett* 22, 109–11.
[65] R.S. Guzzon, et al., Integrated InP-InGaAsP Tunable Coupled Ring Optical Bandpass Filters with Zero Insertion Loss. (2011) *Opt Express* 19, 7816–26.
[66] W. Liu, et al., A Fully Reconfigurable Photonic Integrated Signal Processor. (2016) *Nat Photonics* 10, 190–5.
[67] C. Roeloffzen, et al., Integrated Optical Beamformers. *Optical Fiber Communication Conference*. 22–26 March, Los Angeles, CA, 2015
[68] L. Zhuang, et al., Single-chip Ring Resonator-based 1×8 Optical Beam Forming Network in CMOS-compatible Waveguide Technology. (2007) *IEEE Photon Technol Lett* 19, 1130–2.
[69] L. Zhuang, et al., On-chip Microwave Photonic Beamformer Circuits Operating with Phase Modulation and Direct Detection. (2014) *Opt Express* 22, 17079–91.
[70] M. Burla, et al., Multiwavelength Integrated Optical Beamformer Based on Wavelength División Multiplexing for 2-D Phased Array Antennas. (2014) *J Lightwave Technol* 32, 3509–20.
[71] L. Zhuang, et al., Integrated Microwave Photonic Splitter with Reconfigurable Amplitude, Phase, and Delay Offsets. (2015) *Opt Lett* 40, 5618–21.
[72] J. Wang, et al., Reconfigurable Radio-frequency Arbitrary Waveforms Synthesized in a Silicon Photonic Chip. (2015) *Nat Commun* 6, 5957.
[73] A. Agarwal, et al., Fully Programmable Ring-resonator-based Integrated Photonic Circuit for Phase Coherent Applications. (2006) *IEEE J Lightwave Tech* 24, 77–87.
[74] P. Seth and S. Roy, Ultrafast All-optical Flip-flops, Simultaneous Comparator-decoder and Reconfigurable Logic Unit with Silicon Microring Resonator Switches. (2014) *IEEE J Sel Top Quantum Electron* 20, 5900308.
[75] P. Sethi and S. Roy, All-optical Ultrafast Switching in 2 x 2 Silicon Microring Resonators and its Application to Reconfigurable DEMUX/MUX and Reversible Logic Gates. (2014) *IEEE J Lightwave Tech* 32, 2173–80.
[76] K. Bergman, Silicon Photonics for High Performance Interconnection Networks. *Optical Fiber Communication Conference*. 11–15 March San Diego, CA, 2018.
[77] R. Soref, Integrated-photonic Switching Structures. (2018) *Appl Phys Lett Photonics* 3, 021101.
[78] E. Knill, R. Laflamme and G.J. Milburn, A Scheme for Efficient Quantum Computation with Linear Optics. (2001) *Nature* 409, 46–52.

[79] P. Kok et al., Linear Optical Quantum Computing with Photonic Qubits. (2007) *Rev Mod Phys* 79, 135–74.
[80] J.L. O'Brien, A. Furusawa and J. Vučković, Photonic Quantum Technologies. (2009) *Nat Photonics* 3, 687–95.
[81] M.G. Thompson, A. Politi, J.C. Matthews and J.L. O'Brien. Integrated Waveguide Circuits for Optical Quantum Computing. (2011) *IET Circuits Devices Syst* 5, 94–102.
[82] A. Politi, J. Matthews, M.G. Thompson and J.L. O'Brien. Integrated Quantum Photonics. (2009) *IEEE J Select Top Quantum Electron* 15, 1673–84.
[83] A. Politi, et al., Silica-on-silicon Waveguide Quantum Circuits. (2008) *Science* 320, 646–9.
[84] K. Kieling, J.L. O'Brien and J. Eisert. On Photonic Controlled Phase Gates. (2010) *New J Phys* 12, 013003.
[85] J.B. Spring, et al., Boson Sampling on a Photonic Chip. (2013) *Science* 339, 798–801.
[86] M.A. Broome, et al., Photonic Boson Sampling in a Tunable Circuit. (2013) *Science* 339, 794–8.
[87] A. Crespi, et al., Integrated Multimode Interferometers with Arbitrary Designs for Photonic Boson Sampling. (2013) *Nat Photonics* 7, 545–9.
[88] B.P. Lanyon, et al., Towards Quantum Chemistry on a Quantum Computer. (2010) *Nat Chem* 2, 106–11.
[89] N. Harris, et al., Large-scale Quantum Photonic Circuits in Silicon. (2016) *Nanophotonics* 5, 456–68.
[90] Y. Shen, et al., Software-defined Networking Control Plane for Seamless Integration of Multiple Silicon Photonic Switches in Datacom Networks. (2018) *Opt Express* 26, 10914–29.
[91] C. Sun, et al., Single-chip Microprocessor that Communicates Directly Using Light. (2015) *Nature* 528, 534–8.
[92] H.G. De Chatellus, L.R. Cortés and J. Azaña, Optical Real-time Fourier Transformation with Kilohertz Resolutions. (2016) *Optica* 3, 1–8.
[93] L. Zhuang, et al., Novel Microwave Photonic Fractional Hilbert Transformer Using a Ring Resonator-based Optical All-pass Filter. (2012) *Opt Express* 20, 26499–510.
[94] C. Taddei, et al., Waveguide Filter-based On-chip Differentiator for Microwave Photonic Signal Processing. *IEEE International Topical Meeting on Microwave Photonics*. 28–3 October, Alexandria, VA, 2013.
[95] M. Ferrera, et al., On-chip CMOS-compatible All-optical Integrator. (2010) *Nat Commun* 1, 29.
[96] Y. Shen, et al., Deep Learning with Coherent Nanophotonic Circuits. (2017) *Nat Photon* 11, 441–7.
[97] H.T. Pengy, et al., Neuromorphic Photonic Integrated Circuits. *IEEE J Sel Topics Quant Electron*. In press.
[98] M.C. Estevez, M. Alvarez and L. Lechuga, Integrated Optical Devices for Lab-on-a-Chip Biosensing Applications. (2012) *Laser Photon Rev* 6, 463–87.
[99] R. Heideman, M. Hoekman and E. Schreuder, TriPleX-based Integrated Optical Ring Resonators for Lab-on-a-Chip and Environmental Detection. (2012) *IEEE J Sel Topics Quant Electron* 18, 1583–96.
[100] T. Ozawa, et al., Synthetic Dimensions in Integrated Photonics: From Optical Isolation to Four-dimensional Quantum Hall Physics. (2016) *Phys Rev A* 93, 043827.
[101] G. Harari, et al., Topological Insulator Laser: Theory. (2018) *Science* DOI: 10.1126/science.aar4003.
[102] M.A. Bandres, et al., Topological Insulator Laser: Experiments. (2018) *Science* DOI: 10.1126/science.aar400.

2

Basic Building Blocks and Techniques

As mentioned in Chapter 1, programmable integrated photonics (PIP) relies on designing suitable basic building blocks (BBBs) able to carry elementary signal processing operations and interconnection hardware architectures offering very high spatial regularity. The most popular BBBs proposed so far are based on elementary 2×2 tunable photonics coupling components capable of providing independent setting of the power coupling ratio. Additionally, they are based on the phase shift experienced by the incoming signals from two input waveguides in their transition to the two output waveguide ports of the device. This chapter deals primarily with these components. First, it considers the basic matrix methods required to describe their operation, either standalone or in combination with others to form more complex structures. Next, it describes the main technology approaches for the implementation of BBBs, including 3 dB Mach–Zehnder tunable couplers, directional couplers and beamsplitters, followed by how these BBBs are employed to build up a tunable basic unit (TBU), which is the elementary constituent of waveguide mesh circuits. It concludes by describing the devices and techniques relevant in multiport devices, and discusses the equivalence between mode conversion and linear optics matrix transformations and the universal linear combiner

2.1 Scattering and Transfer Matrices

Throught the book we will see that most programmable photonic circuits use elementary tunable photonic coupling elements as BBBs, e.g. partially reflecting mirrors and waveguide couplers [1]–[22]. These devices are described via a linear set of equations relating incident, transmitted and reflected waves or mode field amplitudes [1–3]. In turn, these relationships can be cast into different matrix representations, the most popular of which are the scattering and the transfer matrices. We can use these representations as a basic computation tool because they can be employed to analyse more complex circuit structures via suitable matrix operations. This section reviews these operations and provides their main basic properties and application scenarios.

Programmable Integrated Photonics. José Capmany and Daniel Pérez, Oxford University Press (2020). © José Capmany and Daniel Pérez.
DOI: 10.1093/oso/9780198844402.001.0001

2.1.1 Scattering Matrices

2.1.1.1 Definition

The two port optical system (Figure 2.1(a)) is defined where either a plane wave, a beam of finite size or an optical mode from a waveguide port is incident on the system both at the input (left) and output (right) planes and generates a corresponding reflected wave, beam or mode.

If we consider one polarisation, then the amplitudes of the incident and reflected waves, beams or modes at the input and output planes (four amplitudes in total) provide a full description of the action of the optical system. In each plane we denote the incident wave amplitudes by a_i and the reflected waves by b_i so that the electric and magnetic fields for a transverse electric (TE) wave, beam or mode in a given plane (considering the axes orientations in Figure 2.1(a)) are expressed as:

$$\begin{aligned} \mathbf{E}_i &= \sqrt{\tfrac{2}{C_i}}\left[a_i e^{-j\beta z} + b_i e^{j\beta z}\right]\hat{y} \\ \mathbf{H}_i &= \sqrt{2C_i}\left[a_i e^{-j\beta z} - b_i e^{j\beta z}\right]\hat{x} \end{aligned} \tag{2.1}$$

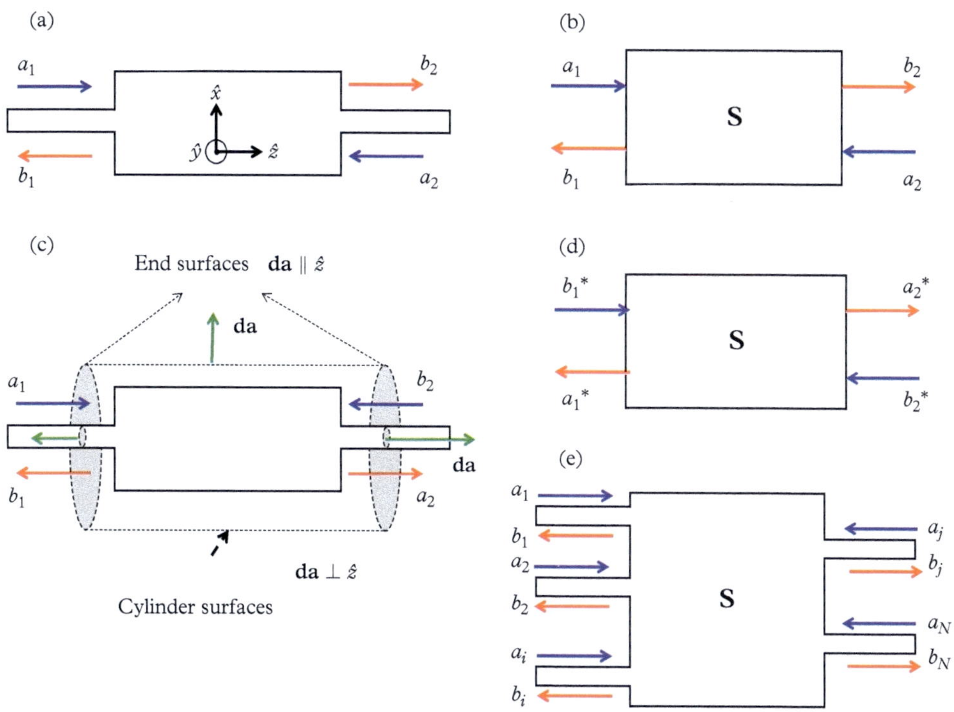

Figure 2.1 *(a) Two port optical system with indication of the in-going and out-going field amplitudes. (b) Black-box representation of the 2x2 optical system. (c) Two port optical system enclosed by a cylindrical volume, (d) Black-box representation of the 2x2 optical system under time reversal operation. (e) NxN optical system with indication of the in-going and out-going field amplitudes.*

where C_i is related to the medium refractive index in plane **i** and β stands for the propagation constant. Note that a_i and b_i can be dependent on the transverse coordinates x and y in the case of beams and modes. Additionally, they are normalised amplitudes so the power per unit area in a given plane is

$$P_i = \frac{1}{2}\mathrm{Re}\left[\mathbf{E}_i \times \mathbf{H}_i^*\right].\hat{z} = |a_i|^2 - |b_i|^2 \tag{2.2}$$

where * denotes the complex conjugate. We can define a linear system relating the reflected and incident wave amplitudes so that only two amplitudes are independently specifiable. For example (Figure 2.1 (b)), if we choose a_1 and a_2 as independent variables, then the system can be described by:

$$b_1 = S_{11}a_1 + S_{12}a_2 \tag{2.3}$$

$$b_2 = S_{21}a_1 + S_{22}a_2. \tag{2.4}$$

We can cast equations (2.11) and (2.12) into a matrix form by defining the vectors

$$\mathbf{a} \equiv \begin{pmatrix} a_1 \\ a_2 \end{pmatrix}; \quad \mathbf{b} \equiv \begin{pmatrix} b_1 \\ b_2 \end{pmatrix} \tag{2.5}$$

and the matrix

$$\mathbf{S} \equiv \begin{pmatrix} S_{11} & S_{12} \\ S_{21} & S_{22} \end{pmatrix}. \tag{2.6}$$

Note that diagonal elements S_{ii} are the reflection coefficients at port i. Thus, (2.11) and (2.12) can be expressed as

$$\mathbf{b} = \mathbf{Sa}. \tag{2.7}$$

2.1.1.2 Reciprocity principle

The reciprocity principle provides a relationship between solutions of Maxwell equations, which, in turn, impose conditions fulfilled by the elements of scattering matrices. We provide here a short derivation [1].

Let us consider a solution $\mathbf{E}^{(1)}, \mathbf{H}^{(1)}$ of the Maxwell equations for an isotropic medium with dielectric constant ε and magnetic permeability μ:

$$\nabla \times \mathbf{E}^{(1)} = -j\omega\mu\mathbf{H}^{(1)}, \tag{2.8}$$

$$\nabla \times \mathbf{H}^{(1)} = j\omega\varepsilon\mathbf{E}^{(1)}. \tag{2.9}$$

We consider now another solution $\mathbf{E}^{(2)}, \mathbf{H}^{(2)}$:

$$\nabla \times \mathbf{E}^{(2)} = -j\omega\mu\mathbf{H}^{(2)}, \tag{2.10}$$

$$\nabla \times \mathbf{H}^{(2)} = j\omega\varepsilon\mathbf{E}^{(2)}. \tag{2.11}$$

We now dot multiply (2.8) by $\mathbf{H}^{(2)}$ and (2.9) by $\mathbf{E}^{(2)}$ and then add both resulting equations:

$$\left(\nabla \times \mathbf{E}^{(1)}\right).\mathbf{H}^{(2)} + \left(\nabla \times \mathbf{H}^{(1)}\right).\mathbf{E}^{(2)} = -j\omega\left[\mu\mathbf{H}^{(1)}.\mathbf{H}^{(2)} - \varepsilon\mathbf{E}^{(1)}.\mathbf{E}^{(2)}\right]. \tag{2.12}$$

We now dot multiply (2.10) by $\mathbf{H}^{(1)}$ and (2.11) by $\mathbf{E}^{(1)}$ and then add both resulting equations:

$$\left(\nabla \times \mathbf{E}^{(2)}\right).\mathbf{H}^{(1)} + \left(\nabla \times \mathbf{H}^{(2)}\right).\mathbf{E}^{(1)} = -j\omega\left[\mu\mathbf{H}^{(1)}.\mathbf{H}^{(2)} - \varepsilon\mathbf{E}^{(1)}.\mathbf{E}^{(2)}\right]. \tag{2.13}$$

Subtracting (2.12) and (2.13) and applying the mixed product properties to the left-hand side member of the equation yields

$$\nabla.\left(\mathbf{E}^{(1)} \times \mathbf{H}^{(2)} - \mathbf{E}^{(2)} \times \mathbf{H}^{(1)}\right) = 0. \tag{2.14}$$

If we now integrate (2.14) over a given volume V enclosed by a surface S, by virtue of Gauss's theorem we have:

$$\oint_S \left(\mathbf{E}^{(1)} \times \mathbf{H}^{(2)}\right).d\mathbf{a} = \oint_S \left(\mathbf{E}^{(2)} \times \mathbf{H}^{(1)}\right).d\mathbf{a}, \tag{2.15}$$

which is known as the *reciprocity theorem*. This principle imposes several conditions over the coefficients of the scattering matrix. First, let us rewrite Eqn. (2.15) in terms of the amplitudes **a** and **b**. For that purpose, we enclose the optical system as shown in Figure 2.1(c) with a cylindrical surface, and with end planes at the input and output surfaces. At the cylinder surface, Eqn. (2.15) vanishes as $\mathbf{E} \times \mathbf{H}$ is orthogonal to d**a**—hence, the only contributions to Eqn. (2.15) are the integrals over the input/output circles. In each circle we choose two counter-propagating solutions of the electromagnetic field. Thus, for the input plane 1 we consider the following fields:

$$\begin{aligned}
\mathbf{E}_1^{(1)} &= \sqrt{\tfrac{2}{C_1}}\left[a_1^{(1)}e^{-j\beta z} + b_1^{(1)}e^{j\beta z}\right]\hat{y} \\
\mathbf{H}_1^{(1)} &= \sqrt{2C_1}\left[a_1^{(1)}e^{-j\beta z} - b_1^{(1)}e^{j\beta z}\right]\hat{x} \\
\mathbf{E}_1^{(2)} &= \sqrt{\tfrac{2}{C_1}}\left[a_1^{(2)}e^{+j\beta z} + b_1^{(2)}e^{-j\beta z}\right]\hat{y} \\
\mathbf{H}_1^{(2)} &= \sqrt{2C_1}\left[a_1^{(2)}e^{+j\beta z} - b_1^{(2)}e^{-j\beta z}\right]\hat{x}
\end{aligned} \tag{2.16}$$

while for the output plane 2 we have:

$$\begin{aligned}
\mathbf{E}_2^{(1)} &= \sqrt{\tfrac{2}{C_2}}\left[a_2^{(1)}e^{j\beta z} + b_2^{(1)}e^{-j\beta z}\right]\hat{y} \\
\mathbf{H}_2^{(1)} &= \sqrt{2C_2}\left[a_2^{(1)}e^{j\beta z} - b_2^{(1)}e^{-j\beta z}\right]\hat{x} \\
\mathbf{E}_2^{(2)} &= \sqrt{\tfrac{2}{C_2}}\left[a_2^{(2)}e^{-j\beta z} + b_2^{(2)}e^{j\beta z}\right]\hat{y} \\
\mathbf{H}_2^{(2)} &= \sqrt{2C_2}\left[a_2^{(2)}e^{-j\beta z} - b_2^{(2)}e^{j\beta z}\right]\hat{x}
\end{aligned} \tag{2.17}$$

By introducing (2.16) and (2.17) after some algebra into (2.8) we find that:

$$\left[\mathbf{b}^{(1)}\right]^T.\mathbf{a}^{(2)} = \left[\mathbf{a}^{(1)}\right]^T.\mathbf{b}^{(2)} \tag{2.18}$$

where the superscript T stands for the transpose matrix. If we now make use of (2.7) and the definition of transpose matrix, we have

$$[\mathbf{b}]^T = [\mathbf{a}]^T\mathbf{S}^T \tag{2.19}$$

Hence,

$$\left[\mathbf{a}^{(1)}\right]^T\mathbf{S}^T\mathbf{a}^{(2)} = \left[\mathbf{a}^{(1)}\right]^T\mathbf{S}\mathbf{a}^{(2)} \tag{2.20}$$

which can only hold if:

$$\begin{aligned}
\mathbf{S}^T &= \mathbf{S} \\
S_{12} &= S_{21}.
\end{aligned} \tag{2.21}$$

In other words, the scattering matrix of a linear reciprocal 2×2 system is symmetric.

2.1.1.3 Power conservation

Further simplifications in the scattering matrix can be obtained if the system under consideration is lossless (i.e. it conserves the power) [1–5]. In this case, the net power flowing to the system must be zero. If we consider (2.2) and apply it to the input and output planes of the 2 × 2 device, then

$$|a_2|^2 - |b_2|^2 + |a_1|^2 - |b_1|^2 = 0, \tag{2.22}$$

which can be written as

$$\mathbf{a}^+\left[\mathbf{I} - \mathbf{S}^+\mathbf{S}\right]\mathbf{a} = 0 \tag{2.23}$$

where $\mathbf{I}$ is the identity matrix and + represents the Hermitian transpose $\mathbf{S}_{ij}^+ = \mathbf{S}_{ji}^*$. Equation (2.23) requires that [1]

$$\mathbf{S}^+\mathbf{S} = \mathbf{I} \Rightarrow \mathbf{S}^+ = \mathbf{S}^{-1} \tag{2.24}$$

In other words, the scattering matrix of a lossless 2 × 2 system is *unitary*, which, according to (2.6), implies the following relationships among its coefficients

$$\begin{aligned} |S_{11}|^2 + |S_{21}|^2 &= 1 \\ |S_{22}|^2 + |S_{12}|^2 &= 1 \\ S_{11}^* S_{12} + S_{21}^* S_{22} &= 0 \end{aligned} \tag{2.25}$$

The scattering matrix of a 2 × 2 system has four complex components. The reciprocity principle eliminates one, leaving three (i.e. six real numbers) as parameters, and (2.25) subject to (2.21), provides three independent real equations that describe a lossless reciprocal 2 × 2 component.

2.1.1.4 Time reversal

A final simplification of the scattering matrix can be obtained if time reversibility can be assumed in the 2 × 2 component. In the time domain, Maxwell equations for a medium with no current sources read [1], [2]

$$\nabla \times \vec{E} = -\mu \frac{\partial \vec{H}}{\partial t} \tag{2.26}$$

$$\nabla \times \vec{H} = \varepsilon \frac{\partial \vec{E}}{\partial t} \tag{2.27}$$

and these become unchanged under a time reversal replacement of t by $-t$ if $-\vec{H}$ is replaced by $-\vec{H}$. Similarly, for isotropic dispersive media, if we take the complex conjugates of (2.8) and (2.9) for a real valued frequency ω we get

$$\nabla \times \mathbf{E}^* = j\omega\mu\mathbf{H}^* \tag{2.28}$$

$$\nabla \times \mathbf{H}^* = -j\omega\varepsilon\mathbf{E}^* \tag{2.29}$$

The form of these becomes identical to (2.8) and (2.9) provided that the fields are now $\mathbf{E}^*$ and $-\mathbf{H}^*$ and $\varepsilon^*(\omega) = \varepsilon(\omega)$, $\mu^*(\omega) = \mu(\omega)$. For a mode travelling in the z direction, the complex conjugation changes the $\exp[-j\beta z]$ variation into $\exp[j\beta z]$. Since the time variation is $\exp[j\omega t]$, the complex conjugation is equivalent to a backward travelling wave.

The time reversibility condition leads to constraints in the scattering matrix coefficients that are similar, but not identical, to those imposed by power conservation. If we take the complex conjugate of (2.7)

$$\mathbf{b}^* = \mathbf{S}^*\mathbf{a}^* \tag{2.30}$$

then $\mathbf{b}^*$ can be interpreted as the time reversed amplitudes of the reflected waves, i.e. incident waves (Figure 2.1(d)), so we can write the equivalence $\mathbf{b}^* \Rightarrow \mathbf{a}$ and similarly $\mathbf{a}^* \Rightarrow \mathbf{b}$. Therefore, from Eqn. (2.30) we get

$$\mathbf{a} = \mathbf{S}^*\mathbf{b} \Rightarrow \mathbf{b} = \left[\mathbf{S}^*\right]^{-1}\mathbf{a} \tag{2.31}$$

and taking Eqn. (2.7) into account as well, this leads to the result

$$\mathbf{S}^* = \mathbf{S}^{-1} \tag{2.32}$$

Note that this condition is different from Eqn. (2.24), which is derived for power conservation. However, if we assume that time reversibility and power conservation hold, then Eqn. (2.24) and Eqn. (2.32) lead to

$$\mathbf{S}^* = \mathbf{S}^+ \tag{2.33}$$

In other words, $\mathbf{S}$ is a symmetric matrix.

2.1.1.5 Extension to N x N systems

The scattering matrix formalism can be extended to an $N \times N$ optical system (Figure 2.1(e)). The basic transformation given by Eqn. (2.7) still holds, provided that Eqn. (2.5) and Eqn. (2.6) are changed to

$$\mathbf{a} \equiv \begin{pmatrix} a_1 \\ a_2 \\ \vdots \\ a_N \end{pmatrix}; \quad \mathbf{b} \equiv \begin{pmatrix} b_1 \\ b_2 \\ \vdots \\ b_N \end{pmatrix} \tag{2.34}$$

and:

$$\mathbf{S} \equiv \begin{bmatrix} S_{11} & S_{12} & \cdots & S_{1j} & \cdots & S_{1N} \\ S_{21} & S_{22} & \cdots & S_{2j} & \cdots & S_{2N} \\ \vdots & \vdots & \vdots & \vdots & \vdots & \vdots \\ S_{i1} & S_{i2} & \cdots & S_{ij} & & S_{iN} \\ \vdots & \vdots & \vdots & \vdots & \vdots & \vdots \\ S_{N1} & S_{N2} & \cdots & S_{Nj} & \cdots & S_{NN} \end{bmatrix} \tag{2.35}$$

It is straightforward to check that the simplifications related to reciprocity, power conservation and time reversal still hold for the $N \times N$ scattering matrix. For instance, the reciprocity condition process Eqns. (2.16)–(2.21) can be applied to a 2×2 reduction linking a pair of ports i, j, leading to

$$S_{ij} = S_{ji} \tag{2.36}$$

The power conservation equation now reads

$$\sum_{i=1}^{N} \left(|a_i|^2 - |b_i|^2 \right) = 0 \tag{2.37}$$

which, according to Eqn. (2.7), Eqn. (2.34) and Eqn. (2.35) leads to Eqn. (2.23), and hence to Eqn. (2.24). The process to obtain the restriction imposed by time reversibility follows exactly the same line of argument as for 2×2 matrices.

2.1.2 Transfer Matrices

Two port components can also be described by means of transfer matrices [2–5]. Here, and referring to Figure 2.1(a), we choose a_1 and b_1 as independent variables and the linear relationships between the field amplitudes are now cast as

$$b_2 = T_{11}a_1 + T_{12}b_1 \tag{2.38}$$

$$a_2 = T_{21}a_1 + T_{12}b_1 \tag{2.39}$$

hence defining the transfer matrix as

$$\mathbf{T} \equiv \begin{pmatrix} T_{11} & T_{12} \\ T_{21} & T_{22} \end{pmatrix} \tag{2.40}$$

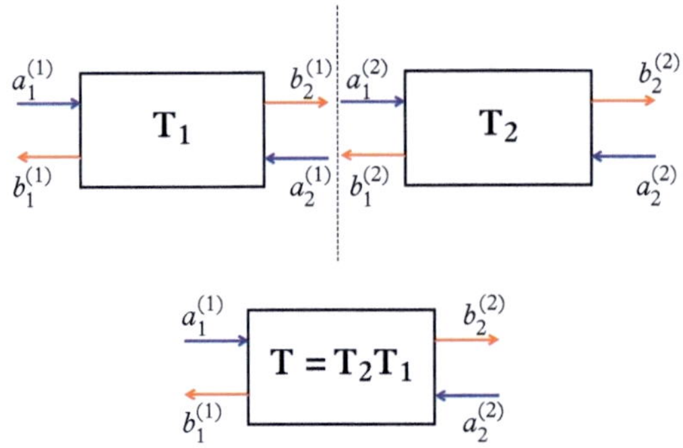

Figure 2.2 *Transfer matrix of a cascade of two 2x2 components.*

Then Eqn. (2.38) and Eqn. (2.39) can be expressed as

$$\begin{pmatrix} b_2 \\ a_2 \end{pmatrix} = \mathbf{T} \begin{pmatrix} b_1 \\ a_1 \end{pmatrix} \tag{2.41}$$

It is straightforward to find the values of the T_{ij} coefficients from the S_{ij} coefficients:

$$\begin{aligned} T_{11} &= -\tfrac{(S_{11}S_{22}-S_{12}S_{21})}{S_{12}}; T_{12} = \tfrac{S_{22}}{S_{21}} \\ T_{21} &= -\tfrac{S_{11}}{S_{21}};\ T_{22} = \tfrac{1}{S_{21}} \end{aligned} \tag{2.42}$$

and from Eqn. (2.42) it can be shown that $\det[\mathbf{T}] = 1$ for reciprocal components.

The transmission matrix is suited for the analysis of cascaded two-port components with forward and backward propagation signals (e.g. cavities and recirculating filters; see Figure 2.2). As shown, the output of the first two-port component 1 is the input of the two-port component 1. Thus,

$$\begin{aligned} \begin{pmatrix} b_2^{(1)} \\ a_2^{(1)} \end{pmatrix} &= T_1 \begin{pmatrix} b_1^{(1)} \\ a_1^{(1)} \end{pmatrix}, \\ \begin{pmatrix} b_2^{(2)} \\ a_2^{(2)} \end{pmatrix} &= T_2 \begin{pmatrix} b_1^{(2)} \\ a_1^{(2)} \end{pmatrix} = T_2 \begin{pmatrix} b_2^{(1)} \\ a_2^{(1)} \end{pmatrix} = T_2 T_1 \begin{pmatrix} b_1^{(1)} \\ a_1^{(1)} \end{pmatrix} \end{aligned} \tag{2.43}$$

Hence, the **T** matrix of a cascade of two ports is the matrix product of their individual **T** matrices and this result can be generalised to a cascade of N two ports.

2.2 Optical Waveguides

Chapter 1 discussed how integrated optical waveguides can have different geometries and refractive index profiles [2]–[3], [6]–[7]; their main role is to propagate stable electromagnetic field patterns known as guided modes. Throughout this book we use the term 'mode' in a very general way as a solution to the Maxwell equations, and we will not consider here other alternatives to guided modes, e.g. radiation and leaky modes.

In general, the solution of Maxwell equations Eqn. (2.8) and Eqn. (2.9) for an integrated waveguide at a given frequency ω can be expressed as a superposition of guided modes, each characterised by a stable transversal field profile and a propagation constant [2], [3].

In practice, a good approximation is to consider only transversal electric (TE) and transversal magnetic (TM) modes. Figure 2.3(a) shows the top view of an integrated optical waveguide, including the axes convention 1, where the TE modes are characterised by having only one non-zero electric field component E_x, H_y being proportional to E_x and $|H_z|<<|H_y|$.

TM modes are characterised by having a single non-zero magnetic field component H_x, and by E_y being proportional to H_x and $|E_z|<<|E_y|$. In general, we can write the field in the waveguide for TE modes as

$$\begin{aligned}\mathbf{E} &= \sum_{\mu} a_\mu \mathbf{e}_\mu e^{-j\beta_\mu z} \\ \mathbf{H} &= \sum_{\mu} a_\mu \mathbf{h}_\mu e^{-j\beta_\mu z}\end{aligned} \tag{2.44}$$

Here, μ is the index representing the mode number, a_μ the mode field amplitude, β_μ is the propagation constant of mode μ and $\mathbf{e}_\mu$, $\mathbf{h}_\mu$ the field patters of mode μ, which can be expressed as:

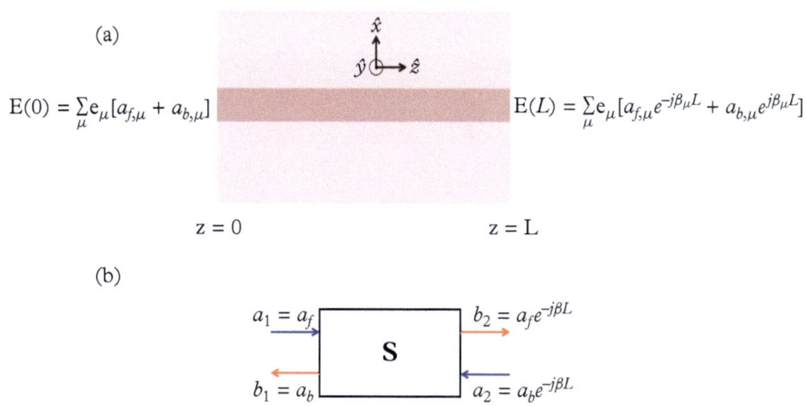

Figure 2.3 *(a) Top-view representation of a photonic integrated waveguide of length L with input and ouputput fields corresponding to the different propagated modes. (b) Black-box representation of the integrated waveguide action over the fundamental* TE_{01} *mode.*

$$\begin{aligned}\mathbf{e}_\mu &= E_\mu(x,y)\,\hat{e}_\mu \\ \mathbf{h}_\mu &= H_\mu(x,y)\,\hat{h}_\mu\end{aligned} \tag{2.45}$$

Where $E_\mu(x,y)$, $H_\mu(x,y)$ represent the field stable transversal profiles of the mode and $\hat{e}_\mu, \hat{h}_\mu$ the polarisation vectors of the electric and magnetic fields, respectively. The actual shape of $E_\mu(x,y)$, $H_\mu(x,y)$ usually requires a numerical calculation using 2D and/or 3D solvers. Power conservation requires that the integral of the real part of the complex Poynting vector over the cross-section S_c of the waveguide structure is independent of distance z along the structure

$$\frac{d}{dz}\int_{S_c} d\mathbf{a}.\left(\mathbf{E}\times\mathbf{H}^* + \mathbf{E}^*\times\mathbf{H}\right) = 0 \tag{2.46}$$

Considering (2.44) and (2.45) and substituting into (2.46) yields, after some algebra,

$$\begin{aligned}&-\sum_\mu\sum_\nu j\left(\beta_\mu-\beta_\nu\right)a_\mu a_\nu^* \int_{S_c} d\mathbf{a}.\left(\mathbf{e}_\mu\times\mathbf{h}_\nu^* + \mathbf{e}_\nu^*\times\mathbf{h}_\mu\right) = 0 \\ &\int_{S_c} d\mathbf{a}.\left(\mathbf{e}_\mu\times\mathbf{h}_\nu^* + \mathbf{e}_\nu^*\times\mathbf{h}_\mu\right) = 0 \quad \text{if } \beta_\mu \neq \beta_\nu\end{aligned} \tag{2.47}$$

If we now apply the time reversal condition, then

$$\begin{aligned}\mathbf{E} &= \sum_\mu a_\mu \mathbf{e}_\mu^* e^{j\beta_\mu z} \\ \mathbf{H} &= \sum_\mu a_\mu \mathbf{h}_\mu^* e^{j\beta_\nu z}\end{aligned} \tag{2.48}$$

which results in a backward propagating solution. Hence, we can apply the reciprocity principle (2.7) to the fields given by Eqn. (2.44) and Eqn. (2.48) in a cylindrical surface closed by two planes at $z = 0$ and z to get

$$\begin{aligned}&-\sum_\mu\sum_\nu a_\mu a_\nu^*\left[e^{j(\beta_\mu-\beta_\nu)}-1\right]\int_{S_c} d\mathbf{a}.\left(\mathbf{e}_\mu\times\mathbf{h}_\nu^* - \mathbf{e}_\nu^*\times\mathbf{h}_\mu\right) = 0 \\ &\int_{S_c} d\mathbf{a}.\left(\mathbf{e}_\mu\times\mathbf{h}_\nu^* + \mathbf{e}_\nu^*\times\mathbf{h}_\mu\right) = 0 \quad \text{if } \beta_\mu \neq \beta_\nu\end{aligned} \tag{2.49}$$

Combining Eqn. (2.49) and Eqn. (2.47), we get the orthogonality condition for propagating modes

$$\int_{S_c} d\mathbf{a}.\left(\mathbf{e}_\mu\times\mathbf{h}_\nu^*\right) = 0 \quad \text{if } \beta_\mu \neq \beta_\nu \tag{2.50}$$

Since, for TE modes $\mathbf{e}_\mu \perp \mathbf{h}_\nu$ and $H_\nu(x,y) \propto E_\nu(x,y)$ we finally have

$$\int_{S_c} E_\mu(x,y)\, E_\nu^*(x,y)\, dxdy = 0 \quad \text{if } \mu \neq \nu \tag{2.51}$$

To compute the scattering matrix for a length z of an integrated optical waveguide it is convenient to distinguish between forward and backward waves; thus, we rewrite equation (2.44) as

$$\begin{aligned} \mathbf{E} &= \sum_\mu \mathbf{e}_\mu \left[a_{f,\mu} e^{-j\beta_\mu z} + a_{b,\mu} e^{j\beta_\mu z}\right] \\ \mathbf{H} &= \sum_\mu \mathbf{h}_\mu \left[a_{f,\mu} e^{-j\beta_\mu z} - a_{b,\mu} e^{j\beta_\mu z}\right] \end{aligned} \tag{2.52}$$

In the case of a single mode (i.e. TE_{01}) propagation waveguide of length L (Figure 2.3(b)) we have ($\mu = 1$), but we drop the subscript in this case to provide

$$\begin{aligned} b_2 &= a_f e^{-j\beta L} = a_1 e^{-j\beta L} \\ b_1 &= a_b e^{-j\beta L} = a_2 e^{-j\beta L} \end{aligned} \tag{2.53}$$

Hence, the scattering matrix for describing the waveguide action over this mode reads

$$\mathbf{S} = \begin{pmatrix} 0 & e^{-j\beta L} \\ e^{-j\beta L} & 0 \end{pmatrix} \tag{2.54}$$

The matrix can be extended as well to consider N propagating modes if needed, but throughout the rest of the book we are mainly concerned with single mode waveguides.

From the general point of view, two solutions centred at different wavelengths will also be considered as different modes—even if they belong to the same class of spatial profile—as they will also be orthogonal. For instance, let us consider a waveguide propagating TE_{01} modes at different frequencies $\omega_k = \omega_o \pm k\Delta\omega$ centred around ω_o:

$$\begin{aligned} \mathbf{E} &= \sum_k a_{\omega_k} \mathbf{e}_{\omega_k} e^{j(\omega_k t - \beta_{\omega_k} z)} \\ \mathbf{H} &= \sum_k a_{\omega_k} \mathbf{h}_{\omega_k} e^{j(\omega_k t - \beta_{\omega_k} z)} \end{aligned} \tag{2.55}$$

Following a similar approach to that leading to Eqn. (2.51) from Eqn. (2.46) we get

$$\int_{S_c} d\mathbf{a}.\left(\mathbf{e}_{\omega_k} \times \mathbf{h}_{\omega_r}^*\right) = 0 \quad \text{if } \beta_{\omega_k} \neq \beta_{\omega_r} \tag{2.56}$$

If the frequencies ω_k are separated far enough from each other so that $\beta_{\omega_k} \neq \beta_{\omega_r}$, then the modes can be considered orthogonal. If this is not the case, the frequencies are conveyed by the same degenerate mode centred at a frequency ω_o and the superposition principle can be applied to their amplitudes, which leads to a *narrowband approximation*

$$\begin{aligned} \mathbf{E} &= \left[\sum_k a_{\omega_k} e^{j\omega_k t}\right] \mathbf{e}_{\omega_o} e^{-j\beta_{\omega_0} z} = S(\omega)\, \mathbf{e}_{\omega_o} e^{-j\beta_{\omega_0} z} \\ \mathbf{H} &= \left[\sum_k a_{\omega_k} e^{j\omega_k t}\right] \mathbf{h}_{\omega_o} e^{-j\beta_{\omega_0} z} = S(\omega)\, \mathbf{h}_{\omega_o} e^{-j\beta_{\omega_0} z} \end{aligned} \tag{2.57}$$

where $S(\omega)$ is the narrowband spectrum centred around ω_o.

2.3 Bulk Beamsplitters

Figure 2.4(a) illustrates the operation of a bulk beamsplitter (BS) [1–5]. Given an incident optical signal, these devices reflect part of it and transmit the rest. In practice, a BS is not employed in integrated programmable photonic circuits, but they are commonly used in their functional description as they play a similar role to integrated waveguide couplers. Our interest here is in the derivation of its scattering matrix as this will be useful to understand the modelling of complex multi-port interferometers discussed in Chapter 3.

The action of the BS can be described by means of its field reflection and transmission coefficients in each face (i.e. r_{11}, t_{21} and r_{22}, t_{12}, respectively). Referring to Figure 2.4(a), we get

$$\begin{aligned} E_1^- &= r_{11} E_1^+ + t_{12} E_2^+ \\ E_2^- &= t_{21} E_1^+ + r_{22} E_2^+ \end{aligned} \tag{2.58}$$

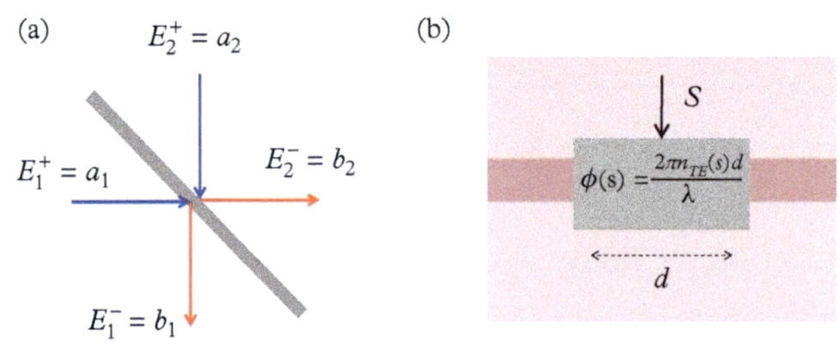

Figure 2.4 *(a) Representation of bulk beamsplitter including input and output fields corresponding to its two surfaces. (b) Schematic layout of an integrated phase shifter where an actuator placed on top of the waveguide generates a change in its refractive index $n_{TE}(s)$ and a subsequent phase shift by means of a control signal s.*

Hence, the scattering matrix is:

$$\mathbf{S} = \begin{pmatrix} r_{11} & t_{12} \\ t_{21} & r_{22} \end{pmatrix} \tag{2.59}$$

Reciprocity requires that **S** is symmetric and thus we have

$$t_{12} = t_{21} = t \tag{2.60}$$

If the BS is lossless, then, according to Eqn. (2.25), Eqn. (2.59) and Eqn. (2.60),

$$\begin{aligned} &|r_{11}|^2 + |t|^2 = 1 \\ &|r_{22}|^2 + |t|^2 = 1 \\ &| r_{11}t \,|\, e^{j(\phi_{12}-\phi_{11})} + | r_{22}t \,|\, e^{j(\phi_{22}-\phi_{21})} = 0 \end{aligned} \tag{2.61}$$

The first two in Eqn. (2.61) imply that

$$|r_{11}|^2 = |r_{22}|^2 = |r|^2 = 1 - |t|^2 \tag{2.62}$$

while the third, together with Eqn. (2.62), imposes a phase shift condition to be fulfilled by the transmission and reflection coefficients

$$\phi_{12} - \phi_{11} - \phi_{22} + \phi_{21} = \pm(2k+1)\pi, \mathrm{k} \in \mathrm{N} \tag{2.63}$$

For the sake of simplicity, we take the arbitrary choice $\phi_{11} = \phi_{22} = 0$ and $k = 0$, with a positive sign. Hence, since Eqn. (2.60) requires that $\phi_{12} = \phi_{21} = \phi$, we have

$$2\phi = \pi \Rightarrow \phi = \frac{\pi}{2} \tag{2.64}$$

Hence, the scattering matrix of the lossless bulk BS is given by:

$$\mathbf{S} = \begin{pmatrix} r & jt \\ jt & r \end{pmatrix}; \quad t = \sqrt{1 - r^2} \tag{2.65}$$

2.4 Phase Shifters

If we recall the scattering matrix of an integrated waveguide of length L given by Eqn. (2.52), we observe that its main action consists in a phase shift of the propagated mode. Actually, the phase shift depends on the waveguide length and the mode propagation constant [2]–[3]. More generally, the propagation constant for the TE fundamental mode β_{TE} can be expressed in terms of an effective refractive mode index as:

$$\beta_{TE} = \frac{2\pi n_{TE}}{\lambda} \tag{2.66}$$

Any change in the value of n_{TE} for a fixed waveguide length will result in a corresponding change in the phase shift (a similar expression holds for the fundamental TM mode). n_{TE} depends upon other factors on the waveguide geometry as well as core and cladding material refractive index n, and this in turn can be changed through different mechanisms, including temperature, current injection, electro-optic effect, mechanical stress, and non-volatile phase actuation based on phase-change materials etc. For each, detailed expressions of these dependences can be found in the literature for the main materials employed in integrated photonics [6]–[9]. Hence, through the application of an external control signal s over a suitable actuating device of length d grown on top of an integrated waveguide (Figure 2.4(b)), we can change the material refractive index in that localised region and therefore the propagation constant of any guided mode in general and the fundamental TE mode in particular. This means that the control signal s can change the phase shift experienced by the propagating signal. This embodiment of an actuator close to the integrated waveguide is known as a tunable phase shifter, and it is a fundamental block of programmable circuits.

Referring to Eqn. (2.52) and Eqn. (2.66) and our previous discussion, the scattering matrix of a tunable phase shifter is given by

$$\mathbf{S} = \begin{pmatrix} 0 & e^{-j\frac{2\pi n_{TE}(s)d}{\lambda}} \\ e^{-j\frac{2\pi n_{TE}(s)d}{\lambda}} & 0 \end{pmatrix} = \begin{pmatrix} 0 & e^{-j\phi(s)} \\ e^{-j\phi(s)} & 0 \end{pmatrix} \tag{2.67}$$

2.5 Integrated Couplers

Integrated couplers are essential building blocks in programmable circuits [11]–[22], as we show in Chapter 3 and Chapter 4. These devices are based on the coupling between field modes propagating within a specific localised region. There are two main types of couplers employed in integrated photonics: the directional coupler (DC) [2]–[3], [6] and the multimode interference (MMI) coupler [23]–[24]. DCs are based on the coupling of the evanescent fields that correspond to the fundamental modes of two integrated waveguides in close proximity. This transfer provides a periodic coupling factor varying from zero to complete cross-coupling along the direction of propagation. They are especially suitable for the implementation of 2×2 devices. MMI couplers are based on the interference between different modes propagating in a multimode region. The MMI coupler is based on the self-imaging principle, a property of multimode waveguides by which an input field profile is reproduced in single or multiple images at periodic intervals along the propagation direction of the guide. MMI devices are thus suitable for implementing $N \times N$ couplers. This section briefly reviews the operating principles and main design equations of both types of couplers.

2.5.1 Directional Couplers

Directional couplers are based on the coupling between the evanescent fields of the co-propagating fundamental modes of two waveguides placed in close proximity (Figure 2.5(a)) [1]–[2]. DCs can be designed so a wave launched in one guide can be partially or completely transferred to the other, and vice versa. In this sense, they behave as an integrated BS. Furthermore, as we show later, this transfer can be controlled by means of an external control signal that enables the so-called tunable directional couplers (TDCs).

2.5.1.1 Co-directional coupling equations

For the description of the DC we consider Figure 2.5(a) and two forward propagating modes a_1 and a_2 of waveguides 1 and 2, respectively (i.e. we only take the first member in Eqn. (2.1)), which, in absence of coupling, have propagating constants β_1 and β_2 and fulfil the equations [1]–[5]

$$\begin{aligned} \frac{da_1}{dz} &= -j\beta_1 a_1 \\ \frac{da_2}{dz} &= -j\beta_2 a_2 \end{aligned} \tag{2.68}$$

Let us assume now that these modes are coupled by means of a weak perturbation, or coupling, which means that we can evaluate the power in the two modes and ignore the coupling. The mode amplitude evolution can now be described by

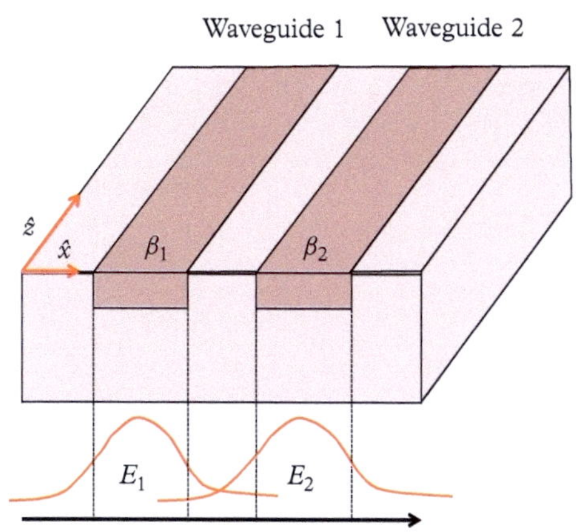

Figure 2.5 *Representation of a directional coupler composed of two waveguides close enough to couple the evanescent fields of their TE_{01} modes.*

$$\left.\begin{array}{l}\frac{da_1}{dz} = -j\beta_1 a_1 + \kappa_{12} a_2 \\ \frac{da_2}{dz} = -j\beta_2 a_2 + \kappa_{21} a_1\end{array}\right\} \Rightarrow \begin{pmatrix}\frac{da_1}{dz} \\ \frac{da_2}{dz}\end{pmatrix} = \begin{pmatrix}-j\beta_1 & \kappa_{12} \\ \kappa_{21} & -j\beta_2\end{pmatrix}\begin{pmatrix}a_1 \\ a_2\end{pmatrix} \tag{2.69}$$

where κ_{12} and κ_{21} are the coupling coefficients from waveguide 2 to waveguide 1, and from waveguide 1 to waveguide 2, respectively. These coefficients are subject to restrictions if we assume power conservation.

Due to the weak coupling (and if we assume that the field amplitudes are normalised so their square moduli directly provide the optical power in each waveguide) the net power is given by

$$P = |a_1|^2 + |a_2|^2 \tag{2.70}$$

Power conservation then implies

$$\frac{dP}{dz} = \frac{d|a_1|^2}{dz} + \frac{d|a_2|^2}{dz} = 0 \tag{2.71}$$

Using Eqn. (2.69) in computing Eqn. (2.71), after some algebra we get

$$\kappa_{12} + \kappa_{21}^* = 0 \tag{2.72}$$

2.5.1.2 Solution of the co-directional coupling equations

To solve Eqn. (2.69) we first need the eigenvalues β of the 2×2 matrix [1]:

$$\begin{vmatrix} j(\beta-\beta_1) & \kappa_{12} \\ \kappa_{21} & j(\beta-\beta_2) \end{vmatrix} = 0 \Rightarrow (\beta-\beta_1)(\beta-\beta_2) + \kappa_{12}\kappa_{21} = 0 \tag{2.73}$$

The solutions for β are given by

$$\beta = \frac{(\beta_1+\beta_2)}{2} \pm \sqrt{\left(\frac{\beta_1+\beta_2}{2}\right)^2 - \kappa_{12}\kappa_{21}} \tag{2.74}$$

Note that, due to Eqn. (2.72), $\kappa_{12}\kappa_{21} = -|\kappa_{21}|^2$, and thus β is always real. Here the solution of the system of coupled differential equations (Eqn. 2.69) is straightforward once we obtain the eigenvectors that correspond to (Eqn. 2.74). Further, if we assume as initial conditions that the waves $a_1(0)$ and $a_2(0)$ are launched at $z = 0$, then [1], [3], [6]

$$
\begin{aligned}
a_1(z) &= e^{-j\frac{(\beta_1+\beta_2)z}{2}}\left[a_1(0)\left(\cos\beta_o z + j\tfrac{\Delta\beta}{2\beta_o}\sin\beta_o z\right) + a_2(0)\tfrac{\kappa_{12}\sin\beta_o z}{\beta_o}\right]\\
a_2(z) &= e^{-j\frac{(\beta_1+\beta_2)z}{2}}\left[a_1(0)\tfrac{\kappa_{21}\sin\beta_o z}{\beta_o} + a_2(0)\left(\cos\beta_o z - j\tfrac{\Delta\beta}{2\beta_o}\sin\beta_o z\right)\right]\\
\Delta\beta &= \beta_2 - \beta_1\\
\beta &= \sqrt{\left(\tfrac{\Delta\beta}{2}\right)^2 - \kappa_{12}\kappa_{21}}
\end{aligned}
\tag{2.75}
$$

Equation (2.75) defines a partial 2×2 scattering matrix for the DC. As in Figure 2.6(a), this matrix is given by

$$
\begin{aligned}
\begin{pmatrix} b_3 \\ b_4 \end{pmatrix} &= e^{-j\frac{(\beta_1+\beta_2)z}{2}} \begin{pmatrix} A(z) & B(z) \\ C(z) & D(z) \end{pmatrix} \begin{pmatrix} a_1 \\ a_2 \end{pmatrix}\\
A(z) &= \left(\cos\beta_o z + j\tfrac{\Delta\beta}{2\beta_o}\sin\beta_o z\right)\\
B(z) &= \tfrac{\kappa_{12}\sin\beta_o z}{\beta_o}\\
C(z) &= \tfrac{\kappa_{21}\sin\beta_o z}{\beta_o}\\
D(z) &= \left(\cos\beta_o z - j\tfrac{\Delta\beta}{2\beta_o}\sin\beta_o z\right)
\end{aligned}
\tag{2.76}
$$

Now, due to the reciprocity principle $B(z) = C(z)$, and hence $\kappa_{12} = \kappa_{21}$, but since Eqn. (2.72) must be fulfilled, then $\kappa_{12} = \mathrm{Im}[\kappa_{21}]$; i.e. both coupling coefficients must be purely imaginary, thus $\kappa_{12} = \kappa_{21} = j\kappa$. Now, by using the properties of the scattering matrix of a lossless system

$$
|A(z)|^2 + |C(z)|^2 = 1 \tag{2.77}
$$

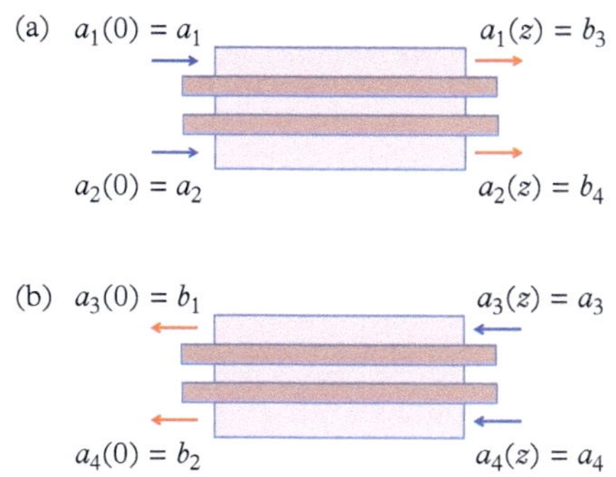

Figure 2.6 *Top view of a two-waveguide 2x2 directional coupler with indication of in-going and out-going fields. (a) Forward propagation, (b) Backward propagation.*

and calling

$$|C(z)|^2 = \left(\frac{\kappa \sin \beta_o z}{\beta_o}\right)^2 = K \tag{2.78}$$

we finally have

$$\begin{pmatrix} b_3 \\ b_4 \end{pmatrix} = e^{-j\frac{(\beta_1+\beta_2)z}{2}} \begin{pmatrix} \sqrt{1-K}e^{j\varphi} & j\sqrt{K} \\ j\sqrt{K} & \sqrt{1-K}e^{-j\varphi} \end{pmatrix} \begin{pmatrix} a_1 \\ a_2 \end{pmatrix}$$
$$\varphi = \arctan\left(\frac{\Delta\beta}{2\beta_o}\tan\beta_o z\right) \tag{2.79}$$

For the complete scattering matrix, we must take into account the relationships established by the propagation of backward modes (Figure 2.6(b)). Here, we get

$$\begin{pmatrix} b_1 \\ b_2 \end{pmatrix} = e^{-j\frac{(\beta_1+\beta_2)z}{2}} \begin{pmatrix} \sqrt{1-K}e^{j\varphi} & j\sqrt{K} \\ j\sqrt{K} & \sqrt{1-K}e^{-j\varphi} \end{pmatrix} \begin{pmatrix} a_3 \\ a_4 \end{pmatrix}$$
$$\varphi = \arctan\left(\frac{\Delta\beta}{2\beta_o}\tan\beta_o z\right) \tag{2.80}$$

and merging Eqn. (2.79) and Eqn. (2.80) we have

$$\begin{pmatrix} b_1 \\ b_2 \\ b_3 \\ b_4 \end{pmatrix} = e^{-j\frac{(\beta_1+\beta_2)z}{2}} \begin{pmatrix} 0 & 0 & \sqrt{1-K}e^{j\varphi} & j\sqrt{K} \\ 0 & 0 & j\sqrt{K} & \sqrt{1-K}e^{-j\varphi} \\ \sqrt{1-K}e^{j\varphi} & j\sqrt{K} & 0 & 0 \\ j\sqrt{K} & \sqrt{1-K}e^{-j\varphi} & 0 & 0 \end{pmatrix} \begin{pmatrix} a_1 \\ a_2 \\ a_3 \\ a_4 \end{pmatrix}$$
$$\varphi = \arctan\left(\frac{\Delta\beta}{2\beta_o}\tan\beta_o z\right) \tag{2.81}$$

2.5.1.3 *Symmetric directional coupler (DC)*

If the two waveguides that implement the DC are identical, which is encountered frequently in practice, the device operates in symmetric mode. Then

$$\begin{aligned} K &= \sin^2(\kappa z) = \sin^2\theta(z) \\ 1-K &= \cos^2(\kappa z) = \cos^2\theta(z) \end{aligned} \tag{2.82}$$

and the scattering matrix of the coupler becomes

$$\begin{pmatrix} b_1 \\ b_2 \\ b_3 \\ b_4 \end{pmatrix} = e^{-j\phi(z)} \begin{pmatrix} 0 & 0 & \cos\theta(z) & j\sin\theta(z) \\ 0 & 0 & j\sin\theta(z) & \cos\theta(z) \\ \cos\theta(z) & j\sin\theta(z) & 0 & 0 \\ j\sin\theta(z) & \cos\theta(z) & 0 & 0 \end{pmatrix} \begin{pmatrix} a_1 \\ a_2 \\ a_3 \\ a_4 \end{pmatrix} \tag{2.83}$$
$$\phi(z) = \beta_1 z = \beta_2 z$$

It is understood that the 4×4 scattering matrix is composed of two uncoupled matrices, one for each direction of propagation. Thus, it is customary, once a given direction is selected, to employ the reduced 2×2 version, rather than the full 4×4 formalism.

For instance, and rewriting Eqn. (2.79), we have

$$\begin{pmatrix} b_3 \\ b_4 \end{pmatrix} = e^{-j\phi} \begin{pmatrix} \cos\theta(z) & j\sin\theta(z) \\ j\sin\theta(z) & \cos\theta(z) \end{pmatrix} \begin{pmatrix} a_1 \\ a_2 \end{pmatrix} \tag{2.84}$$

Complete power transfer between waveguides requires

$$|\sin\theta(z)| = 1 \Rightarrow \theta(z) = (2k+1)\frac{\pi}{2} \Rightarrow z_k (2k+1)\frac{\pi}{2\kappa} \quad k = \{0, 1, 2..\} \tag{2.85}$$

A coupler with a length given by Eqn. (2.85) is said to be in the *cross* state. Alternatively, no power transfer between waveguides is obtained if

$$|\sin\theta(z)| = 0 \Rightarrow \theta(z) = k\pi \Rightarrow z_k = \frac{k\pi}{\kappa}; \; k = \{1, 2, 3...\}. \tag{2.86}$$

Here, the coupler is said to be in the *bar* state. For length values other than these, the device operates as a *variable coupler* with coupling constant values satisfying $0 < K < 1$. Note as well that the DC impresses a common phase shift value.

For practical reasons, we need to use the smallest possible values for the length; hence, in Eqn. (2.85) and Eqn. (2.86) we select the smallest values for k. Doing this yields $l_c \leq z \leq 2l_c$ to implement a full-range coupler, where $l_c = \pi/2\kappa$ is known as the coupling length. The physical size will depend on the value of the coupling coefficients, which are given in general by [1]–[2]

$$\begin{aligned} \kappa_{21} &= -\frac{j\omega}{4} \int_{CS_2} \left[n^2(x,y) - n_2^2\right] E_{TE,1}(x,y)\, E_{TE,2}^*(x,y)\, dxdy \\ \kappa_{12} &= -\frac{j\omega}{4} \int_{CS_1} \left[n^2(x,y) - n_1^2\right] E_{TE,2}(x,y)\, E_{TE,2}^*(x,y)\, dxdy \end{aligned} \tag{2.87}$$

where ω represents the light wave frequency, $n(x,y)$ the perturbed refractive index in one waveguide due to the close proximity of a second waveguide and n_1, n_2 the core refractive indices of waveguide 1 and 2, respectively. $E_{TE,1}(x,y)$ and $E_{TE,2}(x,y)$ are the transverse field profiles of TE modes in waveguide 1 and 2, respectively.

2.5.2 Multimode Interference (MMI) Couplers

An MMI is implemented by means of step-index multimode waveguide (Figure 2.7(a)) [23]–[25], and the structure has a length L, a width W and a refractive index n, with no input/output waveguide ports shown (we address the incorporation of these ports further on). In principle, the 3D structure can be reduced to a 2D slab waveguide by means of the effective index method, so the transversal field dependencies on y of the guided modes can be discarded.

2.5.2.1 *Slab multimode waveguide zone operation*

Let us consider an arbitrary input light field distribution given by $f(x)$, which is decomposed into the eigenmodes of the coupler. In principle, for perfect decomposition, this requires an infinite number of guided modes. However, in practice, a strong confining waveguide is usually assumed and only a limited number of low-order guided modes is required to achieve the decomposition with a negligible degree of error. Figure 2.7(b) shows the optical field distributions corresponding to the lower order modes of the

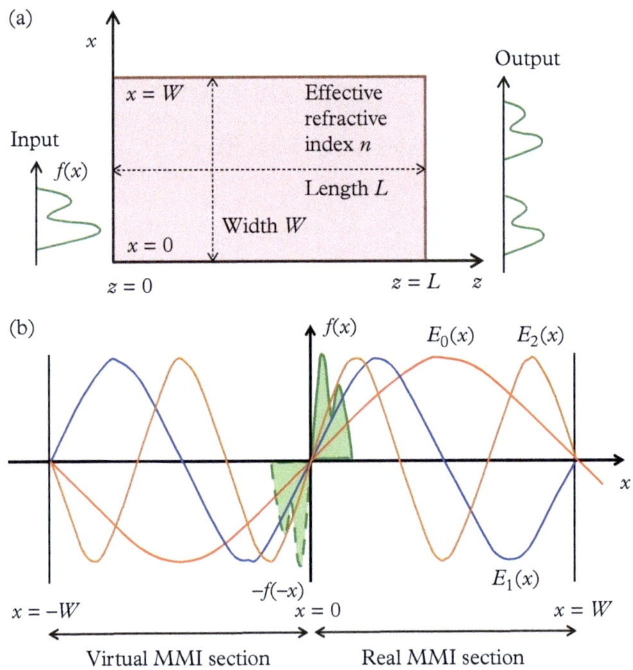

Figure 2.7 *(a) General layout of a multimode interference coupler implemented by means of a 2D slab waveguide (input/output ports are not shown). (b) Real and virtual MMI sections including the input field f(x) distribution and the decomposition into lower order guided modes* $E_i(x)$. *f(x) and* $E_i(x)$ *in the real MMI section are antisymmetrically extended into the virtual section and the resulting functions are then periodic (period 2W) in the x axis.*

multimode waveguide. For mathematical convenience, it also shows a virtual section of equal width W, including the extended asymmetrical eigenmodes and input field distribution.

The overall functions containing the mode profiles of the real and the virtual MMI sections are the periodically extended (with a period of $2W$) to the whole x-axis to allow Fourier decomposition. In this way, the guided eigenmodes of the MMI section can be expressed as

$$E_i(x) = \sin\left[\frac{(i+1)\pi x}{W}\right];\ i = 0, 1, 2, . \tag{2.88}$$

Since strong confinement is assumed, the propagation constants of the guided modes β_i can be calculated since, according to Eqn. (2.88), their transversal propagation constants are given by

$$k_{x,i} = \frac{(i+1)\pi}{W};\ i = 0, 1, 2, \tag{2.89}$$

Furthermore, since $\beta_i^2 = n^2 k_o^2 - k_{x,i}^2$ with $k_o = 2\pi/\lambda$, then applying the paraxial approximation

$$\beta_i = nk_o\sqrt{1 - \left(k_{x,i}/nk_o\right)^2} \approx nk_o - \frac{k_{x,i}^2}{2nk_o} \tag{2.90}$$

and taking into account Eqn. (2.89), we get

$$\begin{aligned} \beta_o &= nk_o - \tfrac{\pi^2}{2nk_o W} \\ \beta_i - \beta_0 &= -\tfrac{i(i+2)\pi^2}{2nk_o W} \end{aligned} \tag{2.91}$$

Defining the beat length as [23]

$$L_c = \frac{\pi}{\beta_0 - \beta_1} = \frac{2W^2 nk_o}{3\pi} \tag{2.92}$$

we finally get

$$\beta_i = \beta_0 - \frac{i(i+2)\pi}{3L_c} \tag{2.93}$$

We now extend the input light field distribution, defined in the real MMI section ($0 < x < W$ in Figure 2.7(b)) to the virtual section ($-W < x < 0$) by forming the antisymmetric function $u(x) = f(x) - f(-x)$ (NB: $u(-x) = f(-x) - f(x) = -u(x)$). The next step is to extend $u(x)$ over the whole x-axis with periodicity $2W$:

$$f_{in}(x) = \sum_{k=-\infty}^{\infty} u(x - k2W) \tag{2.94}$$

Since $f_{in}(x)$ is periodic with period $2W$ it can be expressed in terms of a Fourier series of eigenmodes given by Eqn. (2.88):

$$\begin{aligned} f_{in}(x) &= \sum_{i=0}^{\infty} c_i E_i(x) = \sum_{i=0}^{\infty} c_i \sin\left[\tfrac{(i+1)\pi x}{W}\right] \\ c_i &= -\tfrac{1}{W}\int_{-W}^{0} f(x)\sin\left[\tfrac{(i+1)\pi x}{W}\right] + \tfrac{1}{W}\int_{0}^{W} f(x)\sin\left[\tfrac{(i+1)\pi x}{W}\right] = \\ &= \tfrac{2}{W}\int_{0}^{W} f(x)\sin\left[\tfrac{(i+1)\pi x}{W}\right] dx \end{aligned} \tag{2.95}$$

The output signal from an MMI region of length z can be expressed as

$$f_{out}(x) = \sum_{i=0}^{\infty} c_i E_i(x) \exp\left[-j\beta_o z + j\frac{i(i+2)\pi}{3L_c} z\right] \tag{2.96}$$

2.5.2.2 *Image formation*

When the MMI region length is $z = L_N^M = 3(M/N)L_c$, where M and N are co-prime integers the device produces N-fold images of the extended input field $f_{in}(x)$ at its output. In this case, Eqn. (2.96) is

$$f_{out}(x) = \sum_{i=0}^{\infty} c_i E_i(x) \exp\left[-j\beta_o L_M^N + j\pi\frac{M}{N} i(i+2)\right] \tag{2.97}$$

and it can be shown [23] that

$$\sum_{i=0}^{\infty} c_i E_i(x) \exp\left[-j\beta_o L_M^N + j\pi\frac{M}{N} i(i+2)\right] = \frac{1}{C}\sum_{q=0}^{N-1} f_{in}(x - x_q)\exp\left[j\vartheta_q\right] \tag{2.98}$$

Hence, the output field is the sum of N images of the input field with equal amplitudes numbered by $q = 0, 1, 2, \ldots N-1$. The positions, phases and common amplitude of these images are given by

$$\begin{aligned} x_q &= (2q - N)\tfrac{M}{N} W \\ \vartheta_q &= q(N-q)\tfrac{M}{N}\pi \\ C &= \exp\left[j\beta_0 L_N^M + j\pi\tfrac{M}{N}\right]\sum_{q=0}^{N-1} \exp\left[j\pi\tfrac{M}{N} q(N-q)\right] \end{aligned} \tag{2.99}$$

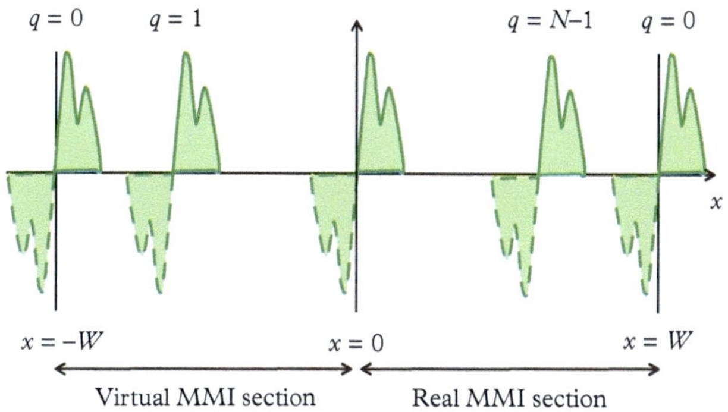

Figure 2.8 *Action of a MMI device of length $L_N^{M=1}$. The real MMI section contains N images of the real input function f(x) at the output. Some images are inverted and some are upright.*

Note that $|C| = 1/\sqrt{N}$. Figure 2.8 illustrates the action described by Eqns. (2.97)–(2.99) for the case of $M = 1$, which, for reasons that are explained further on, is the most interesting case in practice. The images of the extended input function are labelled with their corresponding value of q.

We see that the MMI device of length $L_N^{M=1}$ produces N-fold images of the extended input function $f_{in}(x)$ at its outputs. These correspond to $2N$ images of the real input function $f(x)$ in the range $-W < x < W$. Half of these images are inverted, while the other half are not. Hence, the real MMI section ($0 < x < W$) includes N images of the real input function $f(x)$, again inverted or upright.

2.5.2.3 Practical MMI couplers

When considering the design of an MMI coupler we must take into account that waveguides will be required to inject/collect the input/output signals at the input/output planes of the device. The description for the MMI region provided in Section 2.5.2.2 is based on the case of arbitrary input/output light distributions and, therefore, it does not take into account that input/output waveguides restrict the excited and collected light distributions to limited ranges in the input/output planes. For a proper design, first, the images of the input field must not overlap at the output plane and, second, due to footprint restrictions, the MMI region must be as short as possible; thus, $M = 1$ is the preferred value.

To design the MMI coupler we place an access waveguide with a spatially restricted input light distribution at an arbitrary position in the input plane. Equation (2.99) is then employed to obtain the locations of the N-fold outputs. Then the positions of the other input waveguides rendering the same output position are obtained simply by using the MMI coupler in reverse (see Figure 2.9(a) and 2.9(b) for N even and N odd respectively, noting that the port numbering is reversed in the output plane).

For a given value of W and a fixed number N we have an additional design parameter a, limited by $0 < a < W/N$, for the input and output waveguide positions. If we label the

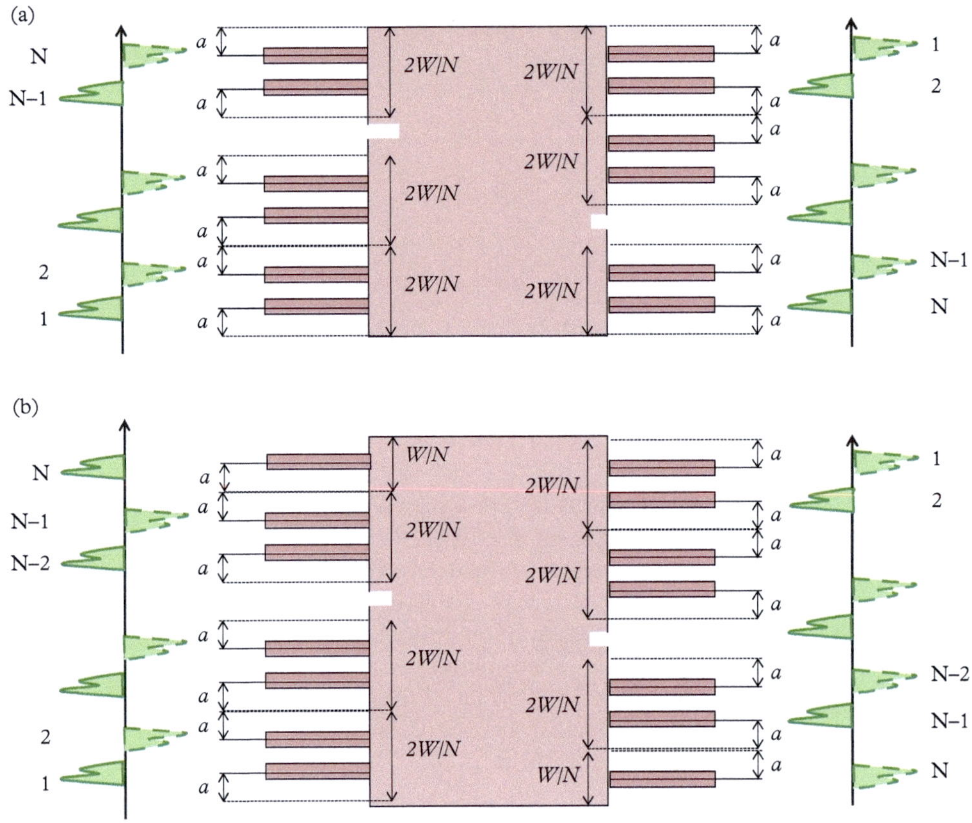

Figure 2.9 *NxN MMI coupler with access guides (a) N even, (b) N odd (after [23]).*

input and output ports using the indices r and s, respectively, then the phases for imaging input port r into output port s are given by [23], [24]

$$\begin{aligned} \varphi_{rs} &= \varphi_o + \pi + \tfrac{\pi}{4N}(s-r)(2N-s+r) \quad r+s \text{ even} \\ \varphi_{rs} &= \varphi_o + \tfrac{\pi}{4N}(s+r-1)(2N-s-r+1) \quad r+s \text{ odd} \\ \varphi_o &= -\beta_o L_N^{M=1} - \tfrac{\pi}{N} - \tfrac{\pi}{4}(N-1) \end{aligned} \tag{2.100}$$

One of the most common applications of the MMI device is in the design of 3-dB 2×2 couplers. In this case:

$$\begin{aligned} \varphi_{11} &= \varphi_{22} = \varphi_o + \pi \\ \varphi_{12} &= \varphi_{21} = \varphi_o + \tfrac{\pi}{2} \\ \varphi_o &= -\beta_o L_N^{M=1} - \tfrac{3\pi}{4} \end{aligned} \tag{2.101}$$

and

$$|C| = \sqrt{2} \tag{2.102}$$

which leads to the following scattering matrix (NB: the inversion in the output port numbering with respect to the input port numbering)

$$S = \frac{je^{j\varphi_o}}{\sqrt{2}} \begin{pmatrix} 0 & 0 & 1 & j \\ 0 & 0 & j & 1 \\ 1 & j & 0 & 0 \\ j & 1 & 0 & 0 \end{pmatrix} \tag{2.103}$$

Again, as with DCs, the 4 × 4 scattering matrix is composed of two uncoupled matrices, one for each direction of propagation. Hence, once a given direction is selected, we can employ the reduced 2 × 2 transfer matrix version, rather than the full 4 × 4 scattering matrix formalism.

$$S = \frac{je^{j\varphi_o}}{\sqrt{2}} \begin{pmatrix} 1 & j \\ j & 1 \end{pmatrix} \tag{2.104}$$

Instances of 2 × 2 MMI couplers with arbitrary splitting ratios have been reported using restricted interference (i.e. a selection of waveguide modes of the MMI zone are excited) [26] and butterfly configurations [27].

2.5.3 Tuneable Couplers (TCs)

A 2 × 2 tunable coupler where the independent setting of the power splitting ratio and a common phase shift can be established by means of applying an external control signal is an essential constituent of the TBUs upon which complex programmable integrated photonic circuits can be built. This section describes the two main approaches that are currently available once the operation of both directional and MMI couplers has been introduced.

2.5.3.1 The 3dB Mach–Zehnder interferometer (MZI) tunable coupler

A tunable coupler can be implemented by using two 3-dB couplers to form a balanced Mach–Zehnder interferometer (MZI) [25], where each arm incorporates a tunable phase shifter (Figure 2.10(a)). If we consider MMIs for the 3-dB couplers, the transmission matrix results in

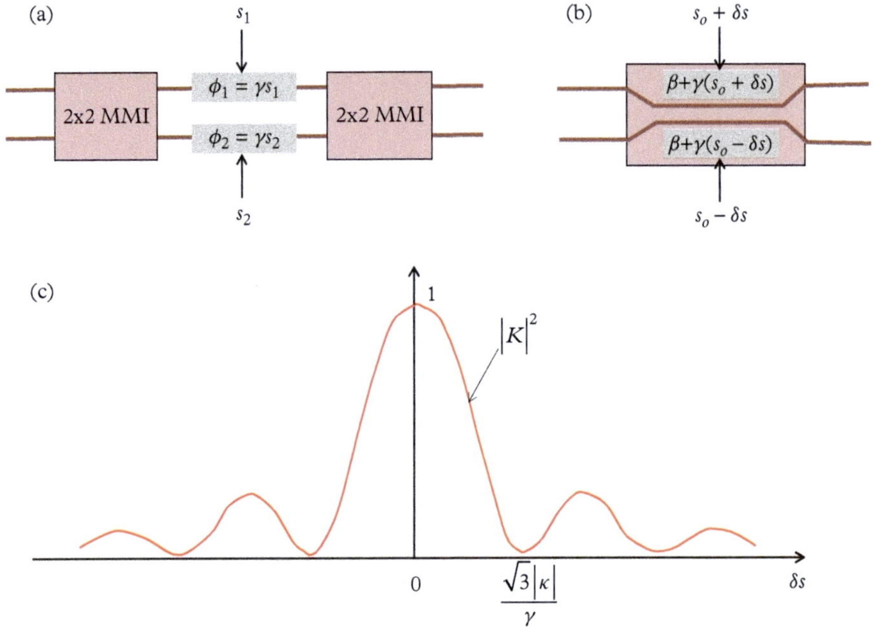

Figure 2.10 *Layouts for the implementation of tunable 2 X 2 couplers. (a) The 3dB MZI coupler with its control signals, (b) Tunable directional coupler (TDC) with its control signals with access guides. (c) Coupling constant tuning of a TDC versus the value of the differential control signal.*

$$
\begin{aligned}
T &= \frac{je^{j\varphi_o}}{\sqrt{2}}\frac{je^{j\varphi_o}}{\sqrt{2}}\begin{pmatrix}1 & j\\ j & 1\end{pmatrix}\begin{pmatrix}e^{-j\phi_1} & 0\\ 0 & e^{-j\phi_2}\end{pmatrix}\begin{pmatrix}1 & j\\ j & 1\end{pmatrix} = \\
&= \frac{e^{j(2\varphi_o+\pi)}}{2}\begin{pmatrix}1 & j\\ j & 1\end{pmatrix}\begin{pmatrix}e^{-j\phi_1} & je^{-j\phi_1}\\ je^{-j\phi_2} & e^{-j\phi_2}\end{pmatrix} = \\
&= \frac{e^{j(2\varphi_o+\pi)}}{2}\begin{pmatrix}e^{-j\phi_1} - e^{-j\phi_2} & je^{-j\phi_1} + je^{-j\phi_2}\\ je^{-j\phi_1} + je^{-j\phi_2} & -e^{-j\phi_1} + e^{-j\phi_2}\end{pmatrix} = \\
&= e^{j\left(2\varphi_o - \frac{\phi_1+\phi_2-3\pi}{2}\right)}\begin{pmatrix}\sin\left(\frac{\phi_2-\phi_1}{2}\right) & \cos\left(\frac{\phi_2-\phi_1}{2}\right) = \\ \cos\left(\frac{\phi_2-\phi_1}{2}\right) & -\sin\left(\frac{\phi_2-\phi_1}{2}\right)\end{pmatrix}
\end{aligned} \tag{2.105}
$$

If we assume that the transduction mechanism leading to the phase shifting in each arm can be described by $\phi_i = \gamma s_i$, $i = 1{,}2$, where γ is the transduction coefficient, then we get

$$
\begin{aligned}
\Delta &= 2\varphi_o - \frac{\phi_1+\phi_2}{2} = 2\varphi_o + \frac{\gamma(s_1+s_2)}{2}\\
\theta &= \frac{\phi_2-\phi_1}{2} = \frac{\gamma(s_2-s_1)}{2}
\end{aligned} \tag{2.106}
$$

which leads to

$$\mathbf{S} = -je^{j\Delta}\begin{pmatrix} \sin\theta & \cos\theta \\ \cos\theta & -\sin\theta \end{pmatrix} \tag{2.107}$$

Equation (2.107) represents a tunable coupler where the coupling constant $K = cos^2\theta$ and the overall phase shift Δ can be independently tuned using the two control signals s_1 and s_2, according to the system of two equations and two unknowns given by Eqn. (2.106).

2.5.3.2 The tunable directional coupler (TDC)

According to Eqns. (2.76)–(2.79), the coupling constant and overall phase shift of a DC can be independently tuned by means of two control signals that induce a propagation constant mismatch between the two waveguides in the coupler (Figure 2.10(b)). We assume that

$$\begin{aligned} \beta_1 &= \beta + \gamma\,(s_o + \delta s) \\ \beta_2 &= \beta + \gamma\,(s_o - \delta s) \end{aligned} \tag{2.108}$$

where s_o is a common tunable bias signal and δs is the differential signal applied to the transducers. We further assume that, under no applied control signals, the coupler length is designed so that the device is in a cross-state (i.e. $z = \pi/2\,|\kappa|$). Under these conditions the differential signal change needed to achieve the transition to the bar state is [3], [6]

$$\delta s = \frac{\sqrt{3}\,|\kappa|}{\gamma} \tag{2.109}$$

Hence, the complete transition ($0 < \mathrm{K} < 1$) is achieved for ($0 < \delta s < \sqrt{3}\,|\kappa|\,/\gamma$) (Figure 2.10(c)). Moreover, the overall phase shift is given by

$$\Delta = \left(\frac{\beta_1 + \beta_2}{2}\right) = \beta + \gamma s_o \tag{2.110}$$

However, note that a residual phase shift is experienced by signals that do not change the propagation waveguide. This phase shift is given by

$$\varphi = \tan^{-1}\left[\frac{1}{\sqrt{1+\left(\frac{|\kappa|}{\gamma\delta s}\right)^2}}\tan\left(\frac{\pi\gamma\delta s}{2\,|\kappa|}\sqrt{1+\left(\frac{|\kappa|}{\gamma\delta s}\right)^2}\right)\right] \tag{2.111}$$

which is not independent of the coupling constant.

2.5.3.3 Tunable basic unit (TBU)

As the next chapters show, many programmable photonic circuits are built using the tunable basic unit (TBU), an extension of a tunable coupler, as the BBB [16]–[22]. The upper part of Figure 2.11 shows the TBU composed by the tunable coupler and its access (input/output) waveguides. The geometry of the latter is a function of the bending radius and varies for each specific circuit topology due to the different allowed angles between waveguide elements.

The TBU length is an important parameter as defined by the basic unit length (BUL)

$$BUL = L_{access} + L_{Tunable\text{-}Coupler} \tag{2.112}$$

where L_{access} is the overall length of the access waveguide segment and $L_{Tunable-Coupler}$ is the length of the tunable coupler. Although different structures of tunable couplers can be considered in principle (balanced and unbalanced 2×2 DCs, 3×3 and 3-dB MZIs) [17], here mainly we consider TBUs composed of a 3-dB balanced MZI structure loaded with a heater on each arm (the lower part of Figure 2.11). Nevertheless, the treatment is not restricted to this case and can be employed for alternative TBU structures and tuning mechanisms.

Referring to the right part of Figure 2.11, the TBU can implement three different states: the *cross state* switch (light path connects in_1 to out_2 and in_2 to out_1); the *bar state* switch (light path connects in_1 to out_1 and in_2 to out_2); and the *tunable splitter*.

For a balanced MZI loaded with heaters on both arms, the splitting ratio is obtained by increasing the effective index due to the Joule effect in the upper or lower arm,

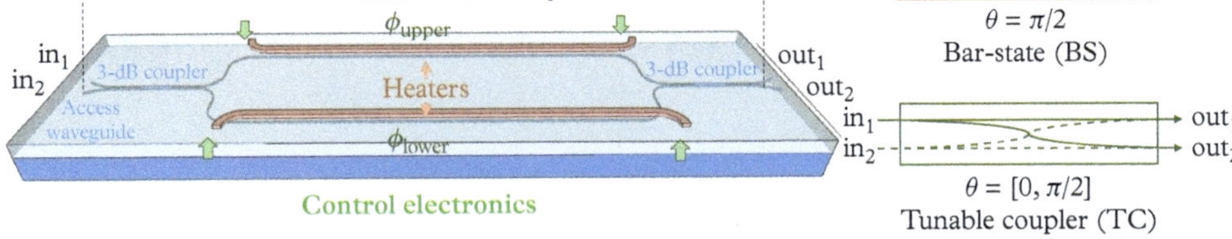

Figure 2.11 *(a) (Upper) Labelled schematic of a general tunable coupler acting as the basic building block of a programmable circuit. The Basic Unit Length (BUL) is illustrated as the sum of the tunable coupler length and the arc length of the access waveguides. (Lower) A particular case of a TBU built upon an integrated 3 dB balanced MZI-based tunable coupler. (b) Signal flow for the different TBU configuration states.*

which produces a ϕ_{upper} and ϕ_{lower} phase shift, respectively. Once set, a common drive in both heaters will provide a common phase shift that leads to independent control of the amplitude ratio and phase. The device matrix under ideal lossless conditions is defined by Eqn. (2.106). However, in practice a loss coefficient α is multiplied to the 2×2 matrix, and α includes the effect of propagation losses in the access waveguides, the tunable coupler waveguide and the insertion losses for both 3-dB couplers. Other TBU designs commonly encountered in practice include an additional phase shifter either before or after the 3-dB MZI tunable coupler (Figure 2.12). These are usually employed in multiport interferometers and provide an additional independent phase shift to either a given row or a given column of the overall scattering matrix.

Referring to the layouts displayed in Figure 2.12(a) and Figure 2.12(b), the final scattering matrices are given by:

$$\mathbf{S} = -je^{j\Delta}\begin{pmatrix} e^{j\varphi}\sin\theta & e^{j\varphi}\cos\theta \\ \cos\theta & -\sin\theta \end{pmatrix} \tag{2.113}$$

and

$$\mathbf{S} = -je^{j\Delta}\begin{pmatrix} e^{j\varphi}\sin\theta & \cos\theta \\ e^{j\varphi}\cos\theta & -\sin\theta \end{pmatrix} \tag{2.114}$$

respectively.

Whatever design is considered, the TBU performance can be characterised in terms of the optical power by its insertion losses (IL) and optical crosstalk parameters (CT_{Bar} and CT_{Cross}):

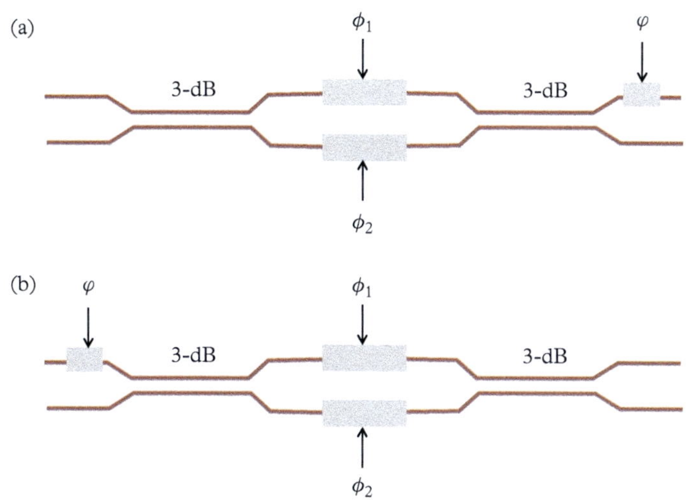

Figure 2.12 *Tunable basic units containing an extra tunable phase shifter in addition to the tunable 3dB MZI coupler (a) Phase shifter after the coupler, (b) Phase shifter before the coupler.*

$$IL(dB) = 20\log_{10}(\alpha) = 10\log_{10}\left(\frac{|a_{out1}|^2+|a_{out2}|^2}{|a_{in1}|^2+|a_{in2}|^2}\right),$$
$$CT_{Bar/Cross}(dB)\Big|_{\substack{|a_{in1}|^2=1 \\ |a_{in2}|^2=0}} = 10\log_{10}\left(\frac{|a_{out2,1}|^2}{|a_{out1,2}|^2}\right). \quad (2.115)$$

These parameters provide valuable information regarding the losses of the tunable units. Since a programmable photonic circuit usually requires the connection of a number of TBUs in cascade, the overall IL of a certain synthetised device will be the sum of the ILs corresponding to the units across which the light has travelled. As an example, if we assume overall device losses of 10 dB and IL of 0.2 dB for each TBU, then the longest path will be limited to fifty units. In the case of the optical crosstalk, the figure determines any signal leak that might cause optical reflections and undesired interferometric paths coupled to the desired photonic circuit to be programmed. Chapter 6 provides an extended analysis on the impact of non-ideal components.

2.6 Linear Optic Components

Section 2.1 showed the main matrix methods employed in characterising photonic components. These methods implicitly assume that signals at the output plane are a set of monochromatic modes that are the transmitted or reflected versions of the input modes. The methods then provide a means to characterise the transfer between the field amplitudes of these input and output modes.

This section takes into account situations where the input and output waves to the photonic device might be of different nature [28]–[30]; Figure 2.13 provides a few examples [29]. Figure 2.13(a) shows an input coupled to an integrated waveguide, where the light focused in the front surface of the device is coupled into the fundamental mode of an integrated waveguide. Figure 2.13(b) shows an optical filter where the input signal is a spatial mode of an optical fibre and the input function space is a set of monochromatic waves, and the device transmits or absorbs waves of different frequencies coupling them to the fundamental mode of an integrated waveguide.

Generally, these systems are characterised by having different input and output mode spaces (H_I and H_O, respectively, as they are usually Hilbert vector spaces) and a linear operator $\mathbf{D}$ relating both. The photonic device carrying the transformation is known as a linear optical component (LOC). Because the input and output functions generally can have arbitrary profiles and can depend on different parameters (space, time, wave vector, etc.), it is especially convenient to resort to the Dirac notation to proceed with the subsequent description of the LOC. The function of operator $\mathbf{D}$ is to take and input function $|\phi_I\rangle$ from H_I and generate a corresponding output function $|\phi_O\rangle$ from H_O:

$$|\phi_O\rangle = \mathbf{D}\,|\phi_I\rangle \quad (2.116)$$

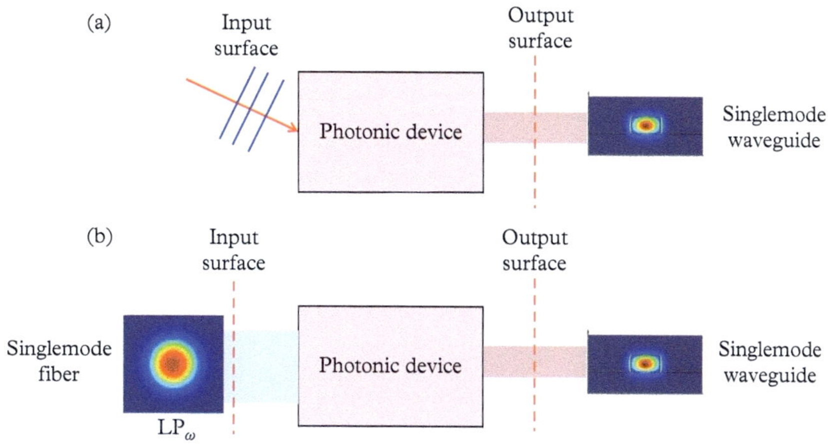

Figure 2.13 *Some examples of photonic devices where input and output modes are of different nature. (a) Input coupler to an integrated waveguide. The input mode space is composed of plane waves or spatial optical beams and the output mode space are the modes of an integrated waveguide. (b) Photonics filter, where the input mode space is composed by the different wavelength profiles of the fundamental mode of an optical fiber and the output mode space is composed of the different wavelength profiles of the TE mode of an integrated waveguide.*

D is similar to the scattering matrix **S** defined in Section 2.1.1 for circuits and waveguides, but it can be considered a generalisation of it, as no restrictions are imposed on the nature of the input and output functions. Furthermore, **D** does not need to be a square matrix since, in practice, we could use a large basis to describe the input wave (e.g. plane waves) and a restricted space to describe the output space (i.e. a limited number of modes of a waveguide). In addition, **D** does not need to be unitary as real photonic devices will be lossy; additionally, the device may scatter part of the input signal into modes that are not included in the output space (e.g. the radiating modes of a waveguide).

In most practical cases encountered in programmable photonic devices, the operator **D** describing its function is bounded, i.e. it renders a finite output for any finite inputs [31]–[40]. Furthermore, we assume that only a finite number of input and output modes are considered for such an operator. Under these conditions, **D** is said to be *compact* and it is always possible to express it under a *singular value decomposition* [29], [41], [42]. This very important result confirms that **D** is expressed as

$$\mathbf{D} = \sum_{k} s_{\mathbf{D},k} \mid \phi_{\mathbf{D}Ok} \rangle \langle \phi_{\mathbf{D}Ik} \mid \tag{2.117}$$

The former relationship is equivalent to write **D** as

$$\mathbf{D} = \mathbf{V}\mathbf{D}_{diag}\mathbf{U}^{\dagger} \tag{2.118}$$

where $\mathbf{U}$ and $\mathbf{V}$ are unitary operators that, in matrix form, have $|\phi_{\mathbf{D}Ik}\rangle$ and $|\phi_{\mathbf{D}Ok}\rangle$ as column vectors, respectively, and $\mathbf{D}_{diag}$ is a diagonal matrix with diagonal complex elements given by $s_{\mathbf{D},k}$, which are the *singular values*. The functions $|\phi_{\mathbf{D}Ik}\rangle$ and $|\phi_{\mathbf{D}Ok}\rangle$ each form orthogonal sets (i.e. modes) in their respective spaces H_I and H_O, and are given by the solutions of the following eigenvalue problems:

$$\begin{aligned} \mathbf{D}^{\dagger}\mathbf{D}\,|\phi_{\mathbf{D}Ik}\rangle &= |s_{\mathbf{D},k}|^2\,|\phi_{\mathbf{D}Ik}\rangle, \\ \mathbf{D}\mathbf{D}^{\dagger}\,|\phi_{\mathbf{D}Ok}\rangle &= |s_{\mathbf{D},k}|^2\,|\phi_{\mathbf{D}Ok}\rangle. \end{aligned} \tag{2.119}$$

Note that both $\mathbf{D}^{\dagger}\mathbf{D}$ and $\mathbf{D}\mathbf{D}^{\dagger}$ are Hermitian and therefore have real eigenvalues, which are the same $|s_{\mathbf{D},k}|^2$. Furthermore, the resulting sets of eigenfunctions $|\phi_{\mathbf{D}Ik}\rangle$ and $|\phi_{\mathbf{D}Ok}\rangle$ are *complete* in the sense that any output function that can be generated in the outer space H_O of the photonic device in response to some input function in the input space H_I that can be expressed as a linear combination of $|\phi_{\mathbf{D}Ok}\rangle$ having non-zero singular values. Additionally, any function that can be generated by the photonic device in the output space can be generated by some function in the input space that is a linear combination of $|\phi_{\mathbf{D}Ik}\rangle$ having non-zero singular values. This important result implies that, since any linear optical device can be written in terms of a linear operator $\mathbf{D}$ between its inputs and outputs and because, in practice, such an operator can always be decomposed using the SVD, then there is a set of orthogonal input modes $|\phi_{\mathbf{D}Ik}\rangle$ that will give rise, one by one, to a set of corresponding orthogonal output modes $|\phi_{\mathbf{D}Ok}\rangle$ with non-zero coupling coefficients $s_{\mathbf{D},k}$. In other words, all optical linear devices are mode converters. In Chapter 4 we use this result to describe the action of multiport interferometers and self-configuring universal LOCs.

2.7 Universal Beam Couplers

In conclusion, we explore the BBBs of programmable photonic circuits by studying the *universal beam coupler* [43]–[45]. This device is a simple one-to-one (input-to-output) mode transformer and is, as shown in Chapter 3, a key component of the general universal LOC, which is based on the principles described in Section 2.6.

2.7.1 Bulk Optics Implementation

Miller's proposed universal beam coupler [43] selects from multiple overlapping modes present at its input and couples each one to a separate port at its output without incurring splitting losses. Figure 2.14(a) depicts an implementation based on bulk BSs that we use to describe the basic device operation.

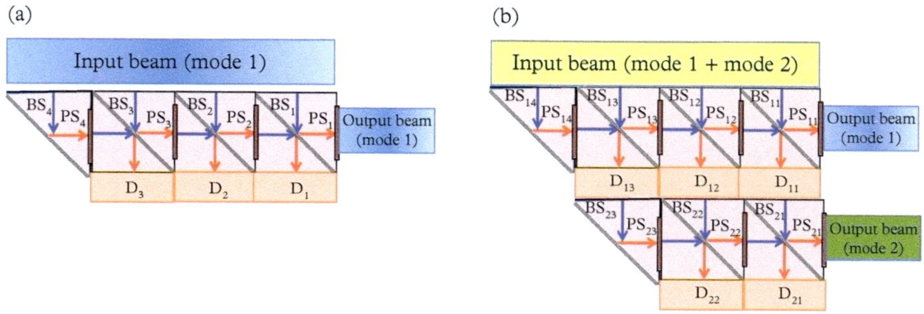

Figure 2.14 *Bulk optics layout of a one (a) and a two (b) mode universal beam coupler with clear indication of the position of each beam splitter BS, Phase shifter PS and Detector D (after [43]).*

We consider an optical beam with a spatially varying profile in the *longitudinal* direction (Figure 2.14(a)). This direction is divided into an arbitrary number of regions (in our case, four) where light incident on each region is then directed to a different BS block. With the exception of the leftmost unit, which contains a complete reflecting mirror and a phase shifter, each block contains a variable BS or mirror characterised by the equations given in Section 2.3, a phase shifter and a detector.

We now assume a monochromatic and single polarisation input beam impinging on the device. Initially, the state of the components in the BS units can be arbitrary, as long as mirror field reflectivities are not zero. The fields incident on the top of the BS blocks ($k = 1,2,3,4$) are given by

$$E_k = E_{ko}e^{j\phi_k} \tag{2.120}$$

where E_{ko} and ϕ_k are real numbers and the mirrors in each beamsplitting unit are characterised by the scattering matrix Eqn. (2.65), each one with a different value of r_k and t_k.

In a first step, the phase shifter PS_4 θ_4 is adjusted to minimise the power in detector D_3. This requires the transmitted and reflected signals at the bottom of BS_3 to be in anti-phase. If we assume that the phase shift from the input of the device to the input surface of BS_3 is a multiple of 2π, then the anti-phase condition is

$$\phi_4 + \theta_4 = \phi_3 + 2m\pi + \frac{\pi}{2} - \pi \tag{2.121}$$

Since m can take any value, we assume for simplicity that $m = 0$; then, the phase shifter PS_4 must be set to

$$\theta_4 = \phi_3 - \phi_4 - \frac{\pi}{2} \tag{2.122}$$

The second step is to tune the reflectivity of BS_3 so the incident power to D_3 is zero. To do so, the transmitted and reflected field magnitudes in BS_3 must be equal:

$$E_{4o} r_3 = E_{3o} t_3 \tag{2.123}$$

However, using Eqn. (2.65) yields

$$r_3 = \frac{E_{3o}}{\sqrt{E_{3o}^2 + E_{4o}^2}}; \quad t_3 = \frac{E_{4o}}{\sqrt{E_{3o}^2 + E_{4o}^2}} \tag{2.124}$$

Hence, considering again the scattering matrix for BS_3 given by Eqn. (2.65) we can write the field entering the phase shifter PS_3 as

$$\begin{aligned} E_{PS_3} &= r_3 E_{3o} e^{j\phi_3} + j t_3 E_{4o} e^{j(\phi_4+\theta_4)} = \\ &= \frac{E_{3o}^2 e^{j\phi_3}}{\sqrt{E_{3o}^2+E_{4o}^2}} + \frac{jE_{4o}^3 e^{j(\phi_4+\theta_4)}}{\sqrt{E_{3o}^2+E_{4o}^2}} = \sqrt{E_{3o}^2 + E_{4o}^2} e^{j\phi_3} \end{aligned} \tag{2.125}$$

where we have taken into account Eqn. (2.122). Note that Eqn. (2.125) has exactly the same form as the field impinging on PS_4, and hence, we can repeat this two-step process for BS_2, which obtains

$$E_{PS_2} = \sqrt{E_{2o}^2 + E_{3o}^2 + E_{4o}^2} e^{j\phi_2} \tag{2.126}$$

Finally, setting θ_1 and r_1 for PS_2 and BS_1, we get

$$E_{PS_1} = \sqrt{E_{1o}^2 + E_{2o}^2 + E_{3o}^2 + E_{4o}^2} e^{j\phi_1} \Rightarrow I_{out} = \left|E_{PS_1}\right|^2 = I_1 + I_3 + I_3 + I_4 \tag{2.127}$$

Therefore, all the input power in the incident mode beam has been coupled to the output.

The device functionality can be extended to operate as a separator for multiple orthogonal input beams or modes [43], [44]. To implement this functionality, we need to insert as many rows of BS units as the number of modes to be separated. Figure 2.14(b) depicts a scheme for a two-mode separator as proposed by Miller [44]. This requires using detectors that are mostly transparent in order to sample only a small amount of the power leaving the first row of BS units. The procedure to set up the device is as follows. We shine the first mode at the device input and we set the values of the elements in row one according to the procedure described so that all the power is directed to output 1 (output beam 1). Then we shine the second input mode (orthogonal to the first) on the device, which will completely transmit it to the second row of BSs, thus becoming its actual input (possibly with some extra longitudinal phase shifting). Miller [29] shows that once an input/output mode configuration is set, then it is not possible to couple power from any other input orthogonal mode beam into that output mode.

The second row of BS units is then similarly programmed to the transmitted beam in order to direct all its power to output 2 (output beam 2). This procedure can be generalised for an arbitrary number N of input orthogonal modes.

2.7.2 Integrated Optics Implementation

The scattering matrices of bulk mirrors Eqn. (2.65) and the tunable 3-dB balanced MZI Eqn. (2.108) are similar, but not identical, which suggests the possibility of implementing the universal beam combiner by means of integrated 3-dB MZIs; Figure 2.15(a) shows an equivalent layout. If the fields incident on the left side input waveguide ports ($k = 1,2,3,4$) are given by Eqn. (2.120), and calling $r_i = \sin\theta_i$ in Eqn. (2.108), we now follow a procedure similar to that in Section 2.7.1, bearing in mind that each 3-dB MZI can provide an independent setting of its equivalent reflectivity, and the overall phase shift impressed to the propagated signal means that, in principle, no extra phase shifting unit should be required.

In a first step, the phase shift of MZI$_4$, Δ_4 is adjusted to minimise the power in detector D$_3$. This requires the transmitted and reflected signals at D$_3$ to be in anti-phase. If we assume that the phase shift from the input port 3 of the device to MZI$_3$ is a multiple of 2π, then the anti-phase condition is:

$$\phi_4 + \Delta_4 - \pi + \Delta_3 = \phi_3 + \Delta_3 + 2(m-1)\pi \Rightarrow \Delta_4 = \phi_3 - \phi_4 + 2m\pi \tag{2.128}$$

which is different from that given by Eqn. (2.121) for the bulk BS configuration. For simplicity, we take the value $m = 0$. The second step is to tune the reflectivity of MZI$_3$ so the incident power to D$_3$ is zero. To do so, the transmitted and reflected field magnitudes in D$_3$ must be equal:

$$E_{4o} r_3 = E_{3o} t_3 \tag{2.129}$$

which yields

$$r_3 = \frac{E_{3o}}{\sqrt{E_{3o}^2 + E_{4o}^2}}; \qquad t_3 = \frac{E_{4o}}{\sqrt{E_{3o}^2 + E_{4o}^2}} \tag{2.130}$$

Hence, if we again consider the scattering matrix given by Eqn. (2.108), we can write the field exciting MZI$_3$ as:

$$\begin{aligned} E_{MZI_3} &= \sqrt{E_{3o}^2 + E_{4o}^2}\, e^{j\phi_3'} \\ \phi_3' &= \phi_3 + \Delta_3 - \tfrac{\pi}{2} \end{aligned} \tag{2.131}$$

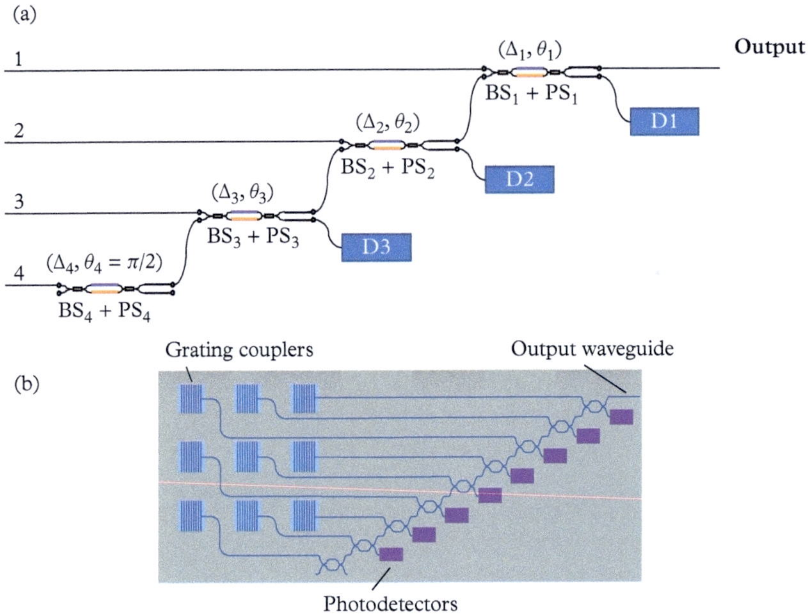

Figure 2.15 *(a) Integrated optics layout of a universal beam coupler equivalent to that of Fig 2.13 (a). (b) Combination of the former device with 2D disposition of grating couplers for coupling continuous spatially varying modes into chip waveguides (after [43]).*

Note now that Eqn. (2.131) has exactly the same form as the field exciting MZI_4; hence, we can repeat this two-step process for MZI_2 obtaining:

$$E_{MZI_2} = \sqrt{E_{2o}^2 + E_{3o}^2 + E_{4o}^2} e^{j\phi'_2} \\ \phi'_2 = \phi_2 + \Delta_2 - \frac{\pi}{2} \tag{2.132}$$

Finally, setting Δ_1 and r_1 for MZI_1, we get:

$$E_{MZI_1} = \sqrt{E_{1o}^2 + E_{2o}^2 + E_{3o}^2 + E_{4o}^2} e^{j\phi'_1} \\ \phi'_1 = \phi_1 + \Delta_1 - \frac{\pi}{2} \tag{2.133}$$

and therefore, as in the bulk optics case, all the input power in the incident mode beam has been coupled to the output.

The integrated configuration can be expanded to implement a multiple mode separator, but we will postpone this extension until Chapter 4, when we discuss its use in the implementation of multiport interferometers.

Miller [43] provides a thorough discussion on the implementation of integrated universal self-couplers in silicon technology and the interested reader is referred to that publication for more detailed information. Practical limitations behind the implementation of this device due to insertion and propagation losses, thermal crosstalk, fluctuations from in the targeted values of the bias signals, etc., are similar to those encountered in more complex programmable circuits; these are thoroughly discussed in Chapter 6. The main limitation of this component resides in the need to insert photonic detectors at selected parts of the circuit to continuously monitor the signal levels in order to verify the conditions for destructive interference. To avoid excessive tapping losses, semi-transparent photodetectors or capacitors are required.

The integrated beam coupler architecture can be combined with efficient types of grating couplers to provide 2D spatially continuous beam coupling into different waveguides (Figure 2.15(b)).

2.8 Chapter Summary

This chapter has covered the main techniques required to analyse and characterise the BBBs employed in programmable photonic integrated circuits. It introduced scattering and transfer matrices to describe devices where input and output signals are conveyed by functions and modes of similar nature and at a later stage presented the operator theory to describe devices where the input and output modes might be of a different nature. It introduced the fundamental properties of the BBBs used in PIP and derived their matrix description, which is needed when considering systems and devices of greater complexity that are implemented by assembling and/or cascading multiple BBB elements. With these tools we are now ready to design and understand the operation of most programmable photonic devices. Chapter 3 considers those with a fixed topology.

REFERENCES

[1] H. Haus, *Waves and Fields in Optoelectronics*, Prentice-Hall, 1984.

[2] A. Yariv, *Photonics: Optical Electronics in Modern Comm*unications, 6th edn, Oxford University Press, 2007.

[3] B.E.A. Saleh and M.C. Teich, *Fundamentals of Photonics*, 2nd edn, John Wiley & Sons, 2011.

[4] K. Iizuka, *Elements of Photonics*, John Wiley & Sons, 2002.

[5] K. Iizuka, *Engineering Optics*, 3rd edn, Springer-Verlag, 2009.

[6] C. Pollock and M. Lipson, *Integrated Photonics*, Kluwer Academic Publishers, 2009.

[7] L. Chrostowski and M. Hochberg, *Silicon Photonics Design: From Devices to Systems*, Cambridge University Press, 2015.

[8] K. Bergman, L.P. Carloni, A. Biberman, J. Chan and G. Hendry, *Photonic Network-on-Chip Design (Integrated Circuits and Systems)*, Springer, 2014.

[9] G. Reed (Ed), *Silicon Photonics: The State of the Art*, John Wiley & Sons, 2008.

[10] A. Peruzzo, et al., Multimode Quantum Interference of Photons in Multiport Integrated Devices. (2011) *Nat Commun* 2, 224.
[11] B.J. Metcalf, et al., Multiphoton Quantum Interference in a Multiport Integrated Photonic Device. (2013) *Nat Commun* 4, 1356.
[12] J. Carolan, et al., Universal Linear Optics. (2015) *Science* 349, 711.
[13] N.C. Harris, et al., Quantum Transport Simulations in a Programmable Nanophotonic Processor. (2017) *Nat Photonics* 11, 447–52.
[14] O. Graydon, Birth of the Programmable Optical Chip. (2016) *Nat Photonics* 10, 1.
[15] L. Zhuang, et al., Programmable Photonic Signal Processor Chip for Radiofrequency Applications. (2015) *Optica* 2, 854–9.
[16] D. Pérez, I. Gasulla, J. Capmany and R.A. Soref, Reconfigurable Lattice Mesh Designs for Programmable Photonic Processors. (2016) *Opt Express* 24, 12093–106.
[17] J. Capmany, I. Gasulla and D. Pérez, Microwave Photonics: The Programmable Processor. (2016) *Nat Photonics* 10, 6–8.
[18] A. Ribeiro, A. Ruocco, L. Vanacker and W. Bogaerts, Demonstration of a 4×4 -Port Universal Linear Circuit. (2016) *Optica* 3, 1348–57.
[19] A. Annoni, et al., Unscrambling Light—Automatically Undoing Strong Mixing between Modes. (2017) *Light Science and Applications* 6, e17110.
[20] L.N. Chen, E. Hall, L. Theogarajan and J. Bowers, Photonic Switching for Data Center Applications. (2011) *IEEE Photon J* 3, 834–44.
[21] D.A.B. Miller, Silicon Photonics: Meshing Optics with Applications. (2017) *Nat Photonics* 11, 403–4.
[22] M. Barchmann, P.A. Besse and H. Melchior, General Self-Imaging Properties in N × N Multimode Interference Couplers Including Phase Relations. (1994) *Appl Opt* 33, 3905–11.
[23] L.H. Soldano and E.C.M. Pennings, Optical Multimode Interference Devices Based on Self-Imaging: Principles and Applications. (1995) *IEEE J Lightwave Technol* 13, 615–27.
[24] C. Madsen and H.J. Zhao, *Optical Filter Design and Analysis: A Signal Processing Approach*, John Wiley & Sons, 1999.
[25] P. Besse, E. Gini, M. Bachmann and H. Melchior, New 1×2 Multi-Mode Interference Couplers with Free Selection of Power Splitting Ratios, *Proc European Conf Opt Commun (ECOC)*. Florence, Italy, 1994.
[26] P. Besse, E. Gini, M. Bachmann and H. Melchior, New 2×2 and 1×3 Multi-Mode Interference Couplers with Free Selection of Power Splitting Ratios. (1996) *J Lightwave Technol* 14 2286–92.
[27] A.E. Willner, J. Wang and H. Huang, Applied Physics. A Different Angle on Light Communications. (2012) *Science* 337(6095), 655–6.
[28] D.A.B. Miller, All Linear Optical Devices are Mode Converters. (2012) *Opt Express* 20, 23985–93.
[29] D.A.B. Miller, How Complicated Must an Optical Component Be?. (2013) *J Opt Soc Am A* 30(2), 238–51.
[30] E. Knill, R. Laflamme and G.J. Milburn, A Scheme for Efficient Quantum Computation with Linear Optics. (2001) *Nature* 409, 46–52.
[31] P. Kok, et al., Linear Optical Quantum Computing with Photonic Qubits. (2007) *Rev Mod Phys* 79, 135–74.
[32] M.G. Thompson, A. Politi, J.C. Matthews and J.L. O'Brien, Integrated Waveguide Circuits for Optical Quantum Computing. (2011) *IET Circuits Devices Syst* 5, 94–102.
[33] A. Politi, et al., Silica-on-Silicon Waveguide Quantum Circuits. (2008) *Science* 320, 646–9.

[34] A. Agarwal, et al., Fully Programmable Ring-Resonator-Based Integrated Photonic Circuit for Phase Coherent Applications. (2006) *IEEE J Lightwave Tech* 24, 77–87.

[35] P. Seth and S. Roy, Ultrafast All-Optical Flip-Flops, Simultaneous Comparator-Decoder and Reconfigurable Logic Unit with Silicon Microring Resonator Switches. (2014) *IEEE J Sel Top Quantum Electron* 20, 5900308.

[36] P. Sethi and S. Roy, All-Optical Ultrafast Switching in 2 x 2 Silicon Microring Resonators and its Application to Reconfigurable DEMUX/MUX and Reversible Logic Gates. (2014) *IEEE J Lightwave Tech* 32, 2173–80.

[37] K. Bergman, Silicon Photonics for High Performance Interconnection Networks. in Optical Fiber Communication Conference. (2018) *OSA Technical Digest* [online], Tu3F.1.

[38] R. Soref, Integrated-Photonic Switching Structures. (201) *Appl Phys Lett Photonics* 3, 021101.

[39] N. Harris et al., Large-Scale Quantum Photonic Circuits in Silicon. (2016) *Nanophotonics* 5, 456–68.

[40] D.A.B. Miller, Communicating with Waves between Volumes—Evaluating Orthogonal Spatial Channels and Limits on Coupling Strengths. (2000) *Appl Opt* 39, 1681–99.

[41] G.W. Hanson and A.B. Yakovlev, *Operator Theory for Electromagnetics* Springer-Verlag, 2002.

[42] D.A.B. Miller, Self-Aligning Universal Beam Coupler. (2013) *Opt Express* 21, 6360–70.

[43] D.A.B. Miller, Self-Configuring Universal Linear Optical Component. (2013) *Photon Res* 1, 1.

[44] D.A.B. Miller, Fundamental Limit for Optical Components. (2007) *J Opt Soc Am B* 24, A1–A17.

3

Reconfigurable Application-Specific Photonic Integrated Circuits

Reconfigurable photonic integrated circuits are the first evolutionary step in programmable photonics, and are application-specific circuits designed to carry a given functionality and feature some degree of flexibility enabled by the possibility of tuning some of its internal parameters by means of appropriate control signals. Another distinctive feature of these circuits resides in the fixed nature of their physical topology and design. The literature reports a vast amount of reconfigurable integrated circuits and it is neither practically possible nor within the scope of this book to cover the topic in detail. For example, reconfigurable circuits for routing and switching matrices, which we do not cover here, can on their own be the subject of an entire book. Rather, this chapter provides some salient and representative examples for other functionalities including filtering, RF phase shifting and time delay lines, beam formers, wave generators and other multifunctional application specific circuits to help the reader understand how these can be emulated, with the more generic programmable photonic circuits discussed in Chapter 4 and Chapter 5.

3.1 Filters

Reconfigurable optical filtering is perhaps one of the functionalities gaining momentum towards implementation in the field of integrated optics. Filtering is a ubiquitous functionality required in many area of photonic systems, e.g. channel selection, noise suppression, signal shaping, etc. The most versatile approach towards the implementation of photonic filters is based on discrete-time signal processing [1], [2], where a number of weighted and delayed samples of the optical signal to be processed are produced in the optical domain and combined prior to detection. In particular, finite impulse response (FIR) [2] filters combine at their output a fixed set of delayed and weighted replicas or taps of the input optical signal, while infinite impulse response (IIR) [2] filters are based on recirculating cavities to provide an infinite number of weighted and delayed replicas of the input optical signal. Taking as an example a FIR configuration, the electronic transfer function is given by

Programmable Integrated Photonics. José Capmany and Daniel Pérez, Oxford University Press (2020). © José Capmany and Daniel Pérez.
DOI: 10.1093/oso/9780198844402.001.0001

$$H(f) = \sum_{k=0}^{N-1} a_k e^{-j2\pi kfT} = \sum_{k=0}^{N-1} |a_k| \, e^{jk\varphi} e^{-j2\pi kfT} \qquad (3.1)$$

where $a_k = |a_k| \, e^{jk\varphi}$ represents the weight of the kth sample, and T the time delay between consecutive samples. Note that Eqn. 3.1 implies the filter is periodic in the frequency domain. Filter reconfigurability is obtained by changing the positions of filter zeros (and poles in the case of IIR filters) in the complex plane; this is achieved by changing the values of the a_k coefficients via suitable control signals. The period, known as the free spectral range (FSR), is given by $FSR = 1/T$. The usual implementation of discrete time filters can follow two approaches (Figure 3.1). In Figure 3.1(a), the delays between consecutive samples are obtained by means of a set of N optical waveguides where the length of the fibre/waveguide in the kth tap is $c(k-1)\,T/n_g$, with c and n_g being the light velocity in the vacuum and the waveguide group index, respectively. This simple scheme allows for spectral tuning if the value of n_g can be changed by means of an

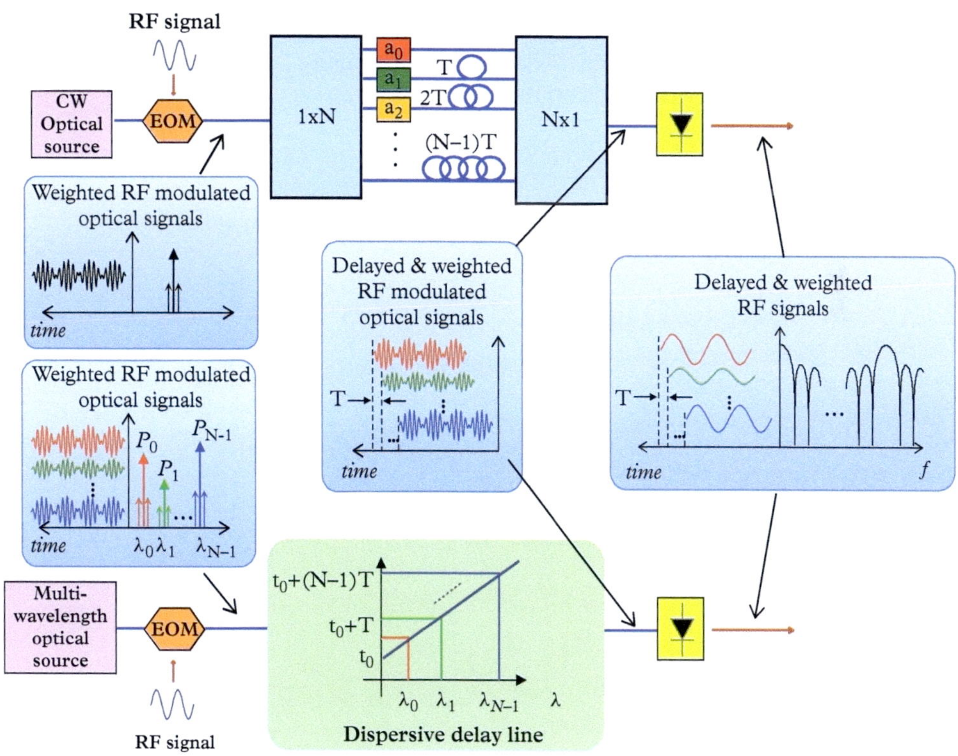

Figure 3.1 *General scheme of a discrete-time FIR photonic filter. (Upper) Traditional approach based on a single optical source in combination with multiple delay lines. (Lower) A more compact approach based on a multi-wavelength optical source combined with a single dispersive element.*

external signal. Note that the most straightforward mechanism involving the change of the T value is not possible once the physical length of the waveguides is set. Figure 3.1 (b) shows an alternative approach, which is based on the combination of a dispersive delay line; furthermore, different optical carriers show that the value of the basic delay T is changed by modifying the wavelength separation among the carriers, thereby allowing tunability [3], [4].

Again, this requires the use of an integrated delay line that can provide enough dispersion, which currently is not possible. While in the first case a_k can be changed by inserting loss/gain and phase shifter devices in the waveguides, the second approach allows us to adjust a_k by changing the optical power emitted by the optical sources [3]. Photonic integrated filters can operate in the incoherent regime, where sample coefficients in Eqn. 3.1 correspond to optical intensities and are thus positive, or in the coherent regime, where the taps in Eqn. 3.1 can be complex-valued. In the former, the basic delay is much greater than the coherence time of the optical source feeding the filter, while in the latter it is much smaller. In practice, the overall majority of proposed schemes work under the coherent regime as it allows the implementation of complex-valued filters.

Filter flexibility in terms of tunability, reconfigurability and selectivity is achieved by acting across the different parameters that characterise the samples in Eqn. 3.1 using the variety of techniques reported in the literature [1]–[4]. Figure 3.2 illustrates the effect of the relevant parameters in Eqn. 3.1 on the filter response.

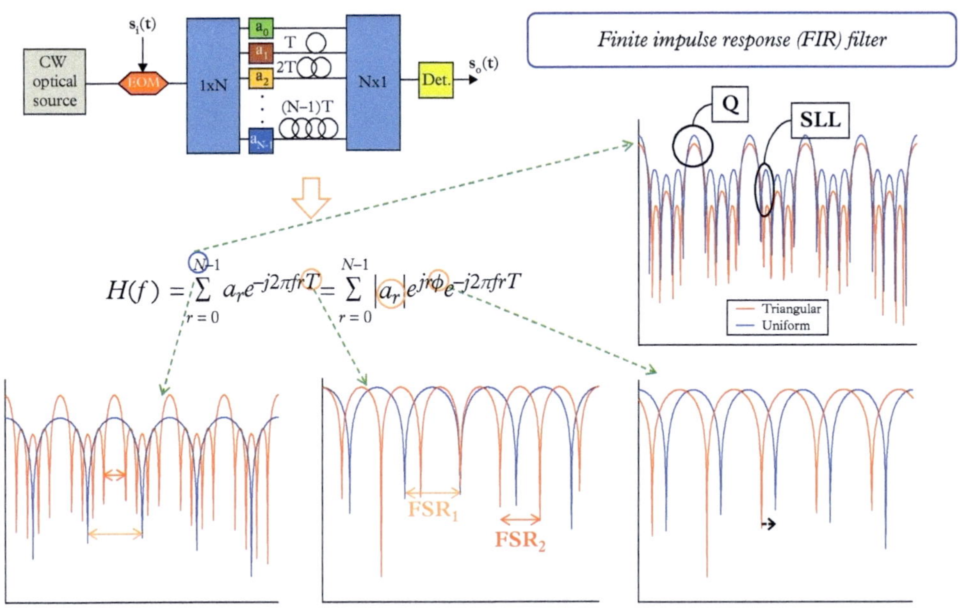

Figure 3.2 *Illustration of the requirements on sample parameters to achieve photonic filter tunability, reconfigurability and selectivity.*

The number of samples dictates whether the filter is either a notch (N = 2) or a bandpass(N > 2) type. As mentioned, T fixes the spectral period, thus changing the T results in compressing or stretching the spectral response. This is a common technique for tuning the notch or bandpass positions of a filter. Figure 3.1(b) shows how fast tuning can be achieved by changing the wavelength separation between adjacent carriers. The phase of the tap coefficients allows the tuning of the spectral response without actually stretching or compressing it. The implementation of phase values depends on the technical approach used and, in the case of coherent filters, it is achieved simply by employing integrated phase shifters.

The possibility of integrating photonic filters in a single chip is highly attractive since this leads to structures with reduced footprint, lower power consumption, higher stability and reliability. Another fundamental advantage is significant cost reduction through a mass production process. Several studies [6]–[26] have reported interesting results in the integration of coherent filters. To date, however, integration has been demonstrated only for selected parts of these subsystems.

Regarding integrated coherent filtering, many of the preliminary approaches reported so far have been based mainly on single cavity ring resonators (RRs). However, a few have also focused on more elaborated designs involving more than one cavity and programmable features. Representative results for single and multiple cavity filters or RRs have been reported [6]–[8], [11], [12]. Norberg and colleagues [9] report the results for a unit cell (Figure 3.3(b)) that can be employed as a basic building block of more complex filters (Figure 3.3(a)).

This unit cell, integrated in InP-InGaAsP is a ring loaded Mach–Zehnder interferometer (MZI). By selectively biasing the semiconductor optical amplifiers (SOAs) and phase modulators (PMs) placed in the arms of the unit cell, filters with a single pole, a single zero or a combination of both can be programmed (Figure 3.3(c)). To understand the operation of the basic unit cell we must use the scattering matrices for the couplers, waveguides and phase shifters derived in Chapter 2. The operation of each SOA is represented by its gain G_i. Also, fractional propagation losses in the waveguides represented by α_i are taken into account. When either $G_1 = 0$ or $G_2 = 0$ the unit cell behaves as a one pole IIR filter with transfer functions given by

$$
\begin{aligned}
|S_{21}|^2_{G_1=0} &= \left| \frac{G_{ff}\alpha_2 G_2}{2 - \alpha_2\alpha_3 G_2 G_3 e^{-j[\beta(L_2+L_3)+\varphi_2]}} \right|^2 \\
|S_{21}|^2_{G_2=0} &= \left| \frac{G_{ff}\alpha_1 G_1}{2 + \alpha_1\alpha_3 G_1 G_3 e^{-j[\beta(L_1+L_3)+\varphi_1]}} \right|^2
\end{aligned}
\tag{3.2}
$$

Where G_{ff} is the gain of the loss-compensating SOAs placed at the input and output of the unit cell, β represents the propagation constant of the fundamental mode in the integrated waveguides; the rest of parameters can be found in Figure 3.3(b). By defining $z^{-1} = e^{-j\beta(L_2+L_3)}$ and $z^{-1} = e^{-j\beta(L_1+L_3)}$ respectively in Eqn. (3.2), we find that the pole locations of these unit cells are given by

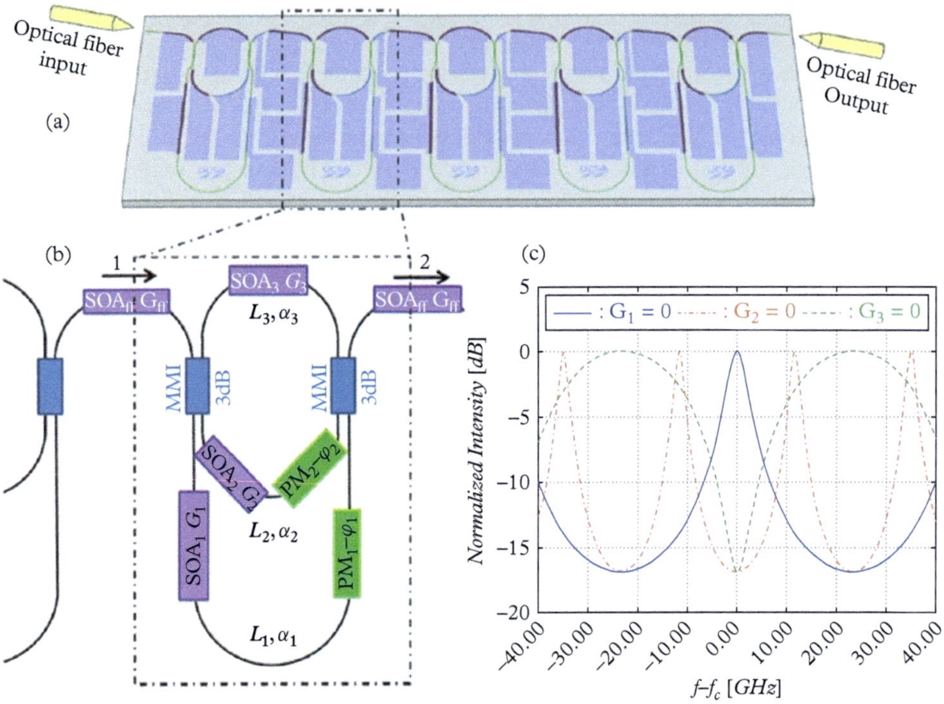

Figure 3.3 *Integrated InP-InGaAsP first-order MWP coherent filter providing one pole and one zero reported in [9]. (a) Cascade configuration of unit cells to implement high-order filters. (b) Unit cell configuration. (c) Examples of measured transfer functions for one pole and one zero configurations.*

$$
\begin{aligned}
z_p\big|_{G_1=0} &= \frac{\alpha_2\alpha_3 G_2 G_3 e^{-j\varphi_2}}{2} \\
z_p\big|_{G_2=0} &= -\frac{\alpha_1\alpha_3 G_1 G_3 e^{-j\varphi_3}}{2}
\end{aligned}
\tag{3.3}
$$

Note that pole magnitude is changed by tuning the operating SOA gains by means of their injection current, while the pole phase is controlled by the transduction mechanism employed to operate the internal optical phase shifters. These will be further discussed in Chapter 6.

When $G_3 = 0$ the unit cell operates as a two-stage (one zero) FIR filter

$$
|S_{21}|^2_{G_3=0} = \frac{G_{ff}^2}{4}\left|1 - \frac{\alpha_2}{\alpha_1} G_1 G_2 e^{-j[\beta(L_2-L_1)+\varphi_2-\varphi_1]}\right|^2 \tag{3.4}
$$

By defining $z^{-1} = e^{-j\beta(L_2-L_1)}$ in Eqn. (3.4), we find that the zero location of the unit cell is given by

$$z_c|_{G_3=0} = \frac{\alpha_2}{\alpha_1} G_1 G_2 e^{-j(\varphi_2-\varphi_1)} \tag{3.5}$$

Again, the magnitude of the zero can be changed by tuning the gain of the operating SOAs while the zero phase is controlled by the internal optical phase shifters. Intermediate SOA biasing will result in a mixed FIR + IIR behaviour, where the cell has one zero and one pole.

The cell concatenation scheme to form complex filters Figure 3.3(a) consists in the uncoupled cascade of a set of N unit cells, where each one can be programmed to independently act as a FIR, IIR or FIR + IIR cell. Hence, the overall transfer function is given by

$$\left|S_{21}^{TOTAL}\right|^2 = \prod_{k=1}^{N} \left|S_{21}^{k}\right|^2 \tag{3.6}$$

Synthesis algorithms developed by Madsen and Zhao [2] can be therefore applied to implement discrete-time optical filtering with arbitrary complexity.

Studies also report [11], [12] hybrid versions incorporating silicon waveguides that combine III–V quantum well layers bonded with low-loss passive silicon waveguides. Low-loss waveguides allow for long loop delays while III–V quantum devices provide active tuning capability. Norberg and colleagues [9] reported results of more complex designs involving two and up to three coupled ring cavities inside the Mach–Zehnder structure [10], which enable second- and third-order filters (Figure 3.4). In particular, and for this design, the frequency tuning range spans around 100 GHz.

Multistage, coupled cavity filters also known as coupled resonant optical waveguides (CROWS) have been reported in SOI technology. Dong and colleagues [13] and Feng and colleagues [14] have reported 1–2 GHz -bandwidth filters with very high extinction ratios (~50 dB). The silicon waveguides employed to construct these filters had

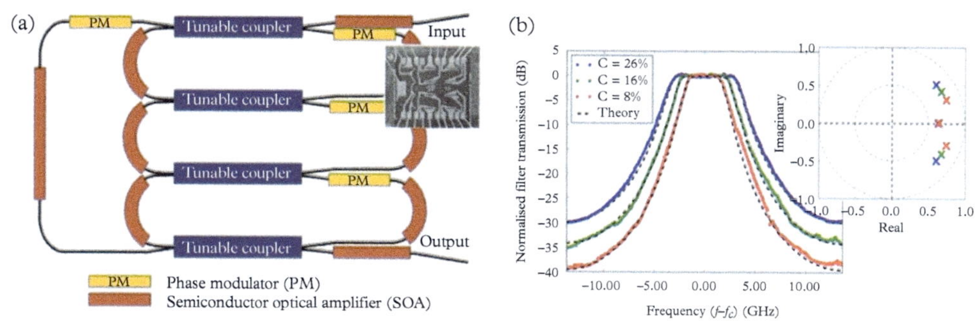

Figure 3.4 *(a) Layout of a multistage InP filter with tunable poles and zeros (after [10]). (b) detail of the resonance shapes for different reconfiguration parameter setups with their corresponding pole positions.*

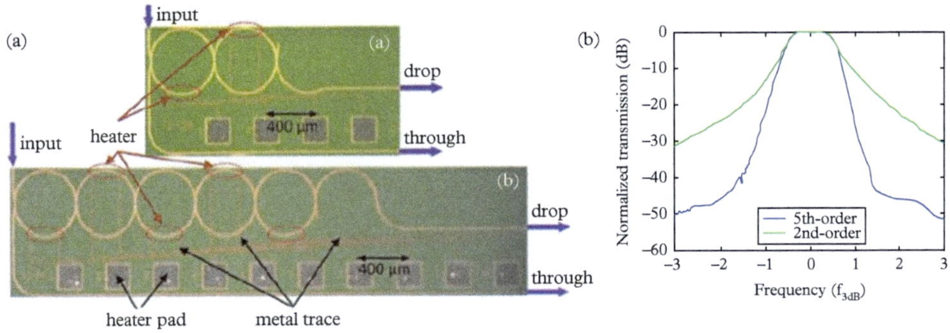

Figure 3.5 *(a) Optical images of fully fabricated 2nd-order (a) and 5th-order CROW filters (after [13]. (b) Measured (drop and through) transfer functions for the filters.*

propagation losses of ~0.5 dB/cm and insertion losses (excluding fibre-to-waveguide coupling) in the range of 2–3.5 dB. Each ring was thermally controlled by means of metal heaters placed on the top of the ring. With a power dissipation of ~72 mW, the ring resonance could be tuned by one free spectral range, resulting in wavelength-tunable optical filters Figure 3.5. Both the second-order and fifth-order RRs were demonstrated.

More recently, filters with a higher degree of sophistication obtained by the cascade of several CROW stages have been reported for channel selection in dense wavelength division multiplexed systems. Figure 3.6 summarizes the results from the study by Romero-García and colleagues [15]. The reported devices were based on three cascaded stages of CROWS loaded on a common bus waveguide. As shown in the layout of Figure 3.6(a) and the fabricated device in Figure 3.6(b), the stages differed in the number of resonators but were implemented with identical unit cells comprised of a matched racetrack resonator layout and a uniform spacing between cells. The different number of resonators in each stage allowed a high rejection in the through port response enabled by the interleaved distribution of zeros. The exact replication of a unique cell avoided the passband ripple and high lobes in the stopband that typically arise in apodized coupled resonator optical waveguide-based filters due to fabrication and coupling induced variations in the effective path length of each resonator. The filters could be reconfigured by means of thermal phase shifters incorporated in each CROW stage (Figure 3.6(c)). The fine tuning enabled the selection of wavelength-division multiplexing (WDM) channels with an increased rejection ratio (Figure 3.7 (d)).

Rasras and colleagues [16,17] demonstrated bandpass and notch filters based on MZI tunable couplers and RRs fabricated in silicon-buried channel waveguides with a width of 2 μm and a propagation loss of 0.25 dB/cm.

Ibrahim and colleagues [18] demonstrated a fully reconfigurable fourth-order optical lattice filter built by cascading identical unit cells consisting of an MZI loaded with an RR (Figure 3.7). The filter was fabricated using a commercial silicon complementary metal oxide semiconductor (CMOS) process and reconfigured by current injection into PIN diodes. The authors demonstrated full control over the single unit cell pole and

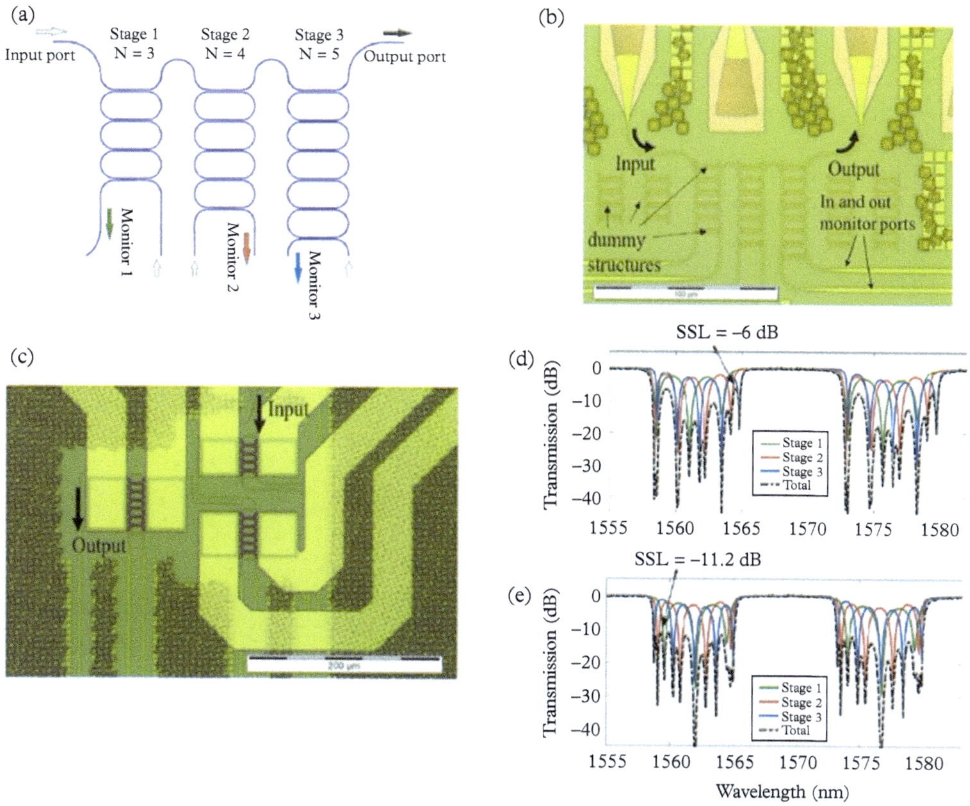

Figure 3.6 *Multiple stage CROW filter in SOI (after [15]) (a) Filter layout, (b) Fabricated device in SOI. (c) A detail of the electrodes driving the phase shifters in each CROW stage. (d) Measurement results of the three individual stages of the filter and combined response before and after tuning with thermal phase shifters.*

zero, thereby switching the unit cell transfer function between a notch filter (FIR) and a band pass filter (IIR). The operation principles and transfer function derivation follow the similar rules as that of the structure reported by Norberg and colleagues [9], with the difference being the absence of gain elements.

High-performance filters have also been demonstrated in a silicon nitride platform. For instance, researchers from Australia's Monash University [19] reported a complex side-coupled integrated spaced sequence of optical resonators (SCISSOR) filter composed of the serial cascade of three RRs of different FSRs (Figures 3.8(a) and 3.8(b)).

A pair of rings with identical FSRs were employed for filter shape synthesis and a third ring with a smaller FSR was used to perform the function as a modulation translator [20], [21] that enables a separate manipulation of the optical carrier phase. A reconfigurable filter was demonstrated using such a processor where all the reconfiguration operations, including filter frequency tuning, bandwidth variation, and switching between bandpass

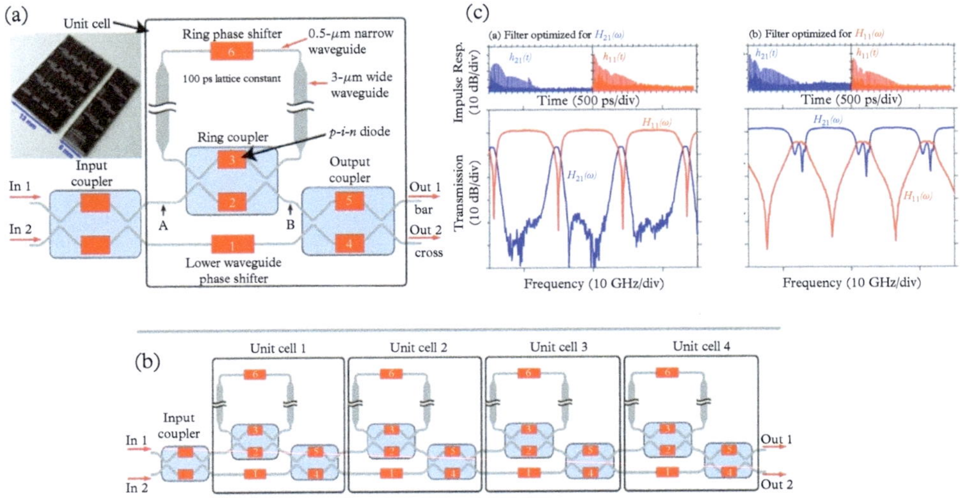

Figure 3.7 *(a) Basic unit cell, (b) Four stage filter layout and (c) Different impulse responses and transfer functions obtained after reconfiguration of the lattice filter reported in [18].*

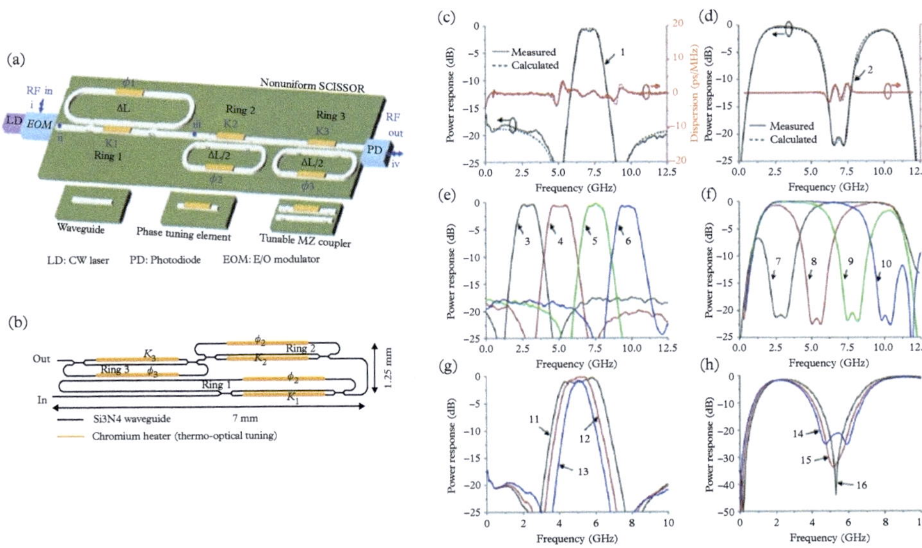

Figure 3.8 *(a), (b) Layout of a Si_3N_4 reconfigurable SCISSOR (after [19]). (c)-(h) Different transfer functions and frequency tunings obtained by reconfiguration. (c) band-pass and (d) band-stop filter response. (e) and (f) Tuning of filter center frequency. (g) and (h) Variations of filter response shape and bandwidth.*

and bandstop filter shapes, were implemented (Figures 3.8(c)–(h)) using the tuning mechanism based on DC-voltage-driven resistor heaters.

Another example of reconfigurable and versatile filters that combines both FIR and IIR sections is the ring-assisted MZI (RAMZI; see Figure 3.9).

It comprises an asymmetric MZI with each arm coupled with an RR. This filter can be subject to different parameter settings resulting in the possibility of implementing either a FIR an IIR or a FIR + IIR response [2]. A team of researchers from LioniX International,

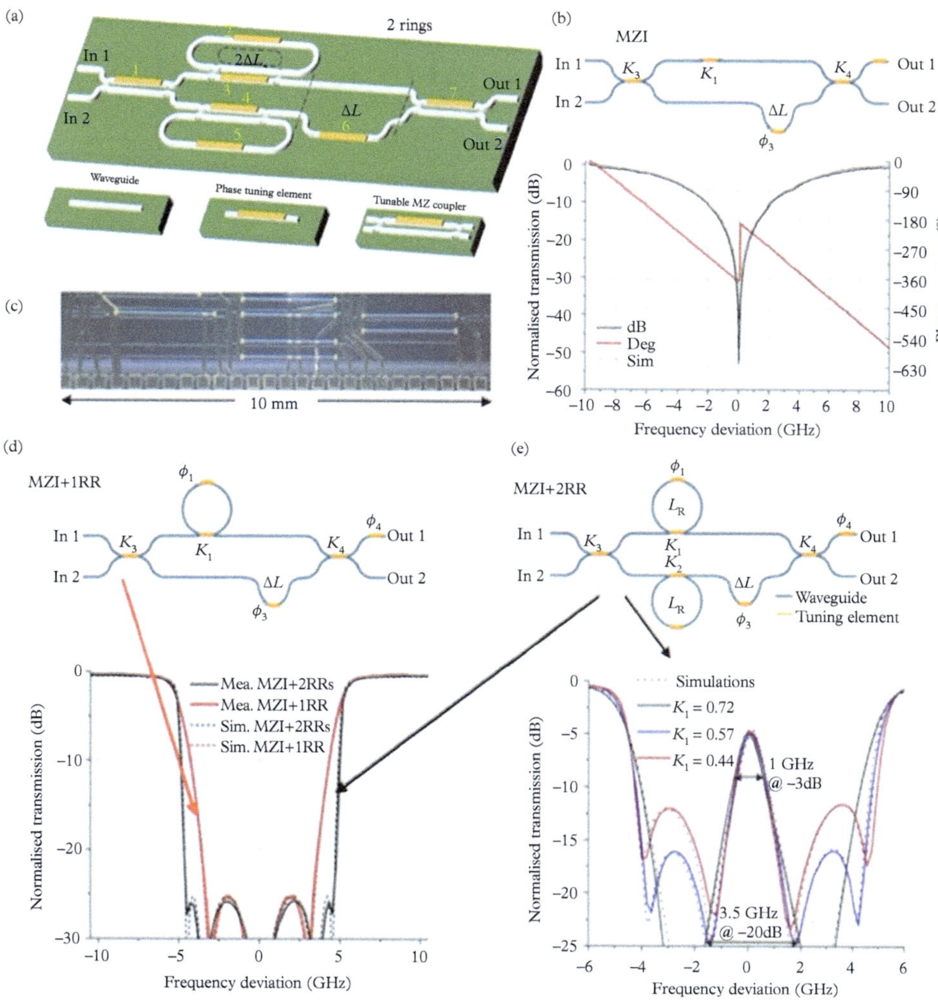

Figure 3.9 *(a) RAMZI filter layout based on a (b) asymmetric Mach-Zehnder interferometer. (c) Photograph of the fabricated device in Si_3N_4 (after [23]). (c), (d) Different filter transfer functions obtained by parameter reconfiguration in one ring and two ring RAMZIs.*

the University of Twente in the Netherlands and Monash University reported a tunable RAMZI circuit [22], [23] employing a total of seven tuning elements, including two tunable couplers at the input and output of the MZI, two tunable couplers of the two RRs and three tunable phase shifters in the ring loops and MZI delay line, all of which were implemented using resistor-based heaters. This RAMZI design enabled a variety of different implementations of optical filters (Figure 3.9). The same team has reported a wide variety of reconfigurable filters for microwave photonics applications [24]. RAMZI filters have also been fabricated in SOI technology [25] and a monolithic version for radiofrequency (RF) filtering has been recently reported in InP [26].

3.2 RF Phase Shifters and Optical Delay Lines

Reconfigurable optical delay lines (ODLs) and wideband tunable phase shifters for RF signals are key elements in many applications such as RF and optical beamforming, signal generation and filtering. Traditionally these operations were carried using fibre-based devices, but this prevents the implementation of compact devices.

3.2.1 Optical Delay Lines

Reconfigurable ODLs are classified into two main categories: variable-length delay lines (VLDL) and variable-propagation-velocity lines (VPVL) based on optical filters. VLDLs can be implemented by means of the so-called switched delay lines (SDLs), where the amount of delay is reconfigured in discrete steps by selectively using an integrated switch a cascade of optical waveguide paths of different lengths by means of photonic switches. SDLs have been proposed in a completely integrated photonics form, which allows increased compactness and improved stability versus thermal and mechanical stress, with reduced delay switching times reaching a few tens of nanoseconds [27]. Higher delay resolution (required for antennae operating at high frequency) can also be improved with an integrated photonics approach. A number of photonic integration platforms have been used, including polymer technology [28]–[30], silica [31]–[34], $LiNbO_3$[35], GaAs [36], [37] and InP [38], [39]. Figure 3.10 illustrates the case of a recent example [40] using ultralow-loss Si_3N_4 substrate featuring a fully integrated 4-bit true-time delay (TTD) line capable of delays above 12 ns, corresponding to about 2.4 m of propagation length, on a chip area of 4.5 cm × 8.5 cm, with waveguide losses as low as 1 dB/m.

VPVL lines can be implemented using optical filters or dispersive optical transmission lines. Their main advantage as compared to VLDLs resides in their capability for continuous delay tuning, which translates into the possibility of continuous steering of the antenna beam direction. Recent efforts have been directed towards VPVL line on-chip implementation using photonic crystals, integrated Bragg gratings and integrated filters. Photonic crystals have been employed as a single fully integrated dispersive delay element. Sancho and colleagues [41] demonstrated the use of a single, very compact, low-loss photonic crystal to implement multiple variable TTDs. The operation is based

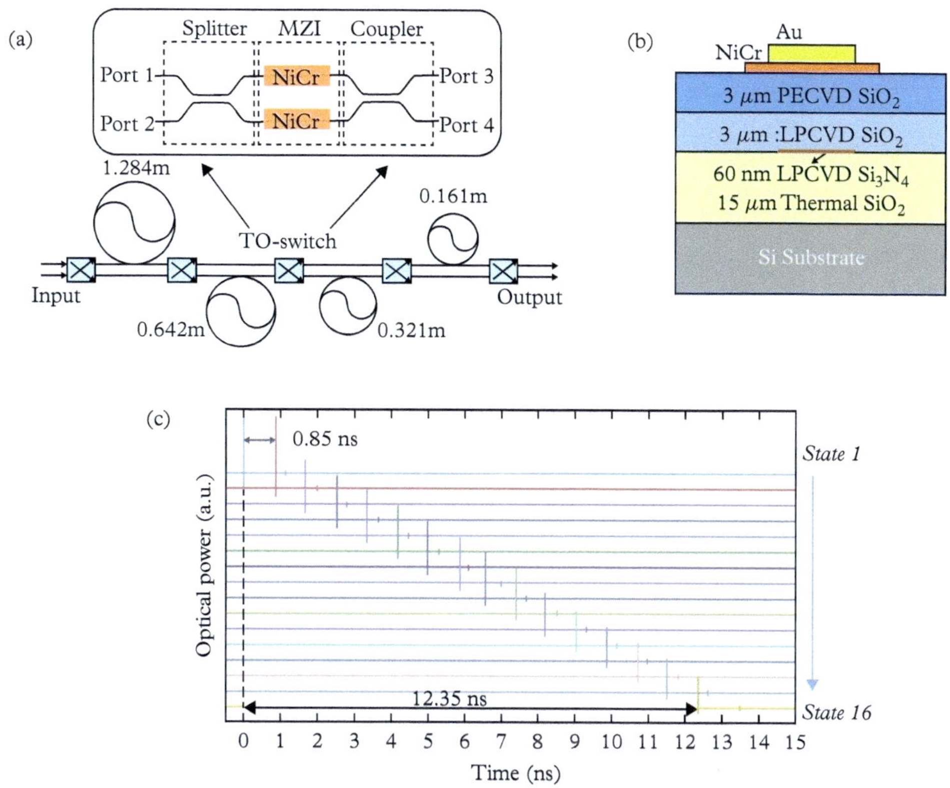

Figure 3.10 *Integrated 4-bit switched optical delay line (after [40]). (a) Circuit layout, (b) Waveguide cross section, (c) Optical output for all delay settings.*

on the slow-light properties of photonic crystals, in which the group delay can be increased to very high values depending on the wavelength demonstrating a 1.5-mm-long device capable of generating delays up to 70 ps with losses below 10 dB over the complete 0–50 GHz band. More recently, Zhang and Yao proposed an electronically reconfigurable Bragg grating device [42] that can operate as an electrically tunable linearly chirped grating providing a dispersive delay line (Figures 3.11(a)–(d)). It consists of multiple series-connected uniform Bragg subgrating sections and a Fabry–Perot (FP) cavity section in the centre of the grating. Each uniform Bragg subgrating section incorporates an independent lateral PN junction (Figure 3.11(b)), and between two neighbouring sections there is an undoped grating to function as an insulator. Distributed electrodes are connected to the independent PN junctions.

By applying a bias voltage to a PN junction, the refractive index of the grating in that particular section could be tuned locally based on a free-carrier plasma dispersion effect (Figure 3.11(c)). Thus, the entire index modulation profile of the grating can be electrically reconfigured by field programming all the bias voltages enabling the grating

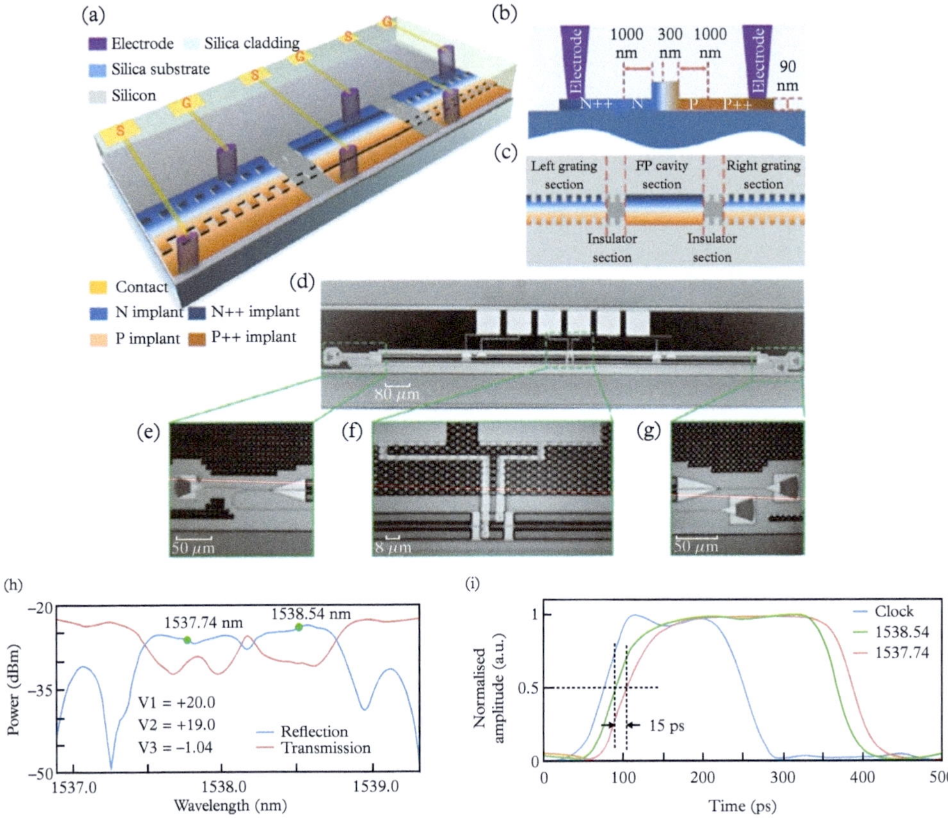

Figure 3.11 *Electrically tunable silicon Bragg grating (after [42]). (a) Perspective view of the grating on a silicon chip. (b) Cross-sectional view of the grating rib waveguide. (c) Top-view of the grating. (d-g) Microscope camera images of the fabricated grating, the input grating coupler and compact Y-branch, the FP cavity section, and the transmission and reflection grating couplers. (h) Reflection and transmission spectra when the device is reconfigured to be a chirped grating. (i) ime-domain measurement when the fabricated grating is reconfigured to be a chirped grating.*

to have diverse spectral characteristics for diverse functionalities. Since the uniform subgratings in the left and right sections can be independently tuned, by shifting the spectral response of one of the two uniform subgratings with different bias voltages, the device can be reconfigured to be a chirped grating.

Figure 3.11(h) presents the measured reflection and transmission spectra of the chirped grating when a maximum reverse bias voltage of +20 V is applied to the left PN junction and a forward bias voltage of −1.04 V is applied to the right PN junction. The 3-dB bandwidth of the spectra is increased to be 1.29 nm, which is larger than the 0.79 nm corresponding to the uniform grating setting. Figure 3.11(i) demonstrates the time delay function where a microwave rectangular pulse with a temporal width of 290 ps is modulated on two optical carriers with different wavelengths and sent to the grating. The two reflected pulses experience two different time delays as they are carried by two different wavelengths, at 1537.74 and 1538.54 nm, respectively. Specifically, the pulse carried by the optical wavelength at 1537.74 nm is reflected by the right subgrating from its centre. The pulse carried by the optical wavelength at 1538.54 nm is reflected by the left subgrating from its centre. The time delay difference is 15 ps. Larger time delays are possible if the length of the grating is increased, and more independent grating sections are incorporated for high-level tuning in terms of group delay and chirp rate.

Integrated optical filters have also been used to implement broadband tunable optical delays. For example, cascaded optical RRs (ORRs) have been demonstrated in silica [31], [34], TriPleX [43], [44] and SOI [45], [46]. This solution combines ORRs comprising tunable phase shifts in the ring loops and tunable couplers. The basic principle is that the ORR can be configured to be an optical all-pass filter [2], [44] (Figure 3.12(a)), featuring a group delay at the resonance frequency, which can be set to be much larger than the round-trip delay of the ring loop and can be controlled via the coupler. This enables the realization of long delays using a small chip area, which can be further increased by cascading multiple RRs in series to form a SCISSOR structure (Figure 3.12(b)). A limitation of this scheme resides in the fact that the delay x bandwidth product is constant and thus, increasing the available delay at resonance implies a reduction in the available bandwidth. This can be overcome by slightly detuning the resonances of the ORRs in the cascade [47] for an eight ORR SCISSOR structure fabricated in Si_3N_4. Figure 3.12(c) provides the basic waveguide structure, individual and SCISSOR layouts and a photograph of one of the fabricated ORRs, including the electrodes placed on top of the ring waveguide to tune its resonance through the thermo-optic effect.

Non-resonant structures based on tunable integrated MZIs have also been recently proposed and experimentally demonstrated [48] to provide continuously tunable delays, which can also be cascaded to provide increased values.

3.2.2 RF Phase Shifters

In addition to delay, phase shifting is also attractive for a number of signal processing applications. Tunable RF phase shifters have been demonstrated using SOI [49], [50] and hybrid SOI and III–V semiconductor RRs [51]. Here, the phase shift is obtained after beating the optical carrier and the single sideband RF modulating signal [52]. Figure 3.13 shows the principle for a structure composed of two ORR in cascade (to achieve full 2π phase excursion [49]). One of the two carriers is placed in coincidence with the resonance of the all-pass transfer function of the resonator where it experiences a different phase-shift compared to the component that is outside the resonance. Tuning is achieved either by moving the resonance of the ring cavity though an internal phase shifter or by tuning the optical carrier.

Both tunable time delay and phase shift operations can be incorporated in a single photonic chip by using the separate carrier tuning (SCT) scheme proposed by Morton and Khurgin [53]. Burla and colleagues [54] have reported and demonstrated the SCT scheme with optical single sideband filtering monolithically integrated in a single chip.

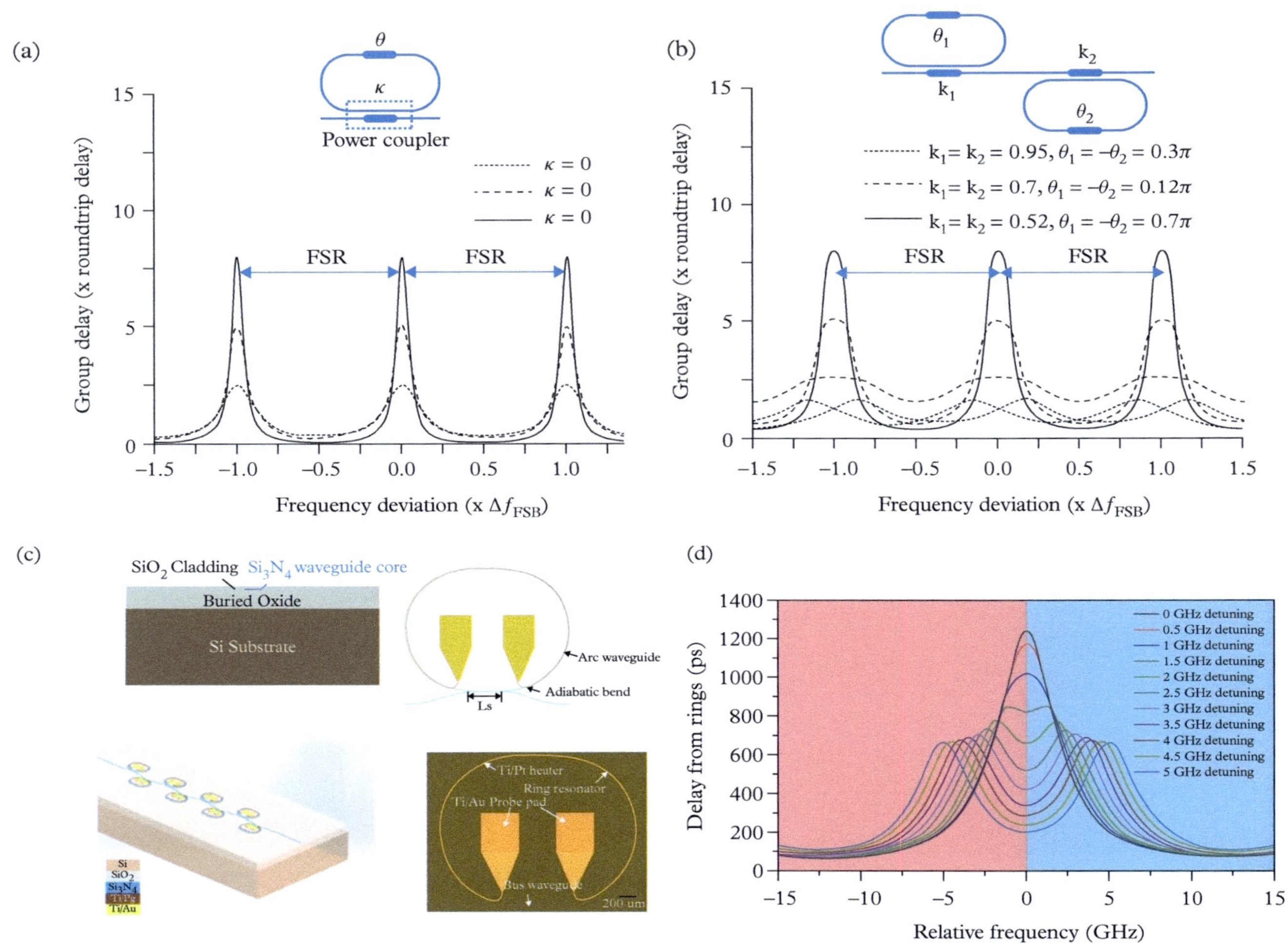

Figure 3.12 *Tunable resonant delay line based on an optical ring resonators acting as an all-pass filter (after [44]). (a) Single ORR structure, (b) Cascade ORR structure. (c) Details from an 8 ORR SCISSOR fabricated in Si3N4 (after [47]). (d) Bandwidth increase by slightly detuning the resonances of the rings I (c) (after [47]).*

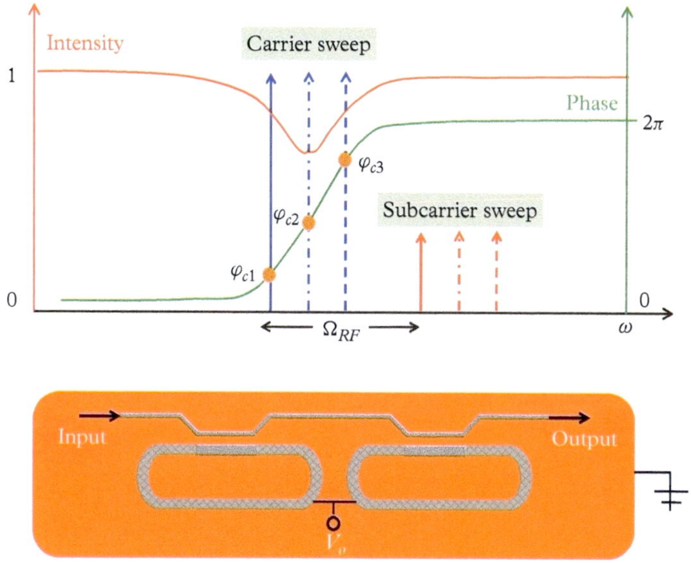

Figure 3.13 *Illustration of the basic principle behind the implementation of a tunable RF phase shifter using a resonant integrated resonator and a single sideband modulated signal (after [52]). The tuning of the chip resonance is here performed using a voltage activated phase-shifter inside the cavities.*

The circuit (Figure 3.14(a)) consists of a reconfigurable optical delay line, a separate carrier tuning unit and an optical sideband filter. The optical sideband filter, a RAMZI, removes one of the radio frequency sidebands of a double-sideband intensity-modulated optical carrier. The ODL and separate carrier tuning unit are individually implemented using a pair of cascaded ORRs. The group delay of the signal sideband is changed by tuning the resonance frequencies and the coupling factor of the ORRs in ODL. This is supplemented by applying an additional phase shift in the [0, 2π] range to the carrier. Burla and colleagues [54] used this module to demonstrate a two-tap microwave photonic filter whose notch position could be shifted by 360° over a bandwidth of 1 GHz. Figures 3.14 (b)–(d) depict the operation of the different modules in a chip.

Waveguide Bragg gratings realized on SOI can also implement tunable ultra-broadband phase shifts in conjunction with optical single-sideband modulation. Morton and Khurgan [53] employed a compact (130 nm long) device to create broadband phase shifts for RF frequencies beyond 900 GHz.

3.3 Beamformers

In phased array antennas a beam is formed and steered to a given angle by adjusting the phase relationship between a given number of radiating elements [2]. This principle of operation works correctly for narrowband signals but, in the case of wideband signals, a

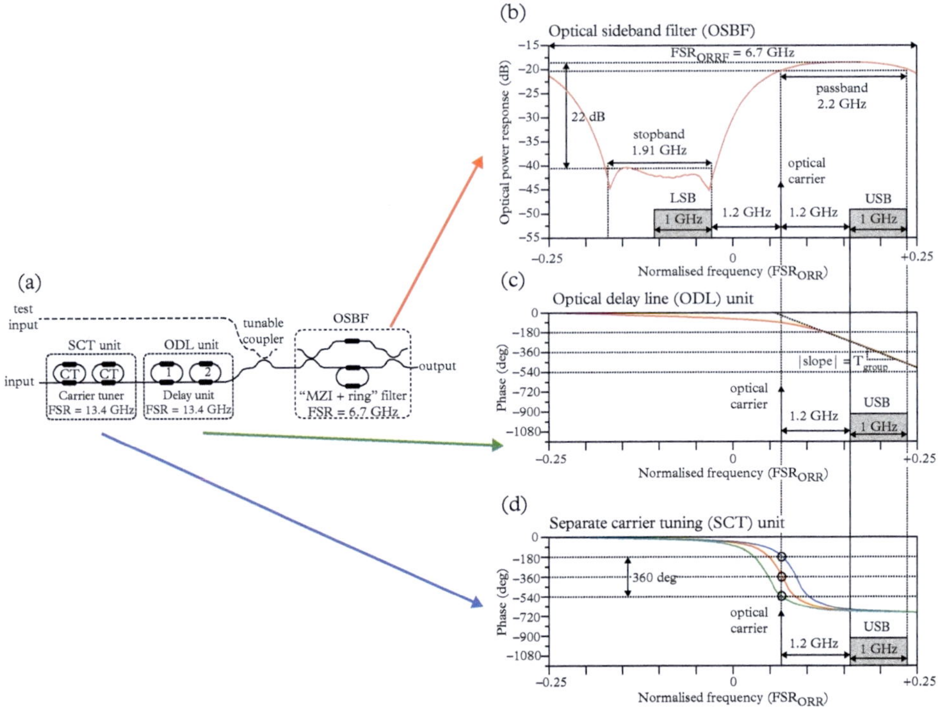

Figure 3.14 *Reconfigurable circuit simultaneously implementing tunable phase shifting and delays using the separate carrier tuning technique (after [54]). (a) Circuit layout. (b) Operation of the tunable SSB filter. (c) Operation of the ODL for the radiofrequency sideband. (d) Operation of the phse shifter for the optical carrier.*

limitation arises. If a constant phase shift is produced from element to element, the beam pointing is different for different frequency components. This deleterious effect, known as beam squint, can be compensated for by using tunable delay lines, rather than by employing phase shifters. Thus, it turns out that both delay lines and phase shifters can be advantageously employed to implement beam steering circuits, each one in a particular context. Attending to the different types of delay lines and phase shifters considered in Section 3.2, there are three categories of integrated photonic beam formers that can be implemented: wideband beam formers based on discretely tunable delay lines TTD [56], wideband beam formers based on continuously tunable delay lines TTD [57]–[64] and narrowband beam formers based on optical phase shifters [65]–[67].

Figure 3.15 shows the configuration and some measured results of a complete tunable broadband beam former based on ORRs arranged in a binary-tree network architecture [63] and fabricated in TriPleX waveguide technology. The beamformer was composed of a four-element array (Figure 3.15(a)), subsequently completed by 1×4 RF phase-shifting stages to feed a final 1×16 linear antenna array (Figure 3.15(b)). Basic

TTDs are obtained using an ODL configuration (Figure 3.14). Delays up to 236 ps were demonstrated over a bandwidth of 2 GHz for Ku-band satellite communication applications [63] (Figure 3.15(c)). The beam former elements could be configured as true delay lines or as pure phase shifters. In the former, it provided squint-free operation (10.7–12.75 GHz), while in the latter, the squint effect was clearly appreciable (Figure 3.15(d)).

Circuit structures with increased sophistication and complexity can also be implemented that leverage on the periodic spectral nature of ring cavity filters [64] and integrated array waveguide gratings [68], [69] to provide multi-wavelength operation.

3.4 Arbitrary Wave Generators and Oscillators

A variety of signal generation techniques have leveraged on the development of reconfigurable integrated photonic circuits for either generating ultra-broad bandwidth RF waves with arbitrary and reconfigurable phase and amplitude or optical frequency comb sources or extremely stable and pure microwave through opto-electronic oscillators (OEO).

3.4.1 Arbitrary Radiofrequency Waveform Generators

Microwave arbitrary waveform generation (AWG) will have very useful applications in 5G communication systems, pulsed radar and advanced instrumentation. The currently available electronic AWG is limited in frequency and bandwidth to around 22 GHz and 6 GHz, respectively. The use of photonic techniques provides a promising solution towards overcoming the main limitations encountered in electronics systems and a variety of techniques have been proposed [70]–[72], including direct space-to-time mapping, spectral shaping and wavelength-to-time mapping and temporal pulse shaping to generate microwave waveforms in the GHz and multiple GHz region. Many proof-of-concept and first demonstrations of these techniques have been implemented through bulky fibreoptic-based systems and clearly an on-chip integrated pulse shaper is thus a desirable solution to overcome the limitations commonly associated with these conventional bulk optics pulse shapers.

Khan and colleagues at Purdue University demonstrated an integrated ultra-broadband arbitrary microwave waveform generator that incorporates a fully programmable spectral shaper fabricated on a silicon photonic chip [73]. The spectral shaper is a reconfigurable filter consisting of eight add/drop micro-ring resonators on a silicon photonics platform. Ultra-compact cross-section (500 nm × 250 nm) silicon nanowires were used to fabricate the rings. The typical bending radius used was 5 μm and the waveguide propagation loss was around 3.5 dB/cm [69]. This spectral shaper programmability is achieved by thermally tuning both the resonant frequencies and the coupling strengths of the micro-ring resonators. A cartoon of the spectral shaper and an optical image showing two channels of RRs with micro-heaters are shown in Figures 3.16(a) and 3.16(b), respectively.

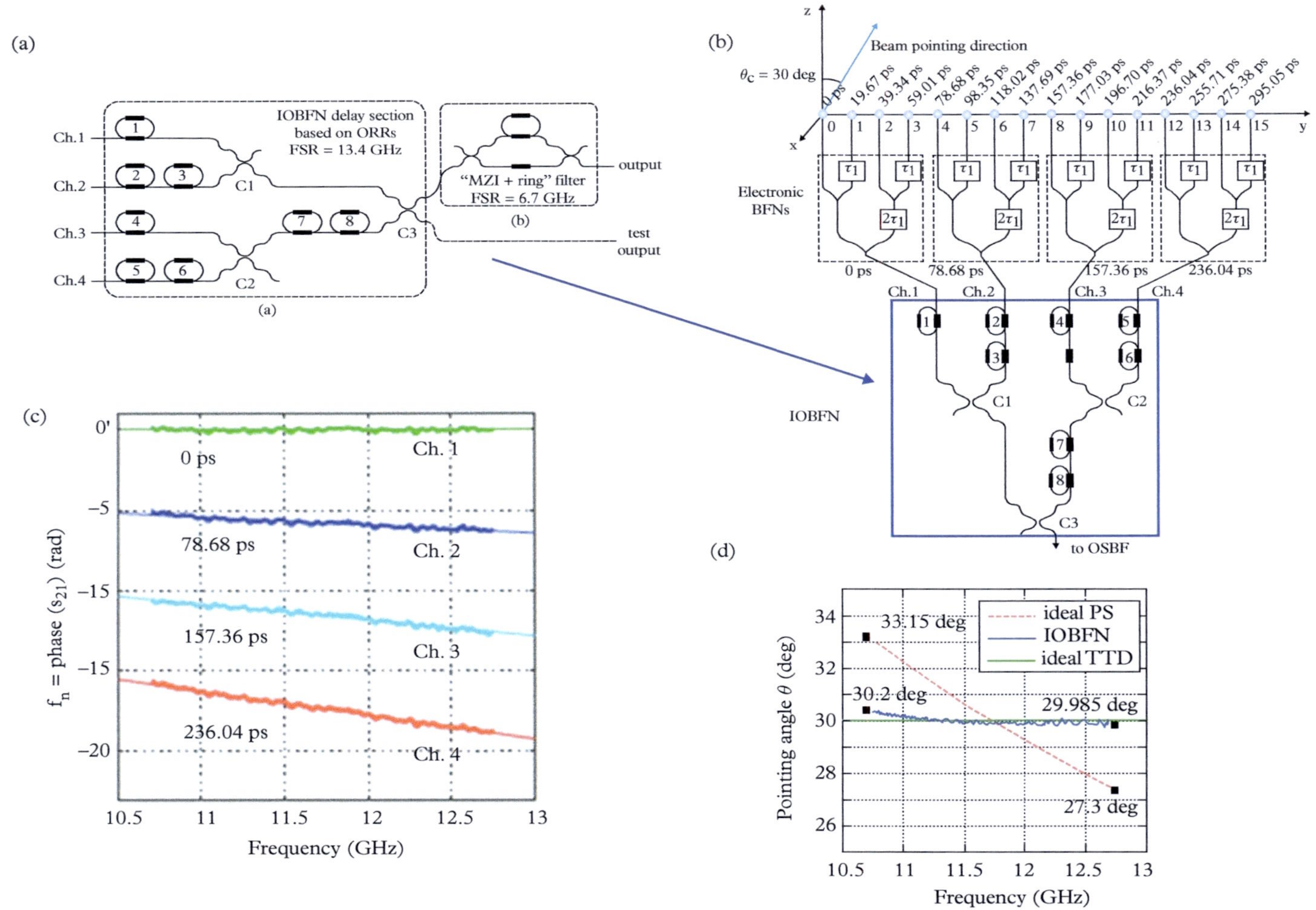

Figure 3.15 *Reconfigurable Delay and phase 1 X 16 beamformer (after [63]). (a) Reconfigurable optical 1 X 4 circuit. (b) Complete beam forming network. (c) Phase shift vs frequency in the four outputs of the optical beam forming stage. (d) Squint and squint free operation in the Ku band for a targeted pointing angle of 30°.*

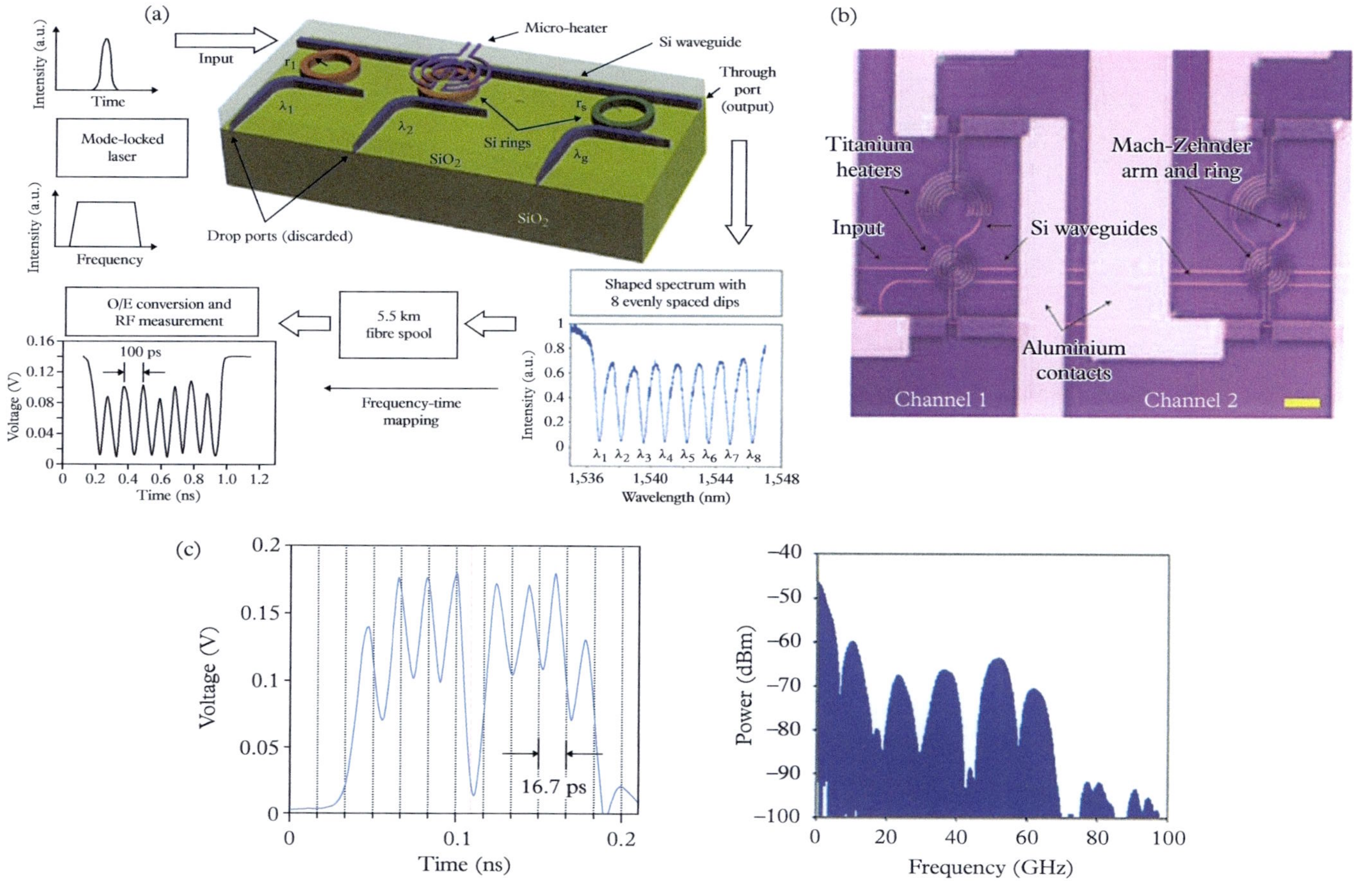

Figure 3.16 *On-chip arbitrary waveform generation using cascaded of 8 microring resonators in SOI platform. (a) Waveform generator system. (b) Optical image of the pulse shaper. (c) Example of waveform generation at 60 GHz (after [73]).*

Figure 3.16(a) shows the principle of the photonic arbitrary microwave waveform generation system implemented here. The spectral shaper is used to modify the spectrum emitted from a mode-locked laser; it then undergoes wavelength-to-time mapping in a dispersive device, which here is a 5.5 km length of optical fibre, before being converted to the electrical domain via a microwave waveform using a high-speed photodetector (PD). By incorporating the spectral shaper, a variety of different waveforms were generated, including those with an apodized amplitude profile, multiple π phase shifts (Fig. 3.16 (c)), two-tone waveforms and frequency-chirped waveforms at the central frequency of 60 GHz.

The main drawback of this design is that it requires a dispersive element for the frequency-to-time mapping operation. This of course can be provided on the flight by an optical fibre link if the signal is to be generated at a centralised site and then delivered to a remote location for radiation. However, if the full signal generation is to be achieved on-chip then the use of the dispersive delay line must be avoided, resulting in the use of alternative techniques described in Section 3.2.1. Figure 3.17 shows the results of a silicon chip based on spectral shaping as well, but here the time delays for each spectral component are provided by separate tunable delay lines implemented with SCISSORS [74].

The AWG layout is shown in Figure 3.17(a). An input ultra-fast pulse is sampled by eight micro-ring add-drop filters of different radii R_i (λ_i denotes the associated resonance wavelength, $i = 1, 2, \ldots, 8$) that download and produce eight replica pulses. Since the wide bandwidth of the input pulse covers several FSRs of the silicon micro-rings, each replica generation micro-ring downloads up to six small slices of the input spectrum. However, in the time domain, each pulse replica actually consists of a series of densely packed, exponentially decaying amplitude pulses. Only the envelopes of the pulse replicas can be detected during optical/electrical conversion, the width of each pulse replica's envelope being inversely proportional to the optical round-trip time in the ring and the ring's power coupling coefficients to bus waveguides. The frequency spacing between the selected channels in the filter bank can be arbitrary and non-uniform, as long as they are not too small to cause interference between channels.

After the pulse replication is performed in each stage for a given wavelength region of the pulse, a selective delay is created by means of an on-chip tuneable delay element comprising a fixed part and a tuneable part (Figure 3.17(b)). The fixed part is a long waveguide with a delay value at multiples of a given set value τ ps, and each channel has τ ps longer delay than the previous one. The tuneable part consists of cascaded micro-ring SCISSOR all-pass filters, which are activated to provide more or less additional delay. A final stage composed of another eight micro-ring add-drop filters of different radii R_i provide the final amplitude for the wavelength sample and combine all the wavelength samples into a common output waveguide bus. Figures 3.17 (c)–(f) provide examples of several AWG waveforms generated by reconfiguring the chip parameters.

Reconfigurable AWGs based on FIR filter architectures have also been reported. For instance, a planar lightwave circuit (PLC) fabricated on silica-on-silicon was reported generating pulse trains at 40 GHz and 80 GHz with flat-top, Gaussian and apodized profiles [75]. The pulse shaper was based on a 12-tap FIR filter performing both phase and amplitude filtering implemented as 12 stages of cascaded MZIs, each with an FSR of 80 GHz.

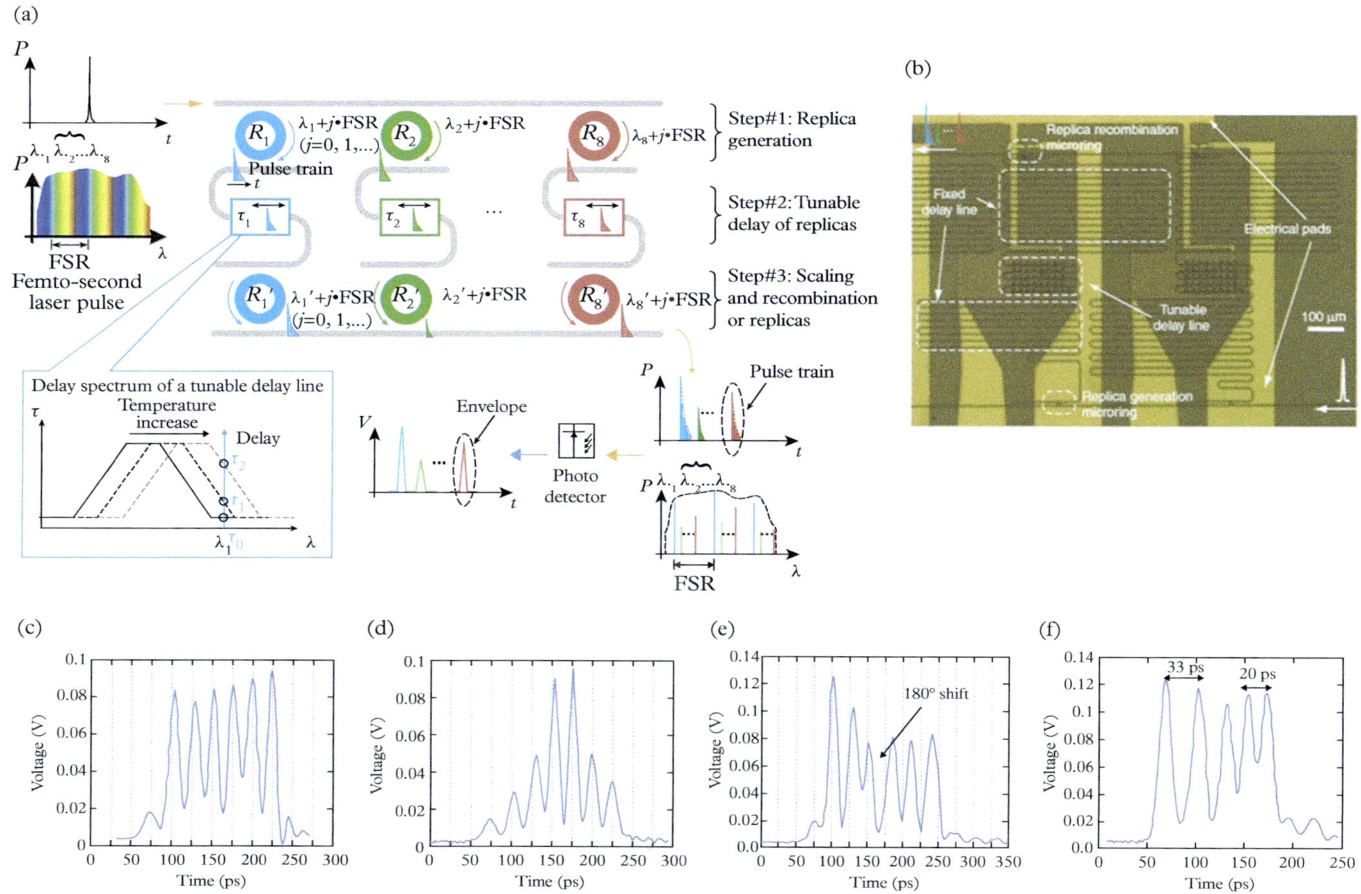

Figure 3.17 *Chip design (a) and fabricated layout (b) for a reconfigurable arbitrary waveform generator, using wavelength to time mapping and selective tunable delay lines based on SCISSORS. (c)-(f) examples of generated waveforms. (c) 40GHz uniform RF burst, (d) 40Ghz apodized RF burst, (e) 40GHz burst with an intermediate π phase shift. (f) Frequency up-chirped burst from 30 to 50 Ghz. (After [74]).*

3.4.2 Opto-electronic Oscillators (OEO)

An OEO [76]–[78] is a microwave generator that provides ultra-pure i.e. (very low phase noise), broadly tunable and stable single mode oscillation by replacing the RF resonator by a very high Q-factor optical cavity, which can be implemented by means of either a low-loss fibre loop or by an integrated waveguide resonant filter. In an OEO, light from a laser is modulated and passed through the optical cavity before reaching a PD (Figure 3.18(a)). The output of the PD is amplified, filtered, adjusted for phase and then fed back to the modulator providing self-sustained oscillation if the overall round trip gain is larger than the loss and the circulating waves can be combined in phase. It therefore works much in the same way as a laser.

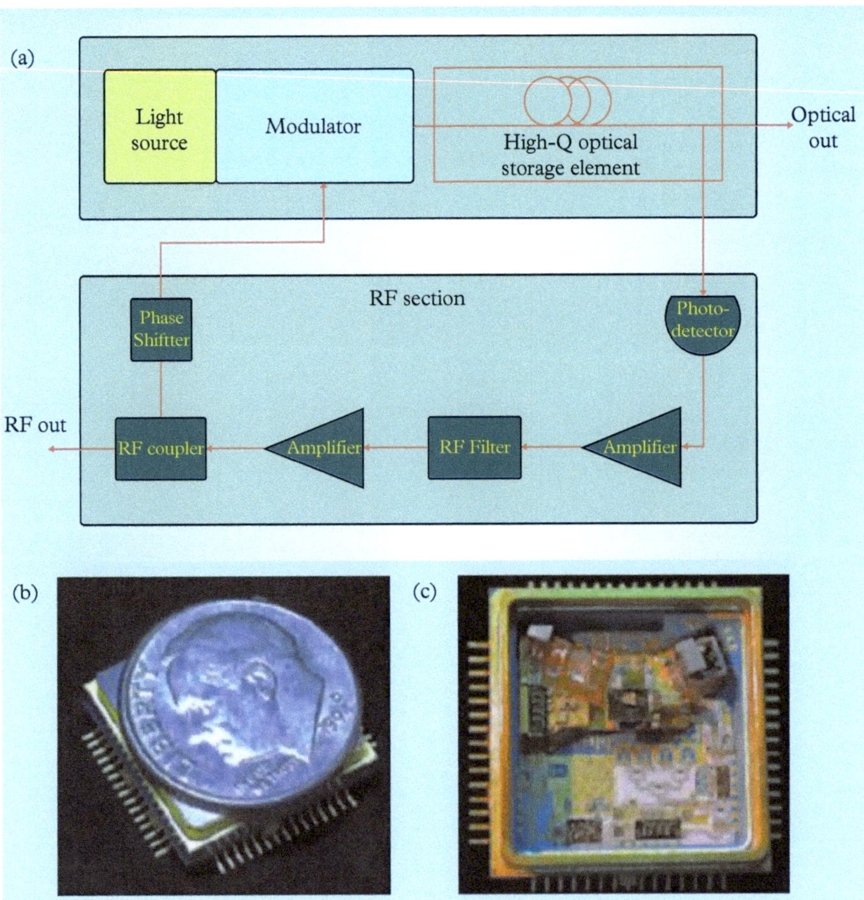

Figure 3.18 *Chip design (a) and fabricated a) Block diagram of a generic OEO. (b) and (c) Miniature OEO based on a lithium niobate WGMR (after [78]).*

OEOs bring many unique advantages, e.g. exceptionally low phase noise and versatility of the output frequency. Such ultra-pure microwaves are indeed needed in a wide range of applications, including time-frequency metrology, frequency synthesis and aerospace engineering.

The spectral purity of the signal in the OEO is directly related to the Q-factor of the loop and therefore, most OEOs utilise a long length of fibre to achieve high spectral purity. A disadvantage of using a fibre loop is the production of super modes that appear in the phase noise spectrum caused by the propagation of waves multiple times around the OEO loop. In addition, a fibre delay line is bulky, so the oscillators are not an optimal solution for the implementation of compact microwave sources. In addition, this bulky delay line element has to be temperature stabilised, an energy-consuming feedback control process. A solution based on discrete integrated components was been proposed, which could also lead to the possibility of full integration of OEOs. It consisted in replacing the optical fibre loop by a high Q-factor cavity implemented by means of a whispering gallery mode resonator (WGMR) [78]. WGMRs ranging in size from a few hundred micrometres to a few millimetres can be fabricated from a wide variety of optically transparent materials reaching Q-factors in the range of 3×10^{11}. OEOs based on high Q-factor WGMRs made from electro-optic materials can provide high performance in a miniaturized package smaller than a coin (Figures 3.18(b) and 3.18(c)) [78], operating in frequency ranges that are 10–40 GHz and featuring instantaneous line widths below 200 Hz.

More recently several attempts have been reported towards the full integration of the photonic part of the OEO in a chip [79], [80]. Zhang and Yao [79] proposed and experimentally demonstrated a silicon photonic integrated OEO that includes a high-speed PM, a thermally tunable micro-disk resonator (MDR), and a high-speed PD (Figure 3.19). The OEOs was able to generate a microwave signal with a frequency tuning range of 3–8 GHz. The phase noise of the generated microwave signal was measured to be around −80 dB/Hz at a 10-KHz offset frequency, which is still far from the values achieved with fibre-based OEOs.

Tang and colleagues [80] reported and experimentally demonstrated an InP-based integrated reconfigurable OEO where the chip contains the active elements as well as all the main RF components. The storage cavity here was implemented by means of a spiral waveguide rather than by an integrated resonator. All of the devices needed in the OEO loop circuit were monolithically integrated on chip within size of 5×6 cm^2. By tuning the injection current, the output frequency of the proposed OEO was tuned from 7.30 GHz to 8.87 GHz featuring a phase noise value of around −91 dBc/Hz@1MHz.

3.4.3 Frequency Combs

The generation of optical combs on a chip [81] allows for the precise measurement of optical frequencies through direct referencing to microwave atomic clocks and the production of multiple taps for high sidelobe rejection signal processors. Recently, the generation of optical frequency combs through the non-linear process of continuous-wave optical-parametric oscillation using micro-scale resonators (or optical micro-combs)

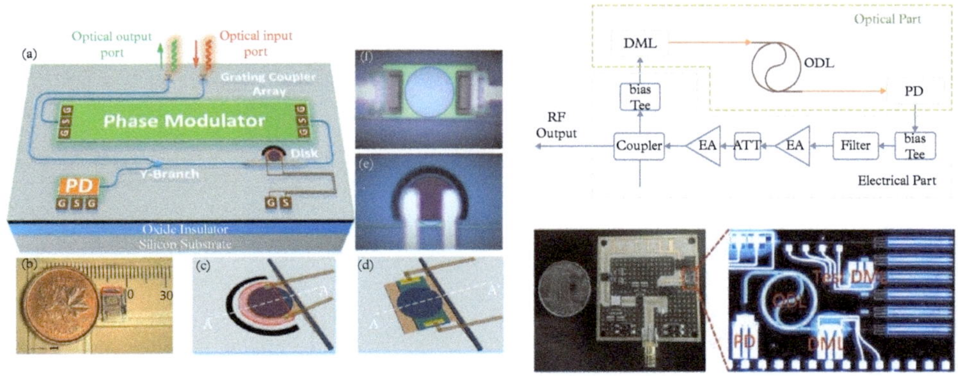

Figure 3.19 *Schematic of the proposed integrated photonics OEO (after [79]). (Left) Perspective view of the proposed OEO; the fabricated chip prototype captured by a microscope camera; perspective view of the MDR with a top-placed micro-heater; perspective view of the MDR with a p-type doped micro-heater; image of the MDR with a top-placed micro-heater; and image of the MDR with a p-type doped micro-heater (after [79]). (Right) Schematic and photographs of the InP OEO, photograph of the fabricated device, photonic part of the OEO. DML: directly modulated laser; ODL: optical delay line; PD: photo detector; EA: electrical amplifier; ATT: attenuator (after [80]).*

has attracted significant interest as these devices have the potential to yield highly compact and frequency-agile comb sources. Several authors have reported the generation of optical frequency combs from highly-robust CMOS-compatible integrated micro-resonator optical parametric oscillators [81], although we will not delve deeper into this approach as it is based on nonlinear effects, which are outside the scope of this book.

Combs have been traditionally based on bulky and high-cost femtosecond lasers with line spacing in the range of 100 MHz–1 GHz. However, their integrated and low-cost counterparts, i.e. integrated semiconductor mode-locked lasers, are limited by their large comb spacing, small number of lines and broad optical linewidth. However, Whang and colleages [82] reported a demonstration of a III–V-on-Si comb laser that can function as a compact, low-cost frequency comb generator after frequency stabilization. Figure 3.20 shows the layout and the emitted spectrum of the integrated comb source under optimum operation conditions.

The use of low-loss passive silicon waveguides enables the integration of a long laser cavity, which enables the laser to be locked in the passive mode at a repetition rate of 1 GHz. The optical spectrum features a 12-nm bandwidth comb structure with 10-dB output optical spectrum. The very small optical mode spacing results in a dense optical comb of over 1,400 equally spaced optical lines with sub-kHz 10-dB radio frequency linewidth. Furthermore, the narrow longitudinal mode linewidth (<400 kHz) indicates notably stable mode-locking.

The main problem with the former approach is the reconfigurability, as it is set by the cavity length and might require the use of external phase filters to tune the repetition frequency. One solution, which can also be integrated on a chip, is to generate the comb spectrum by cascading amplitude and PMs at the output of a continuous-wave laser source [83], driven by specially tailored RF waveforms to generate an optical frequency comb with very high spectral flatness. Wu and colleagues [83] experimentally

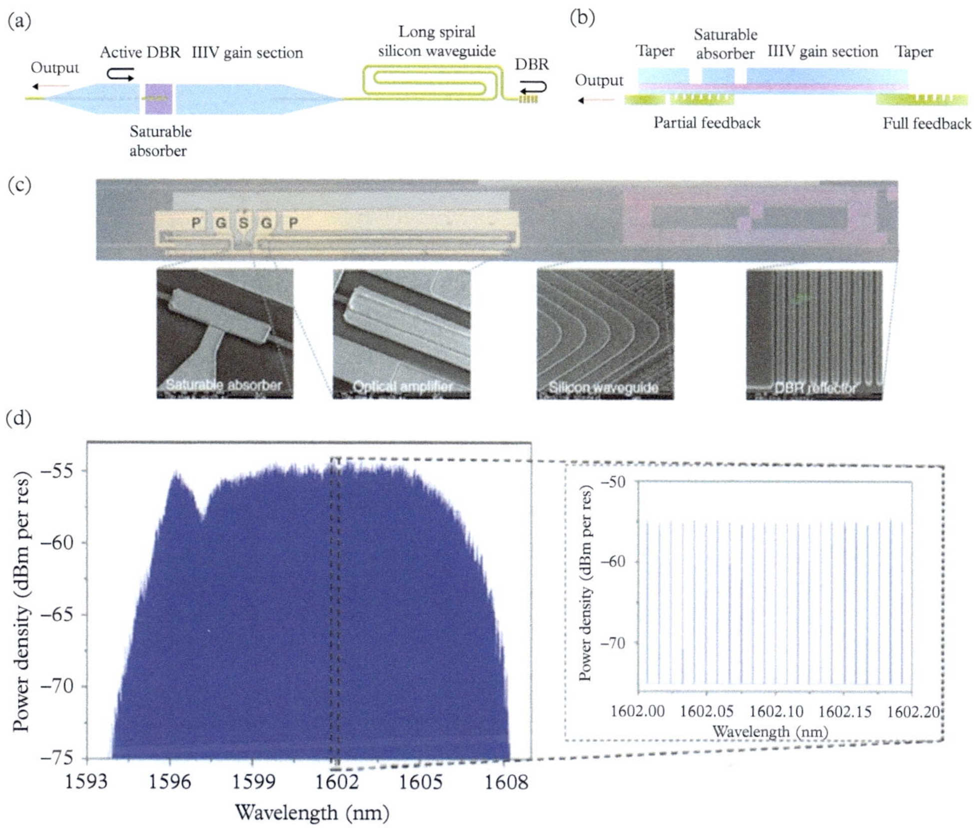

Figure 3.20 *(a) Top and (b) side views of a III-V on Si anti-colliding MLL comb source design (after [82]). (c) Microscope image of the III-V-on-Si MLL. Insets: Scanning electron microscope images of various constituent parts of the laser. (d) Optical comb generated by the passively locked 1 GHz MLL. Inset: a detail of evenly spaced optical modes in the comb.*

demonstrated this technique and showed it leading to an extremely flat comb spectrum with a spectral power variation below 1 dB. The number of comb lines that can be generated is limited by the power-handling capability of the PM, and this can be scaled without compromising the spectral flatness.

3.5 Multifunctional Reconfigurable PICs

Most of the reconfigurable photonic circuits reviewed here are application specific, i.e. they are specifically designed and targeted for a given application. Nevertheless, clearly some circuit structures, e.g. ring cavities, are employed in building circuits that support different functionalities. This leads to the question of whether a given application-specific circuit architecture can be implemented so it can support, even if few, different functionalities.

It is indeed possible to design some circuits from a set of known traditional optical signal processing building blocks, e.g. ring cavities, couplers, delay lines, that can implement a reduced set of functionalities by suitable parameter reconfiguration. Thus, they can be considered as multi-functional but not general or universal in the sense used for the architectures considered in Chapter 4 and Chapter 5.

One example is the multi zero-pole filter reported by Guzzon and colleagues [10] (Figure 3.4(a)); originally designed as a tunable filter, Liu and colleagues recently reported this structure [84] as able to support useful functionalities for time-domain optical signal processing, e.g. differentiation, integration and Hilbert transformation. Figure 3.21 displays a chart showing the basic demonstrated functionalities for this photonic circuit.

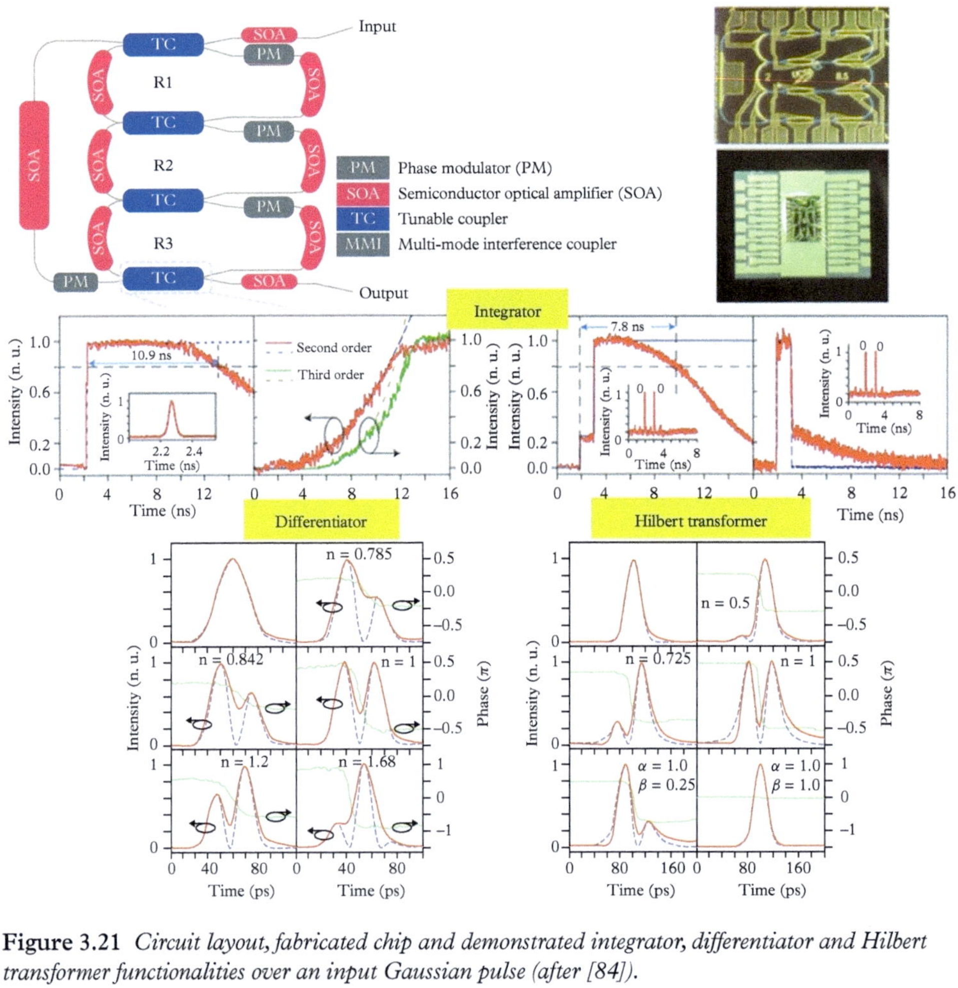

Figure 3.21 *Circuit layout, fabricated chip and demonstrated integrator, differentiator and Hilbert transformer functionalities over an input Gaussian pulse (after [84]).*

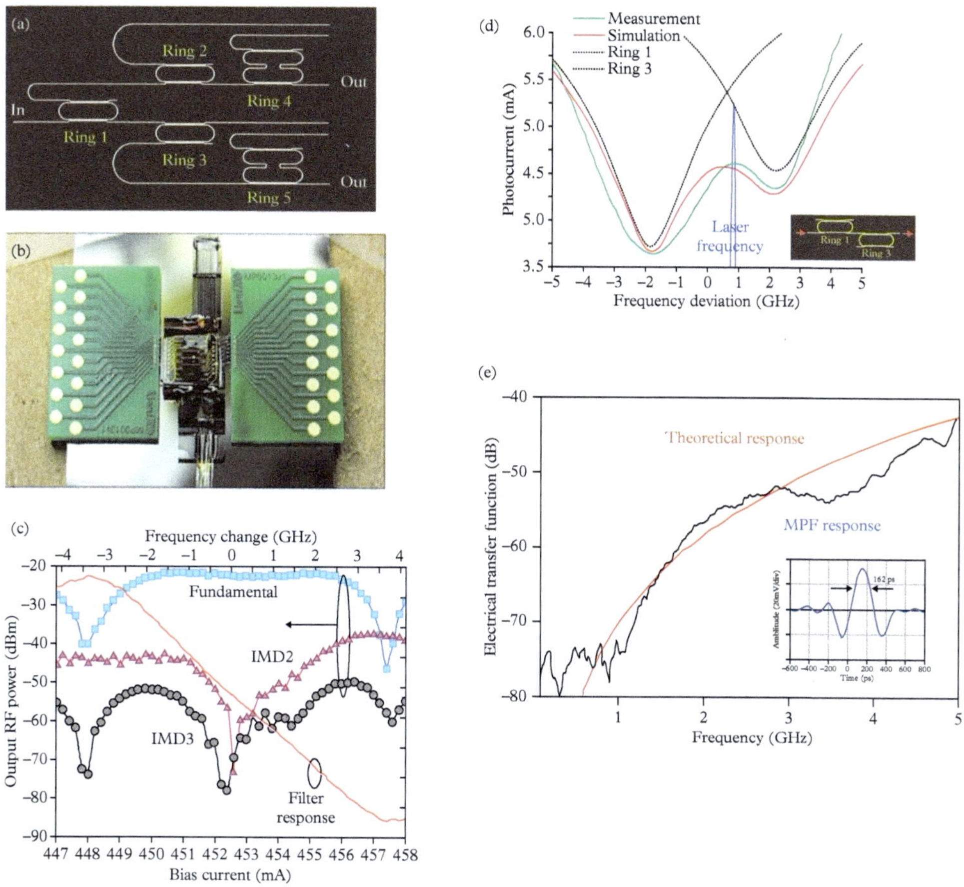

Figure 3.22 *Circuit layout (a), fabricated chip (b) and demonstrated discriminator (c) and doublet pulse generator (d)–(e) of a* Si_3N_4 *multifunctional circuit, after [85] and [86]).*

Each functionality requires a specific setting for the different parameters characterizing the internal blocks (SOAs, PMs and tunable couplers) of the circuit.

Another example is shown in Figure 3.22. This integrated circuit built in Si_3N_4 technology was initially designed to operate as a frequency discriminator for phase-modulated optical communication links [85] and subsequently employed to operate as a programmable ultra-wideband (UWB) doublet pulse generator [86] and an instantaneous frequency measurement system [87]. The full circuit is composed of five rings (Figure 3.22(a)). All of them are required for the operation as a discriminator, while only two are required for operation as UWB doublet generator (the rest were disabled) and only one ring is employed for the frequency measurement.

As a final example we consider now the microwave photonic (MWP) processor [88], which illustrates the transition between multi-purpose reconfigurable specific circuits

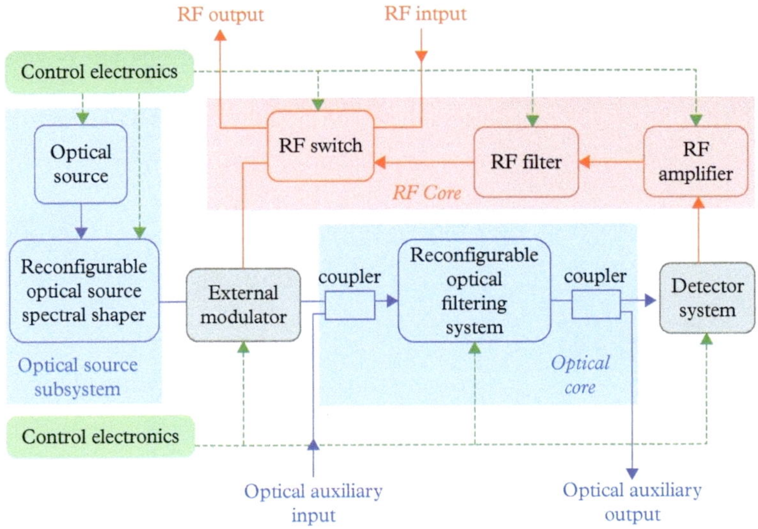

Figure 3.23 *Circuit layout of the MWP transistor (after [89]).*

and the more general and universal architectures discussed in Chapter 4 and Chapter 5. The originating concept of this device can be traced back to the proposal of the MWP transistor [89] (Figure 3.23), which is composed of subsystems, each of which is a collection of connected fixed and variable components.

The transistor configuration is obtained by opening the feedback loop of a generalised OEO [78]. As demonstrated by Capmany and colleagues [89], this architecture is—in principle—capable of performing a few, but important, required functionalities in MWP provided that its optical subsystems can be reconfigured by means of suitable control signals. However, the main limitation of the MWP transistor is that it does not support the reconfiguration of interconnections between its internal optical subsystems. This results in an intermediate performance between ASPIC and a universal processor.

As regards increased flexibility, the former approach can be extended to implement a programmable MWP processor architecture [20], which incorporates optical routing and switching elements (ORSEs) into the MWP transistor layout and replaces the reconfigurable optical filtering system with a more powerful and versatile optical core. The upper part of Figure 3.24 schematizes the concept.

The central element of the programmable optical processor is the optical core, where the main signal processing tasks are carried out in the photonic domain. Ideally, the optical core should be built upon a versatile architecture capable of implementing different functionalities in response to different electronic control signals. In practice, among the reconfigurable photonic circuits proposed in the literature, those that are potential candidates for this core are either based on the cascade of finite (i.e. MZIs) or infinite (i.e. ring cavity) impulse response cells, or a combination of both.

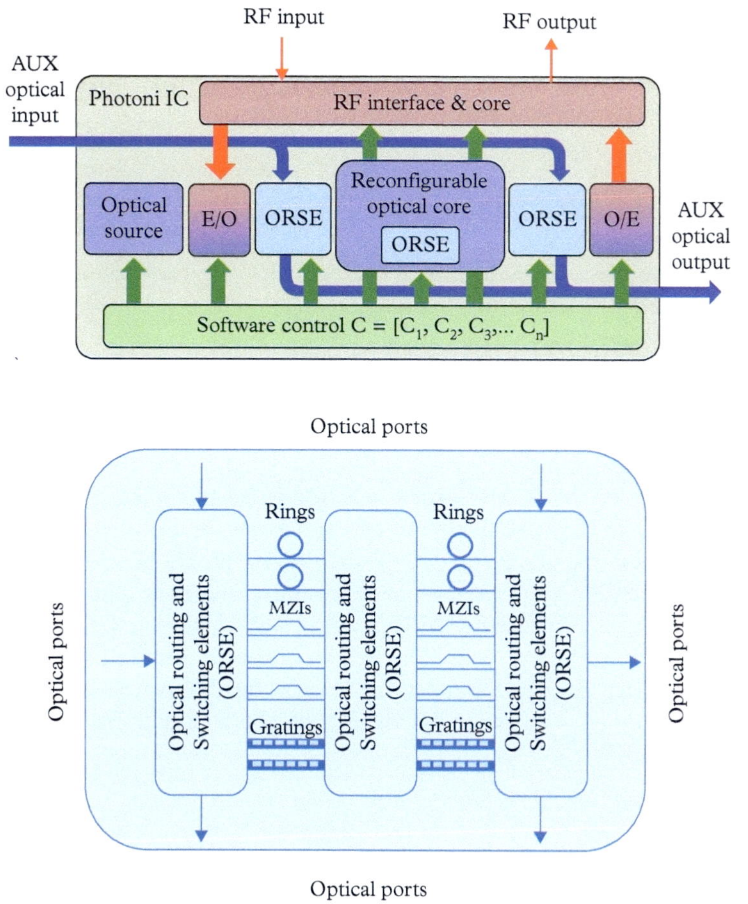

Figure 3.24 *(Upper) Layout of the universal processor (after [88]). (Lower) Particular implementation of the optical core using ad-hoc photonic components such as Mach Zehnder interferometers (MZIs), optical ring resonators (ORRs) and gratings connected by several optical routing and switching elements (ORSEs).*

These structures feature periodic filters and allow for bandwidth reconfiguration and notch tunability by moving the zeros and poles along the z-plane. Thus, to further increase the degree of flexibility of the processor, the former configuration can be extended with a subsystem of switched optical signal processing elements, including different types of fixed and reconfigurable filtering structures and interconnections, as illustrated in the lower part of Figure 3.24. This example includes FIRs, IIRs and dispersive delay lines (DDLs) that can be accessible by the correct programming of the ORSEs that interconnect them. The optical routing and switching elements require medium-count ports depending on the number of component banks. This architecture [88] [89] is still limited in: flexibility, as ad-hoc designed photonic subsystems must be included and interconnected within the optical core; scalability, as fixed size photonic components are the building blocks for more complex circuits; and loss, as considerable numbers of ORSEs need to be traversed to implement complex configurations.

3.6 Chapter Summary

Although reconfigurable photonic integrated circuits are fundamentally application specific, the fact that some features describing their operation can be changed by the application of external control signals qualify them as a first step towards programmable photonic circuits. This chapter has shown most of these circuits have a fixed topology and display a limited range of functionalities. Application-specific circuits do play a significant role in electronics and could potentially serve the same purpose in photonics. Nevertheless, they have a number of drawbacks related to the time needed for their efficient design, coupled with the frequent need for redesign cycles and the rigid nature of their performance in terms of flexibility and scalability. In as much as programmable circuits, micro-processors and field programmable gate arrays have been developed in electronics to complement application-specific circuits, integrated photonics will require a similar approach to complement ASPICS. This is the rationale behind the need for developing the concepts and the technology of generic programmable photonics circuits. Inspired by these similar concepts, but bearing in mind the differences between photonics and electronics, Chapter 4 and Chapter 5 next lay the foundations of programmable photonics.

REFERENCES

[1] J. Capmany, B. Ortega, D. Pastor and S. Sales, Discrete-Time Optical Processing of Microwave Signals. (2005) *J Lightw Technol* 23, 702–23.
[2] C. K. Madsen and J. H. Zhao, *Optical Filter Design and Analysis: A Signal Processing Approach*, Wiley Interscience, 1999.
[3] J. Capmany, B. Ortega and D. Pastor, A Tutorial on Microwave Photonic Filters. (2006) *J Lightw Technol* 24(1), 201–29.
[4] R. A. Minasian, Photonic Signal Processing of Microwave Signals. (2006) *IEEE Trans Microw Theory Tech* 54(2), 832–46.
[5] E. J. Norberg, et al., Programmable Photonic Lattice Filters in InGaAsP-InP. (2010) *Photon Technol Lett* 22, 109–11.
[6] R. S. Guzzon, et al., Monolithically Integrated Programmable Photonic Microwave Filter with Tunable Inter-ring Coupling. *IEEE Topical Meeting on Microwave Photonics*. 5–9 October Montreal, QC, Canada, 2010. pp. 23–6.
[7] E. J. Norberg, R. S. Guzzon, J. S. Parker, L. A. Johansson, and L. A. Coldren, A Monolithic Programmable Optical Filter for RF-signal Processing. *IEEE Topical Meeting on Microwave Photonics*. 5–9 October Montreal, QC, Canada, 2010. pp. 365–8.

[8] E. J. Norberg, R. S. Guzzon, J. S. Parker, and L. A. Coldren, Programmable Photonic Filters from Monolithically Cascaded Filter Stages. *Proceedings Integrated Photonics Research, Silicon and Nanophotonics and Photonics in Switching*, 25–28 July, Monterey, CA, 2010, paper ITuC3.
[9] E. J. Norberg, et al., Programmable Photonic Microwave Filters Monolithically Integrated in InP/InGaAsP. (2011) *J Lightw Technol* 29, 1611–19.
[10] R. S. Guzzon, et al., Integrated InP-InGaAsP Tunable Coupled Ring Optical Bandpass Filters with Zero Insertion Loss. (2011). Opt Express 19, 7816–26.
[11] H.-W. Chen, et al., Hybrid Silicon Tunable Filter Based on a Mach–Zehnder Interferometer and Ring Resonator. *Proceedings of the International Topical Meeting on Microwave Photonics.* 14-16 October, Valencia, Spain, 2009. pp. 1–4.
[12] H.-W. Chen, et al., Integrated Microwave Photonic Filter on a Hybrid Silicon Platform. (2010) *IEEE Trans Microw Theory Tech* 58, 3213–19.
[13] P. Dong, et al., GHz-bandwidth Optical Filters Based on High-order Silicon Ring Resonators. (2010) *Opt Express* 18, 23784–9.
[14] N. N. Feng, et al., Thermally-efficient Reconfigurable Narrowband RF-photonic Filter. (2010) *Opt Express* 18, 24648–53.
[15] S. Romero-García, et al., 'Wideband Multi-stage CROW Filters with Relaxed Fabrication Tolerances. (2018) *Opt Express* 26, 4723–37.
[16] M. Rasras, et al., Demonstration of a Tunable Microwave-photonic Notch Filter Using Low-loss Silicon Ring Resonators. (2009) *J Lightw Technol* 27, 2105–10.
[17] M. Rasras, et al., Demonstration of a Fourth-Order Pole-Zero Optical Filter Integrated Using CMOS Processes. (2007) *J Lightw Technol* 25, 87–92.
[18] S. Ibrahim, et al., Demonstration of a Fast-reconfigurable Silicon CMOS Optical Lattice Filter. (2011) *Opt Express* 19, 13245–56.
[19] L. Zhuang, Flexible RF Filter Using a Non-uniform SCISSOR.' (2016) *Opt Lett* 4, 11118–21.
[20] L. Zhuang, Ring Resonator-based On-chip Modulation Transformer for High-performance Phase-modulated Microwave Photonic Links. (2013) *Opt Express* 21, 25999–6013.
[21] L. Zhuang, et al., Novel Microwave Photonic Fractional Hilbert Transformer Using a Ring Resonator-based Optical All-pass Filter. (2012) *Opt Express* 20, 26499–510.
[22] C. Taddei, et al., Waveguide Filter-based On-chip Differentiator for Microwave Photonic Signal Processing. Paper presented at the *IEEE International Topical Meeting on Microwave Photonics (MWP)*, 28–31 October, Alexandria, VA, 2013. pp. 28–31.
[23] L. Zhuang, et al., CRIT-alternative Narrow-passband Waveguide Filter for Microwave Photonic Signal Processors. (2014) *IEEE Photon Technol Lett* 26, 1034–7.
[24] Y. Xie, et al., Programmable Optical Processor Chips: Toward Photonic RF Filters with DSP-level Flexibility and MHz-band Selectivity. (2017) *Nanophotonics* 7, 421–54.
[25] P. Orlandi, et al., Photonic Integrated Filter with Widely Tunable Bandwidth. (2014) *J Lightwave Technol* 32, 897–907.
[26] J. S. Fandiño, et al., A Monolithic Integrated Photonic Microwave Filter. (2017) *Nat Photon* 11, 124–9.
[27] W. Ng, et al., The first demonstration of an Optically Steered Microwave Phased Array Antenna Using True-time Delay. (1991) *J Lightwave Technol*, 9(9), 1124–31.
[28] Y. Chen, K. Wu, F. Zhao and R. T. Chen, Loss Compensated Photonic True-time Delay for Phased-array Antenna. (2004) *Proc IEEE Antennas Propagation Society Int Symp* 4, 4324–7.
[29] X. Wang, et al., 'Fully-integrated 4-bit True-time Delay Module Using Polymer Optical Switches and Waveguide Delay Lines. Integrated Photonics Research Applications, Optical Society America. 24–26 April, Uncasville, CT, 2006. Paper no. IMF4.
[30] Y. Chen, et al., Reconfigurable True-time Delay for Wideband Phased-array Antennas. (2004) *Proc SPIE* 5363, 125–30.

[31] K. Horikawa, I. Ogawa, H. Ogawa and T. Kitoh, Photonic Switched True-time Delay Beam Forming Network Integrated on Silica Waveguide Circuits. (1995) *IEEE MTT-S Int Microwave Symp Dig.* 16–20 May, Orlando, FL. pp. 65–8.

[32] K. Horikawa, I. Ogawa, T. Kitoh, and H. Ogawa, Silica-based Integrated Planar Lightwave True-time-delay Network for Microwave Antenna Applications. (1996) *Proc Optical Fiber Communications*, pp. 100–1.

[33] J. LeGrange, et al., Demonstration of an Integrated, Tunable High Resolution True Time Delay Line. (2004) *Proc 17th Ann Meeting IEEE Lasers Electro-Optics Soc* 2, 790–1.

[34] M. S. Rasras, et al., Integrated Resonance-enhanced Variable Optical Delay Lines. (2005) *IEEE Photon Technol Lett* 17(4), 834–6.

[35] K. Horikawa, Y. Nakasuga and H. Ogawa, Self-heterodyning Optical Waveguide Beam Forming and Steering Network Integrated on Lithium Niobate Substrate. (1995) *IEEE Trans Microwave Theory Tech* 43(9), 2395–401.

[36] W. W. Ng, et al., GaAs and Silica-based Integrated Optical Time-shift Network for Phased Arrays. (1994) *SPIE Proc* 2155, 114–23.

[37] W. Ng, D. Yap, A. Narayanan and A. Walston, High-precision Detector-switched Monolithic GaAs Time-delay Network for the Optical Control of Phased Arrays. (1994) *IEEE Photon Technol Lett* 6(2), 231–4.

[38] J. Stulemeijer, et al., Compact photonic integrated phase and amplitude controller for phased-array antennas. (1999) *IEEE Photon Technol Lett* 11(1), 122–4.

[39] G. Flamand, et al., InP-based PIC for an Optical Phased-array Antenna at 1.06 μm. (2000) *IEEE Photon Technol Lett* 12(7), 876–8.

[40] L. M. Renan, et al., Integrated Ultra-low-loss 4-bit Tunable Delay for Broadband Phased Array Antenna Applications. (2013) *IEEE Photon Technol Lett* 25(12), 1165–8.

[41] J. Sancho, et al., Integrable Microwave Filter Based on a Photonic Crystal Delay Line. (2012) *Nature Commun* 3, 1075.

[42] W. Zhang and J. Yao, A Fully Reconfigurable Waveguide Bragg Grating for Programmable Photonic Signal Processing. (2018) *Nature Communications* 9, Art. no. 1396.

[43] L. Zhuang, et al., Single-chip Ring Resonator-based 1×8 Optical Beam Forming Network in CMOS-compatible Waveguide Technology. (2007) *IEEE Photon Technol Lett* 19(15), 1130– 2.

[44] L. Zhuang et al., Novel Low-loss Waveguide Delay Lines Using Vernier Ring Resonators for On-chip Multi-λ Microwave Photonic Signal Processors. (2013) *Laser Photon Rev* 7, 994–1002.

[45] J. Cardenas, et al., Wide-bandwidth continuously Tunable Optical Delay Line Using Silicon Micro-Ring Resonators. (2010) *Opt Express* 18(25), 26525–34.

[46] P. A. Morton, J. Cardenas, J. B. Khurgin and M. Lipson, Fast Thermal Switching of Wideband Optical Delay Line with No Long-term Transient. (2012) *IEEE Photon Technol Lett* 24(6), 512– 14.

[47] C. Xiang, et al., Low-Loss Continuously Tunable Optical True Time Delay Based on Si_3N_4 Ring Resonators. (2018) *IEEE J Sel Top Quantum Electron* 24, 5900109.

[48] D. Melati et al., Wideband Integrated Optical Delay Line Based on a Continuously Tunable Mach–Zehnder Interferometer. (2018) *IEEE J Sel Top Quantum Electron* 24, 4400108.

[49] M. Pu, et al., Widely Tunable Microwave Phase Shifter Based on Silicon-on-Insulator Dual-micro-ring Resonator. (2010) *Opt Express* 18, 6172–82.

[50] J. Lloret, et al., Tunable Complex-valued Multi-tap Microwave Photonic Filter Based on Single Silicon-on-Insulator Micro-ring Resonator. (2011) *Opt Express* 19(13), 12402–7.

[51] J. Lloret et al., Broadband Microwave Photonic Fully Tunable Filter Using a Single Heterogeneously Integrated III-V/SOI-Microdisk-based Phase Shifter. (2012) *Opt Express* 20, 10796–806.

[52] D. B. Adams and C. Madsen, A Novel Broadband Photonic RF Phase Shifter. (2008) *J Lightwave Technol* 15(15), 2712–17.
[53] P. A. Morton and J. B. Khurgin, Microwave Photonic Delay Line with Separate Tuning of the Optical Carrier. (2009) *IEEE Photon Technol Lett* 21(22), 1686–8.
[54] M. Burla, et al., On-chip CMOS Compatible Reconfigurable Optical Delay Line with Separate Carrier Tuning for Microwave Photonic Signal Processing. (2011) *Opt Express* 19, 21475–84.
[55] M. Burla, et al., On-chip Programmable Ultra-wideband Microwave Photonic Phase Shifter and True Time Delay Unit. (2014) *Opt Lett* 39(21), 6181–4.
[56] B. Howley, X. Wang, M. Chen and R. T. Chen, Reconfigurable Delay Time Polymer Planar Lightwave Circuit for an X-band Phased-array Antenna Demonstration. (2007) *J Lightw Technol* 25(3), pp. 883–90.
[57] A. Meijerink, et al., Novel Ring Resonator-based Integrated Photonic Beamformer for Broadband Phased Array Receive Antennas–Part I: Design and Performance Analysis. (2010) *J Lightw Technol* 28(1), 3–18.
[58] L. Zhuang, et al., 'Novel Ring Resonator-based Integrated Photonic Beamformer for Broadband Phased Array Receive Antennas–Part II: Experimental Prototype. (2010) *J Lightw Technol* 28(1), 19–31.
[59] D. Marpaung, et al., Towards a Broadband and Squint-free Ku-band Phased Array Antenna System for Airborne Satellite Communications. *Proc 5th European Conf. Antennas Propagation.* 11–15 April, Rome, Italy, 2011. pp. 2623–7.
[60] D. Marpaung, et al., Photonic Integration and Components Development for a Ku-band Phased Array Antenna System. *Proc IEEE Topical Meeting Microwave Photonics Asia Pacific Microwave Photonics.* Oct. 18–21, Singapore, 2011. pp. 458–61.
[61] M. Burla, et al., System Integration and Radiation Pattern Measurements of a Phased Array Antenna Employing an Integrated Photonic Beamformer for Radio Astronomy Applications. (2012) *Appl Opt* 51(7), 789–802.
[62] L. Zhuang, et al., Low-loss, High-index-contrast Si_3N_4/SiO_2 Optical Waveguides for Optical Delay Lines in Microwave Photonics Signal Processing. (2011) *Opt Express* 19(23), 23162–70.
[63] M. Burla, et al., Integrated Photonic Ku-band Beamformer Chip with Continuous Amplitude and Delay Control. (2013) *IEEE Photon Technol Lett* 25(12), 1145–8.
[64] M. Burla, et al., Multiwavelength-integrated Optical Beamformer Based on Wavelength Division Multiplexing for 2-D Phased Array Antennas. (2014) *J Lightw Technol* 32(20), 3509–20.
[65] K. Horikawa, I. Ogawa, T. Kitoh, and H. Ogawa, Silica-based Integrated Planar Lightwave True-time-delay Network for Microwave Antenna Applications. Proc. Optical Fiber Communications. 25 February, San Jose, CA, 1996. pp. 100–1.
[66] K. Horikawa, Y. Nakasuga, and H. Ogawa, Self-heterodyning Optical Waveguide Beam Forming and Steering Network Integrated on Lithium Niobate Substrate. (1995) *IEEE Trans Microwave Theory Tech* 43(9), 2395–401.
[67] G. Grosskopf, et al., Maximum Directivity Beamformer at 60 GHz with Optical Feeder. (2003) *IEEE Trans Antennas Propagat* 51(11), 3040–6.
[68] T. Koonen and Z. Cao, Optically Controlled 2D Radio Beam Steering System. *Proc Asia-Pacific Microwave Photonics Conf Int Topical Meeting Microwave Photonics.* 20–23 October, Sapporo, Japan, 2014, pp. 389–91.
[69] B. Jalali and S. Yegnanarayanan, Optically Controlled Phased-array Antenna Using Wavelength-Selective True Time Delay. *Proc IEEE Int Conf Phased Array Systems Technology.* 21–25 May, Dana Point, CA, 2000, pp. 367–70.

[70] J. Yao, Photonic Generation of Microwave Arbitrary Waveforms. (2011) *Opt Commun* 284, 3723–36.
[71] J. D. McKinney, D. Seo, D. E. Leaird and A. M. Weiner, Photonically Assisted Generation of Arbitrary Millimeter-wave and Microwave Electromagnetic Waveforms via Direct Space-to-Time Optical Pulse Shaping. (2003) *J Lightwave Technol* 21, 3020–8.
[72] I. S. Lin, J. D. McKinney and A. M. Weiner, Photonic Synthesis of Broadband Microwave Arbitrary Waveforms Applicable to Ultra-wideband Communication. (2005) *IEEE Microw Wirel Compon Lett* 15, 226–8.
[73] M. H. Khan, et al., Ultrabroad-bandwidth Arbitrary Radiofrequency Waveform Generation with a Silicon Photonic Chip-based Spectral Shaper. (2010) *Nat Photonics* 4, 117–22.
[74] J. Wang, et al., Reconfigurable Radio-frequency Arbitrary Waveforms Synthesized in a Silicon Photonic Chip. (2015) *Nature Comms* 6, 5957.
[75] P. Samadi, et al., RF Arbitrary Waveform Generation Using Tunable Planar Lightwave Circuits. (2011) *Opt Commun* 284, 3737–41.
[76] X. S. Yao and L. Maleki, Optoelectronic Oscillator for Photonic Systems. (1996) *IEEE J Quantum Electron* 32(7), 1141–9.
[77] X. S. Yao and L. Maleki, Optoelectronic Microwave Oscillator. (1996) *J Opt Soc Amer B* 13(8), 1725–35.
[78] L. Maleki, Sources: The Optoelectronic Oscillator. (2011) *Nature Photonics* 5(12), 728–30.
[79] W. Zhang and J. Yao, Silicon Photonic Integrated Optoelectronic Oscillator for Frequency-Tunable Microwave Generation. (2018) *IEEE J Lightwave Tech* 36, 4655–63.
[80] J. Tang, et al., Integrated Optoelectronic Oscillator. (2018) *Opt Express* 26, 12257–65.
[81] J. Yu, et al., RF Photonics: An Optical Microcombs' Perspective. (2018) *IEEE J Sel T Quantum Electron* 24, 6101020.
[82] Z. Whang, et al., A III-V-on-Si Ultra-dense Comb Laser. (2017) *Light: Science & Applications* 6, e16260.
[83] R. Wu, et al., Generation of Very Flat Optical Frequency Combs from Continuous-wave Lasers Using Cascaded Intensity and Phase Modulators Driven by Tailored Radio Frequency Waveforms. (2010) *Opt Lett* 19, 3234–6.
[84] W. Liu, et al., A Fully Reconfigurable Photonic Integrated Signal Processor. (2016) *Nat Photon* 10, 190–5.
[85] D. Marpaung, C. Roeloffzen, A. Leinse and M. Hoekman, A Photonic Chip-based Frequency Discriminator for a High-performance Microwave Photonic Link. (2010) *Opt Express* 18, 27359–70.
[86] D. Marpaung, L. Chevalier, M. Burla and C. Roeloffzen, Impulse Radio Ultrawideband Pulse Shaper Based on a Programmable Photonic Chip Frequency Discriminator. (2011) *Opt Express* 19, 24838–48.
[87] D. Marpaung, On-chip Photonic-assisted Instantaneous Microwave Frequency Measurement System. (2013) *IEEE Photon Technol Lett* 25(9), 837.
[88] D. Pérez, I. Gasulla and J. Capmany, Software-defined Reconfigurable Microwave Photonics Processor. (2015) *Opt Express* 23, 14640–54.
[89] J. Capmany, J. Mora, P. Muñoz y S. Sales, A Microwave Photonics Transistor. *IEEE International Topical Meeting on Microwave Photonics (MWP)*, 28–31 October, Alexandria, VA. 2013.

4

Integrated Multi-port Interferometers

Integrated multi-port interferometers are the second step in programmable photonics. Initially conceived and designed to implement fixed linear optics unitary transformations between N input and N output ports to support mode transformations in quantum optics, the rapid development of this field and the possibility of applying these structures to other areas stimulated development of devices capable of supporting arbitrary linear transformations by suitable programming and surpassing use in square matrix transformations and unitary operators. This chapter reviews some basic concepts on unitary $N \times N$ matrices and then describes in detail the basic design principles and integrated optics implementations of triangular and rectangular designs for multi-port interferometers. It then considers more specific interferometer designs, which do not provide possible implementation of general linear transformations but are optimised for performing important operations, e.g. the Fourier transform. Lastly, it covers the linear universal component, which integrates both universal beam couplers and integrated multi-port interferometers.

4.1 Basic Concepts on $N \times N$ Unitary Matrices

Multi-port optical interferometers are conceived and designed to implement unitary transformations between N input and N output ports [1]–[4]. It will provide therefore useful to review some salient properties of these transformations. We will assume an $N \times N$ unitary transformation $U(N)$ given by the following matrix

$$U = \begin{pmatrix} u_{11} & u_{12} & \cdots & u_{1N} \\ u_{21} & u_{22} & \cdots & u_{2N} \\ \cdots & \cdots & \cdots & \cdots \\ u_{N1} & u_{N2} & \cdots & u_{NN} \end{pmatrix} \tag{4.1}$$

Because U(N) is unitary, then

$$U^{\dagger} = \begin{pmatrix} u_{11}^{*} & u_{21}^{*} & \cdots & u_{N1}^{*} \\ u_{12}^{*} & u_{22}^{*} & \cdots & u_{N2}^{*} \\ \cdots & \cdots & \cdots & \cdots \\ u_{1N}^{*} & u_{2N}^{*} & \cdots & u_{NN}^{*} \end{pmatrix} = U^{-1} \tag{4.2}$$

Programmable Integrated Photonics. José Capmany and Daniel Pérez, Oxford University Press (2020). © José Capmany and Daniel Pérez.
DOI: 10.1093/oso/9780198844402.001.0001

So

$$U^{\dagger}U = U^{-1}U = I$$
$$UU^{\dagger} = UU^{-1} = I \tag{4.3}$$

Where $\star$ denotes the complex conjugate. We shall now describe four important properties of unitary matrices $U(\mathrm{N})$ that will be relevant in the design of multi-port interferometers.

Property 1: *If all the matrix coefficients in a given row i are zero except u_{ii}, that is $u_{ij} = 0$ for $j \neq i$, then all the matrix coefficients in column i are zero, except u_{ii}, that is, $u_{ji} = 0$ for $j \neq i$. Furthermore, u_{ii} is a modulus one complex number.*

To prove this, we use Eqn. (4.1), Eqn. (4.2) and Eqn. (4.3)

$$\begin{pmatrix} u_{11} & u_{12} & \cdots & u_{1i} & \cdots & u_{1N} \\ u_{21} & u_{22} & \cdots & u_{2i} & \cdots & u_{2N} \\ \cdots & \cdots & \cdots & \cdots & \cdots & \cdots \\ 0 & 0 & \cdots & u_{ii} & \cdots & 0 \\ \cdots & \cdots & \cdots & \cdots & \cdots & \cdots \\ u_{N1} & u_{N2} & \cdots & u_{Ni} & \cdots & u_{NN} \end{pmatrix} \begin{pmatrix} u_{11}^* & u_{21}^* & \cdots & 0 & \cdots & u_{N1}^* \\ u_{12}^* & u_{22}^* & \cdots & 0 & \cdots & u_{N2}^* \\ \cdots & \cdots & \cdots & \cdots & \cdots & \cdots \\ u_{1i}^* & u_{2i}^* & \cdots & u_{ii}^* & \cdots & u_{Ni}^* \\ \cdots & \cdots & \cdots & \cdots & \cdots & \cdots \\ u_{1N}^* & u_{2N}^* & \cdots & 0 & \cdots & u_{NN}^* \end{pmatrix} =$$
$$= \begin{pmatrix} 1 & 0 & \cdots & 0 & \cdots & 0 \\ 0 & 1 & \cdots & 0 & \cdots & 0 \\ \cdots & \cdots & \cdots & 0 & \cdots & \cdots \\ 0 & 0 & \cdots & 1 & \cdots & 0 \\ \cdots & \cdots & \cdots & 0 & \cdots & \cdots \\ 0 & 0 & \cdots & 0 & \cdots & 1 \end{pmatrix} \tag{4.4}$$

If we focus on the results for column i we have

$$|u_{ii}|^2 = 1 \Rightarrow u_{ii} = e^{j\alpha_i}$$
$$u_{ji}u_{ii}^* = 0, \;\; j \neq i \Rightarrow u_{ji} = 0 \tag{4.5}$$

Property 2: *if all the matrix coefficients in a given column i are zero except u_{ii}, that is $u_{ji} = 0$ for $j \neq i$, then all the matrix coefficients in row i are zero, except u_{ii}, that is, $u_{ij} = 0$ for $j \neq i$ and, again, u_{ii} is a modulus one complex number.*

The demonstration of this property follows the same procedure as Property 1.

One matrix of special interest is the *rotation matrix* in the plane defined by rows m and n

$$T_{mn}(\varphi,\theta) = \begin{pmatrix} 1 & 0 & \cdots & 0 & \cdots & 0 & \cdots & 0 \\ 0 & 1 & \cdots & 0 & \cdots & 0 & \cdots & 0 \\ \cdots & \cdots & \ddots & \cdots & \cdots & \cdots & \cdots & \cdots \\ 0 & 0 & \cdots & e^{j\varphi}\cos\theta & \cdots & -\sin\theta & \cdots & 0 \\ \cdots & \cdots & \cdots & \cdots & \ddots & \cdots & \cdots & \cdots \\ 0 & 0 & \cdots & e^{j\varphi}\sin\theta & \cdots & \cos\theta & \cdots & 0 \\ \cdots & \cdots & \cdots & \cdots & \cdots & \cdots & \ddots & \cdots \\ 0 & 0 & \cdots & 0 & \cdots & 0 & \cdots & 1 \end{pmatrix} \boxed{\begin{matrix} 1 \\ 2 \\ \cdots \\ m \\ \cdots \\ n \\ \cdots \\ N \end{matrix}} \tag{4.6}$$

Which, according to Eqn. (2.111), corresponds to a lossless beamsplitter or tunable coupler between inputs m and n and outputs n and m in this order. As can be immediately checked, $T_{mn}(\varphi,\theta)$ is unitary and its inverse is given by

$$T_{mn}^{-1}(\varphi,\theta) = \begin{pmatrix} 1 & 0 & \cdots & 0 & \cdots & 0 & \cdots & 0 \\ 0 & 1 & \cdots & 0 & \cdots & 0 & \cdots & 0 \\ \cdots & \cdots & \ddots & \cdots & \cdots & \cdots & \cdots & \cdots \\ 0 & 0 & \cdots & e^{-j\varphi}\cos\theta & \cdots & e^{-j\varphi}\sin\theta & \cdots & 0 \\ \cdots & \cdots & \cdots & \cdots & \ddots & \cdots & \cdots & \cdots \\ 0 & 0 & \cdots & -\sin\theta & \cdots & \cos\theta & \cdots & 0 \\ \cdots & \cdots & \cdots & \cdots & \cdots & \cdots & \ddots & \cdots \\ 0 & 0 & \cdots & 0 & \cdots & 0 & \cdots & 1 \end{pmatrix} \boxed{\begin{matrix} 1 \\ 2 \\ \cdots \\ m \\ \cdots \\ n \\ \cdots \\ N \end{matrix}} \tag{4.7}$$

Property 3: *The product $UT_{mn}^{-1}(\varphi,\theta)$ is a unitary matrix that leaves all the columns in U unaltered with the exception of columns m and n and we can choose values for (φ,θ) to null whatever (but only one) coefficient in these columns.*

$UT_{mn}^{-1}(\varphi,\theta)$ is the product of two unitary matrices and it is thus unitary. Now, taking into account Eqn. (4.1) and Eqn. (4.7) we obtain

$$UT_{mn}^{-1}(\varphi,\theta) = \begin{pmatrix} u_{11} & u_{12} & \cdots & u_{1N} \\ u_{21} & u_{22} & \cdots & u_{2N} \\ \cdots & \cdots & \ddots & \cdots \\ u_{N1} & u_{N2} & \cdots & u_{NN} \end{pmatrix} \begin{pmatrix} 1 & 0 & \cdots & 0 & \cdots & 0 & \cdots & 0 \\ 0 & 1 & \cdots & 0 & \cdots & 0 & \cdots & 0 \\ \cdots & \cdots & \ddots & \cdots & \cdots & \cdots & \cdots & \cdots \\ 0 & 0 & \cdots & e^{-j\varphi}\cos\theta & \cdots & e^{-j\varphi}\sin\theta & \cdots & 0 \\ \cdots & \cdots & \cdots & \cdots & \ddots & \cdots & \cdots & \cdots \\ 0 & 0 & \cdots & -\sin\theta & \cdots & \cos\theta & \cdots & 0 \\ \cdots & \cdots & \cdots & \cdots & \cdots & \cdots & \ddots & \cdots \\ 0 & 0 & \cdots & 0 & \cdots & 0 & \cdots & 1 \end{pmatrix} =$$

$$= \begin{pmatrix} u_{11} & u_{12} & \cdots & u_{1m}e^{-j\varphi}\cos\theta - u_{1n}\sin\theta & \cdots & u_{1m}e^{-j\varphi}\sin\theta + u_{1n}\cos\theta & \cdots & u_{1N} \\ u_{21} & u_{22} & \cdots & u_{2m}e^{-j\varphi}\cos\theta - u_{2n}\sin\theta & \cdots & u_{2m}e^{-j\varphi}\sin\theta + u_{2n}\cos\theta & \cdots & u_{2N} \\ \cdots & \cdots & \ddots & \cdots & \cdots & \cdots & \cdots & \cdots \\ u_{m1} & u_{m2} & \cdots & u_{mm}e^{-j\varphi}\cos\theta - u_{mn}\sin\theta & \cdots & u_{mm}e^{-j\varphi}\sin\theta + u_{mn}\cos\theta & \cdots & u_{mN} \\ \cdots & \cdots & \cdots & \ddots & \cdots & \cdots & \cdots & \\ u_{n1} & u_{n2} & \cdots & u_{nm}e^{-j\varphi}\cos\theta - u_{nn}\sin\theta & & u_{nm}e^{-j\varphi}\sin\theta + u_{nn}\cos\theta & \cdots & u_{nN} \\ \cdots & \cdots & \cdots & \cdots & \ddots & \cdots & \cdots & \cdots \\ u_{N1} & u_{N2} & \cdots & u_{Nm}e^{-j\varphi}\cos\theta - u_{Nn}\sin\theta & \cdots & u_{Nm}e^{-j\varphi}\sin\theta + u_{Nn}\cos\theta & \cdots & u_{NN} \end{pmatrix} \tag{4.8}$$

For example, if we want to null coefficient (N, n) then

$$u_{Nm}e^{-j\varphi}\sin\theta + u_{Nn}\cos\theta = 0 \tag{4.9}$$

leading to

$$\begin{pmatrix} u_{11} & u_{12} & \cdots & * & \cdots & * & \cdots & u_{1N} \\ u_{21} & u_{22} & \cdots & * & \cdots & * & \cdots & u_{2N} \\ \cdots & \cdots & \ddots & \cdots & \cdots & \cdots & \cdots & \cdots \\ u_{m1} & u_{m2} & \cdots & * & \cdots & * & \cdots & u_{mN} \\ \cdots & \cdots & \cdots & \cdots & \ddots & \cdots & \cdots & \cdots \\ u_{n1} & u_{n2} & \cdots & * & \cdots & * & \cdots & u_{nN} \\ \cdots & \cdots & \cdots & \cdots & \cdots & \cdots & \ddots & \cdots \\ u_{N1} & u_{2} & \cdots & * & \cdots & 0 & \cdots & u_{nN} \end{pmatrix} \tag{4.10}$$

where $*$ denotes non-zero complex valued coefficients.

Property 4: *The product $T_{mn}(\varphi,\theta)\,U$ is a unitary matrix that leaves all the rows in U unaltered with the exception of rows m and n and we can choose values for (φ,θ) to null whatever (but only one) coefficient in these rows.*

$T_{mn}(\varphi,\theta)\,U$ is the product of two unitary matrices and it is thus unitary. Now, taking into account Eqn. (4.1) and Eqn. (4.6) we obtain

$$T_{mn}(\varphi,\theta)\,U = \begin{pmatrix} 1 & 0 & \cdots & 0 & \cdots & 0 & \cdots & 0 \\ 0 & 1 & \cdots & 0 & \cdots & 0 & \cdots & 0 \\ \cdots & \cdots & \ddots & \cdots & \cdots & \cdots & \cdots & \cdots \\ 0 & 0 & \cdots & e^{j\varphi}\cos\theta & \cdots & -\sin\theta & \cdots & 0 \\ \cdots & \cdots & \cdots & \cdots & \ddots & \cdots & \cdots & \cdots \\ 0 & 0 & \cdots & e^{j\varphi}\sin\theta & \cdots & \cos\theta & \cdots & 0 \\ \cdots & \cdots & \cdots & \cdots & \cdots & \cdots & \ddots & \cdots \\ 0 & 0 & \cdots & 0 & \cdots & 0 & \cdots & 1 \end{pmatrix} \begin{pmatrix} u_{11} & u_{12} & \cdots & u_{1N} \\ u_{21} & u_{22} & \cdots & u_{2N} \\ \cdots & \cdots & \cdots & \cdots \\ u_{N1} & u_{N2} & \cdots & u_{NN} \end{pmatrix} = \tag{4.11}$$

$$= \begin{pmatrix} u_{11} & u_{12} & \cdots & u_{1m} & \cdots & u_{1n} & \cdots & u_{1N} \\ u_{21} & u_{22} & \cdots & u_{2m} & \cdots & u_{2n} & \cdots & u_{2N} \\ \cdots & \cdots & \ddots & \cdots & \cdots & \cdots & \cdots & \cdots \\ u_{m1}e^{j\varphi}\cos\theta - u_{n1}\sin\theta & u_{m2}e^{j\varphi}\cos\theta - u_{n2}\sin\theta & \cdots & u_{mm}e^{j\varphi}\cos\theta - u_{nm}\sin\theta & \cdots & u_{mn}e^{j\varphi}\cos\theta - u_{nn}\cos\theta & \cdots & u_{mN}e^{j\varphi}\cos\theta - u_{nN}\sin\theta \\ \cdots & \cdots & \cdots & \cdots & \ddots & \cdots & \cdots & \cdots \\ u_{m1}e^{j\varphi}\sin\theta + u_{n1}\cos\theta & u_{m2}e^{j\varphi}\sin\theta + u_{n2}\cos\theta & \cdots & u_{mm}e^{j\varphi}\sin\theta + u_{nm}\cos\theta & \cdots & u_{mn}e^{j\varphi}\sin\theta + u_{nn}\cos\theta & \cdots & u_{mN}e^{j\varphi}\sin\theta + u_{nN}\cos\theta \\ \cdots & \cdots & \cdots & \cdots & \cdots & \cdots & \ddots & \cdots \\ u_{N1} & u_{N2} & \cdots & u_{Nm} & \cdots & u_{Nn} & \cdots & u_{NN} \end{pmatrix}$$

For example, if we want to null coefficient (n, N) then

$$u_{mN}e^{j\varphi}\sin\theta + u_{nN}\cos\theta = 0 \tag{4.12}$$

Leading to

$$\begin{pmatrix} u_{11} & u_{12} & \cdots & u_{1m} & \cdots & u_{1n} & \cdots & u_{1n} \\ u_{21} & u_{22} & \cdots & u_{2m} & \cdots & u_{2n} & \cdots & u_{2N} \\ \cdots & \cdots & \ddots & \cdots & \cdots & \cdots & \cdots & \cdots \\ * & * & \cdots & * & \cdots & * & \cdots & * \\ \cdots & \cdots & \cdots & \cdots & \ddots & \cdots & \cdots & \cdots \\ * & * & \cdots & * & \cdots & * & \cdots & 0 \\ \cdots & \cdots & \cdots & \cdots & \cdots & \cdots & \ddots & \cdots \\ u_{N1} & u_{N2} & \cdots & u_{Nm} & \cdots & u_{nN} & \cdots & u_{NN} \end{pmatrix} \tag{4.13}$$

Property 5: *If D is a diagonal matrix with $d_{ii} = e^{j\alpha_i}$ then for any $T_{mn}^{-1}(\varphi,\theta)$ one can find a diagonal matrix D' with $d'_{ii} = e^{j\delta_i}$ and a matrix $T_{mn}(\varphi',\theta')$ such that $T_{mn}^{-1}(\varphi,\theta)D = D'T_{mn}(\varphi',\theta')$.*

We construct

$$T_{mn}^{-1}(\varphi,\theta)D = \begin{pmatrix} 1 & 0 & \cdots & 0 & \cdots & 0 & \cdots & 0 \\ 0 & 1 & \cdots & 0 & \cdots & 0 & \cdots & 0 \\ \cdots & \cdots & \ddots & \cdots & \cdots & \cdots & \cdots & \cdots \\ 0 & 0 & \cdots & e^{-j\varphi}\cos\theta & \cdots & e^{-j\varphi}\sin\theta & \cdots & 0 \\ \cdots & \cdots & \cdots & \cdots & \ddots & \cdots & \cdots & \cdots \\ 0 & 0 & \cdots & -\sin\theta & \cdots & \cos\theta & \cdots & 0 \\ \cdots & \cdots & \cdots & \cdots & \cdots & \cdots & \ddots & \cdots \\ 0 & 0 & \cdots & 0 & \cdots & 0 & \cdots & 1 \end{pmatrix} \begin{pmatrix} e^{j\alpha_1} & \cdots & 0 \\ \vdots & \ddots & \vdots \\ 0 & \cdots & e^{j\alpha_N} \end{pmatrix} = \tag{4.14}$$

$$= \begin{pmatrix} e^{j\alpha_1} & 0 & \cdots & 0 & \cdots & 0 & \cdots & 0 \\ 0 & e^{j\alpha_2} & \cdots & 0 & \cdots & 0 & \cdots & 0 \\ \cdots & \cdots & \ddots & \cdots & \cdots & \cdots & \cdots & \cdots \\ 0 & 0 & \cdots & e^{j\alpha_m} e^{-j\varphi} \cos\theta & \cdots & e^{j\alpha_n} e^{-j\varphi} \sin\theta & \cdots & 0 \\ \cdots & \cdots & \cdots & \cdots & \ddots & \cdots & \cdots & \cdots \\ 0 & 0 & \cdots & -e^{j\alpha_m} \sin\theta & \cdots & e^{j\alpha_n} \cos\theta & \cdots & 0 \\ \cdots & \cdots & \cdots & \cdots & \cdots & \cdots & \ddots & \cdots \\ 0 & 0 & \cdots & 0 & \cdots & 0 & \cdots & e^{j\alpha_N} \end{pmatrix}$$

and

$$D' T_{mn}(\varphi',\theta') D = \begin{pmatrix} e^{j\delta_1} & \cdots & 0 \\ \vdots & \ddots & \vdots \\ 0 & \cdots & e^{j\delta_N} \end{pmatrix} \begin{pmatrix} 1 & 0 & \cdots & 0 & \cdots & 0 & \cdots & 0 \\ 0 & 1 & \cdots & 0 & \cdots & 0 & \cdots & 0 \\ \cdots & \cdots & \ddots & \cdots & \cdots & \cdots & \cdots & \cdots \\ 0 & 0 & \cdots & e^{j\varphi'} \cos\theta' & \cdots & -\sin\theta' & \cdots & 0 \\ \cdots & \cdots & \cdots & \cdots & \ddots & \cdots & \cdots & \cdots \\ 0 & 0 & \cdots & e^{j\varphi'} \sin\theta' & \cdots & \cos\theta' & \cdots & 0 \\ \cdots & \cdots & \cdots & \cdots & \cdots & \cdots & \ddots & \cdots \\ 0 & 0 & \cdots & 0 & \cdots & 0 & \cdots & 1 \end{pmatrix} = \tag{4.15}$$

$$= \begin{pmatrix} e^{j\delta_1} & 0 & \cdots & 0 & \cdots & 0 & \cdots & 0 \\ 0 & e^{j\delta_2} & \cdots & 0 & \cdots & 0 & \cdots & 0 \\ \cdots & \cdots & \ddots & \cdots & \cdots & \cdots & \cdots & \cdots \\ 0 & 0 & \cdots & e^{j\delta_m} e^{j\varphi'} \cos\theta' & \cdots & -e^{j\delta_m} \sin\theta' & \cdots & 0 \\ \cdots & \cdots & \cdots & \cdots & \ddots & \cdots & \cdots & \cdots \\ 0 & 0 & \cdots & e^{j\delta_n} e^{j\varphi'} \sin\theta' & \cdots & e^{j\delta_n} \cos\theta' & \cdots & 0 \\ \cdots & \cdots & \cdots & \cdots & \cdots & \cdots & \ddots & \cdots \\ 0 & 0 & \cdots & 0 & \cdots & 0 & \cdots & e^{j\delta_N} \end{pmatrix}$$

Identifying Eqn. (4.14) and Eqn. (4.15) results in the following equations

$$\begin{aligned} &\delta_i = \alpha_i, \quad i \neq m,n \\ &e^{j\alpha_m} e^{-j\varphi} \cos\theta = e^{j\delta_m} e^{j\varphi'} \cos\theta' \\ &e^{j\alpha_n} e^{-j\varphi} \sin\theta = -e^{j\delta_m} \sin\theta' \\ &- e^{j\alpha_m} \sin\theta = e^{j\delta_n} e^{j\varphi'} \sin\theta' \\ &e^{j\alpha_n} \cos\theta = e^{j\delta_n} \cos\theta' \end{aligned} \tag{4.16}$$

Which result either in

$$\begin{aligned} \delta_i &= \alpha_i, \quad i \neq m, n \\ \delta_m &= \alpha_n - \varphi \\ \delta_n &= \alpha_n - \pi \\ \varphi' &= \alpha_m - \alpha_n + \pi \\ \theta' &= \theta - \pi \end{aligned} \tag{4.17}$$

Or

$$\begin{aligned} \delta_i &= \alpha_i, \quad i \neq m \\ \delta_m &= \alpha_n - \varphi - \pi \\ \varphi' &= \alpha_m - \alpha_n + \pi \\ \theta' &= \theta \end{aligned} \tag{4.18}$$

4.2 Triangular Interferometers

4.2.1 Principles and Design

The triangular arrangement for a multi-port programmable interferometer was proposed by Reck and colleagues in 1994 [1]. It is based on the iterative use of Property 3 described in Section 4.1 to make zero all the off-diagonal elements in the unitary matrix U.

To this end and in a first step, U is multiplied from the right by a succession of matrices $T_{Nq}^{-1}(\varphi_q, \theta_q)$ with $q = N-1, \ldots 1$. This product is equivalent to the attachment of successive 2×2 tunable beamsplitters to ports q and N. By making use of Property 3 after each product to null element (q, N) and then property 1, we have

$$U(N)\, T_{NN-1}^{-1}\, T_{NN-2}^{-1} \cdots T_{N1}^{-1} = \begin{pmatrix} * & * & * & * & 0 \\ * & * & * & \cdots & 0 \\ \cdots & \cdots & \cdots & \cdots & 0 \\ * & * & * & * & 0 \\ 0 & 0 & 0 & 0 & e^{j\alpha_N} \end{pmatrix} = \begin{pmatrix} U'(N-1) & \mathbf{0} \\ \mathbf{0} & e^{j\alpha_N} \end{pmatrix} \tag{4.19}$$

Note that the effective dimension of the resulting matrix has been reduced to $N-1$. We can now apply the same procedure based on Property 3, bearing in mind that the elements in the last row will remain zero and thus will not be affected by successive multiplications from the right, to reduce the dimensions to $N-2$, $N-3$ and finally to 1, leading to

$$U(N)T_{NN-1}^{-1}T_{NN-2}^{-1}\cdots T_{N1}^{-1}\cdots T_{32}^{-1}T_{31}^{-1}T_{21}^{-1} = \begin{pmatrix} e^{j\alpha_1} & 0 & 0 & 0 & 0 \\ 0 & e^{j\alpha_2} & 0 & \cdots & 0 \\ \cdots & \cdots & \cdots & \cdots & 0 \\ 0 & 0 & 0 & e^{j\alpha_{N-1}} & 0 \\ 0 & 0 & 0 & 0 & e^{j\alpha_N} \end{pmatrix} = D \tag{4.20}$$

Hence and as D is a diagonal (and unitary) matrix we have

$$U(N)T_{NN-1}^{-1}T_{NN-2}^{-1}\cdots T_{N1}^{-1}\cdots T_{32}^{-1}T_{31}^{-1}T_{21}^{-1}D^{-1} = I \tag{4.21}$$

Hence

$$U^{-1}(N) = T_{NN-1}^{-1}T_{NN-2}^{-1}\cdots T_{N1}^{-1}\cdots T_{32}^{-1}T_{31}^{-1}T_{21}^{-1}D^{-1} \tag{4.22}$$

which finally yields

$$U(N) = DT_{21}T_{31}T_{32}\cdots T_{N-1N-2}T_{N1}\cdots T_{NN-2}T_{NN-1} = D\prod_{i=2}^{N}\left[\prod_{j=1}^{i-1}T_{ij}\right] \tag{4.23}$$

Equation (4.23) provides the desired decomposition of the unitary operator into a set of beamsplitter operations completed by a phase shifting operation. Each beamsplitter relates two inputs and two outputs of the overall structure. Reck and co-workers proposed for $T_{mn}(\varphi_{mn}, \theta_{mn})$ the bulk optics configuration shown in Figure 4.1 (a), which is represented in chip layouts by the symbol shown in Figure 4.1 (b). Figure 4.1 (c) and Figure 4(d) show as an example, the decomposition and the layout for a general 4×4 unitary matrix.

As it can be appreciated, the practical implementation of this method involves a triangular array of beamsplitters, each diagonal row in the triangle reducing the effective dimension of the Hilbert space by one. The maximum number of required beamsplitters to implement U(N) is given by $N(N-1)/2$.

For later comparison with the rectangular design it is useful to define the *depth* of an interferometer [2] as the longest path through the interferometer, enumerated by counting the number of beamsplitters traversed by that path. In the case of the triangular interferometer, this corresponds to the edges of the triangle and is given by $(N-1)+(N-2) = 2N-3$.

The formalism developed for $N \times N$ unitary transformations can be extended to include any $M \times N$ linear (non-unitary) transformation. It has been shown that that any $M \times N$ linear transformation, with $M \leq N$ ($M \geq N$), can be either directly embedded within a $2N \times 2N$ ($2M \times 2M$) unitary transformation [5] to within a scaling factor, or, implemented by 2 separate $N \times N$($M \times M$) interferometers connected to each other via

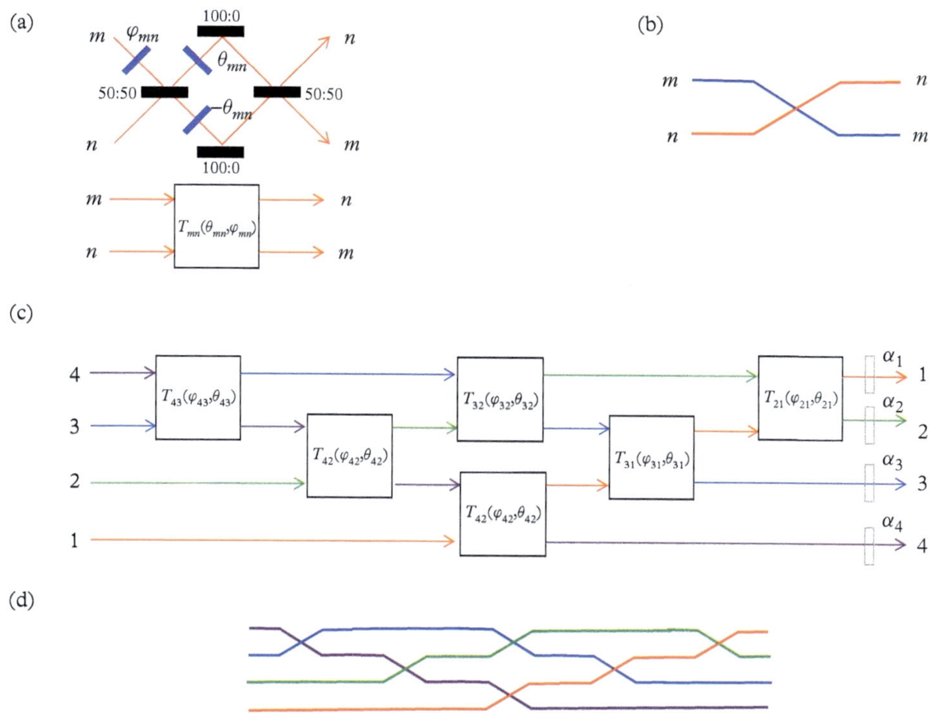

Figure 4.1 *(a) Bulk optics implementation of a tunable beam splitter (after [1]). (b) Symbol representation of the beam splitter layout. (c) Triangular decomposition implementing a general 4x4 unitary transformation and layout (d). Colours represent the propagation of a given input port.*

phase and amplitude modulators [6]. A similar formalism can be developed multiplying U from the left by a succession of matrices T and making use of Property 4.

4.2.2 Integrated Triangular Interferometers

Triangular interferometers can be integrated on a chip as suggested by Miller [6], [7] by replacing the tunable bulk optics beamsplitter shown in Figure 4.1(a) by its equivalent version in integrated optics, which is the tunable basic unit (TBU) shown in Figure 2.12. Up to now, this has been the preferred option for implementing most reconfigurable integrated optical quantum circuits [3], [8]–[11] and some classical applications as well [4], [5], [7], [12]. Figure 4.2 (a)–(b) illustrates this equivalence for the 4×4 unitary transformer discussed above. The triangular interferometer has been reported in several material platforms. We show now some examples although further discussion of these and other implementations will be provided in Chapter 7 and Chapter 8, respectively, in the context of their applications.

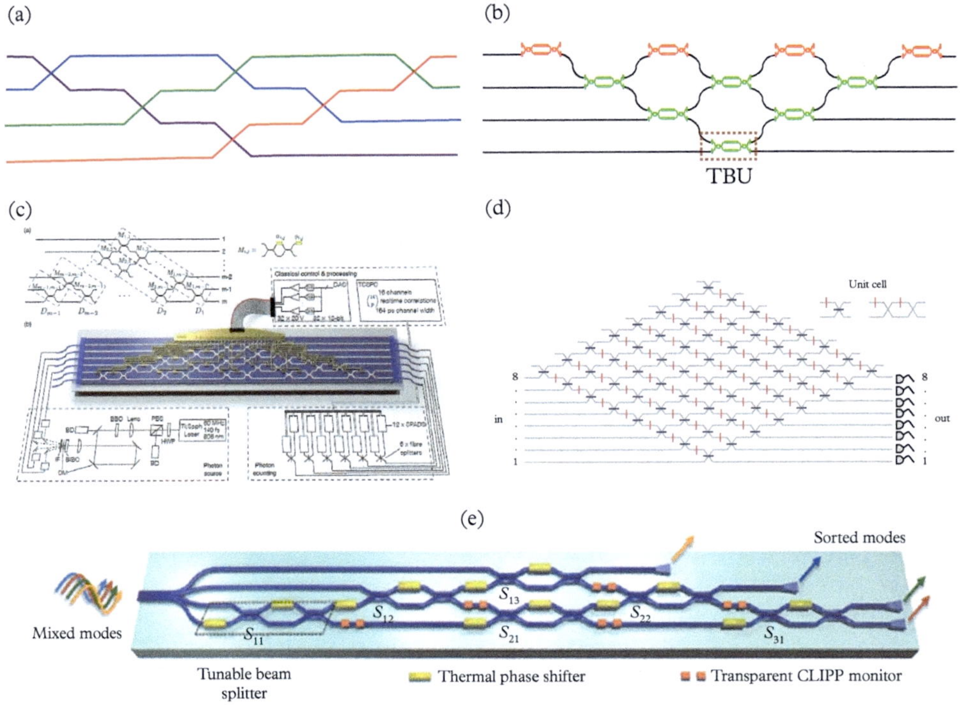

Figure 4.2 *Integrated triangular interferometers. (a)–(b) U(4) layout of Figure 4.1(d) and its implementation using MZI based TBUs. TBUs in red colour operate as phase shifters. (c)–(e) Implementations in different material platforms reported in the literature. (c) U(6) in Silica on Silicon (after [8]), (d) U(8) in Silicon Nitride (after [9]) and (e) U(4) in Silicon (after [12]).*

Figure 4.2(c) shows an integrated circuit implementing a $U(6)$ fabricated in SiO_2 on silicon to implement a boson sampling circuit operating with six photon states [8]. The chip consists of a cascade of 15 Mach–Zehnder interferometers with 30 thermo-optic phase shifters integrated into a single photonic chip that is electrically and optically interfaced for arbitrary setting of all phase shifters. the chip could be programmed to implement different functionalities like heralded quantum logic and entangling gates, boson sampling with verification tests, and six-dimensional complex Hadamards. Figure 4.2(d) shows a silicon nitride chip implementing a programmable $U(8)$ transformation [11]. The chip contains 128 tunable elements and was programmed to implement a variety of photonic quantum information processing primitives, in the form of Hong–Ou–Mandel interference, bosonic coalescence/anti-coalescence and high dimensional single-photon quantum gates. Finally, Figure 4.2(e) shows the layout of Silicon chip implemented in silicon photonics designed to provide programmable mode scrambling and unscrambling functionalities [12]. The chip included six MZIs and twelve phase shifters.

4.3 Rectangular Interferometers

4.3.1 Principles and Design

The rectangular universal $N \times N$ interferometer has been recently proposed [2]. This design is based on an alternative arrangement of $U(2)$ beamsplitters and phase shifters and outperforms the triangular arrangement in a series of figures of merit, including the optical depth, which is half of that of the triangular design and robustness against optical losses.

In the triangular decomposition [1], *successive columns or rows of the* U(N) *matrix are nulled* using either only T_{mn} matrices (multiplied from the left) or only T_{mn}^{-1} matrices (multiplied by the right) respectively, making sure that nulled elements of $U(N)$ are not affected by subsequent operations. The rectangular decomposition [2], proceeds by *nulling successive diagonals* of U(N) by alternating between T_{mn} and T_{mn}^{-1} matrices where $m = n - 1$, in such a way that every nulled diagonal in the matrix corresponds to one diagonal line of beamsplitters through the interferometer. The sequence of T_{mn} and T_{mn}^{-1} matrices must both correspond to the desired order of beamsplitters in the interferometer and guarantee that the nulled elements of U(N) are not affected by subsequent operations. This is illustrated as an example in Figure 4.3 for the case of a $U(5)$ interferometer [2].

The process [2], starts by nulling the bottom left element of $U(5)$ (first diagonal) with a T_{12}^{-1} matrix, which, according to property 3, causes the first two columns of $U(5)$ to mix. This corresponds to adding the top left beamsplitter in the interferometer as shown in the right hand-side of Figure 4.3(b). When then null the second diagonal using a T_{34} matrix followed by a T_{45} matrix, which correspond to the two bottom right beamsplitters in the interferometer (Figure 4.3(b)). According to property 4, T_{34} mixes rows 3 and 4, and T_{45} mixes rows 4 and 5. Note that since both the (4,1) and (5,1) elements of $U(5)$ have been previously nulled, they are not affected by T_{45}. We then proceed through each step to null a successive diagonal of the updated $U(5)$ matrix by alternating between T_{mn}^{-1} and T_{mn} matrices, which corresponds to adding diagonal lines of beamsplitters to the interferometer (Figure 4.3(d)–(e)).

The final expression after the decomposition process is:

$$T_{45}T_{34}T_{23}T_{12}T_{45}T_{34}UT_{12}^{-1}T_{34}^{-1}T_{23}^{-1}T_{12}^{-1} = D \tag{4.24}$$

or:

$$U = T_{34}^{-1}T_{45}^{-1}T_{12}^{-1}T_{23}^{-1}T_{34}^{-1}T_{45}^{-1}DT_{12}T_{23}T_{34}T_{12} \tag{4.25}$$

However, taking property 5 into consideration, we finally have:

$$U = D'T_{34}T_{45}T_{12}T_{23}T_{34}T_{45}T_{12}T_{23}T_{34}T_{12} \tag{4.26}$$

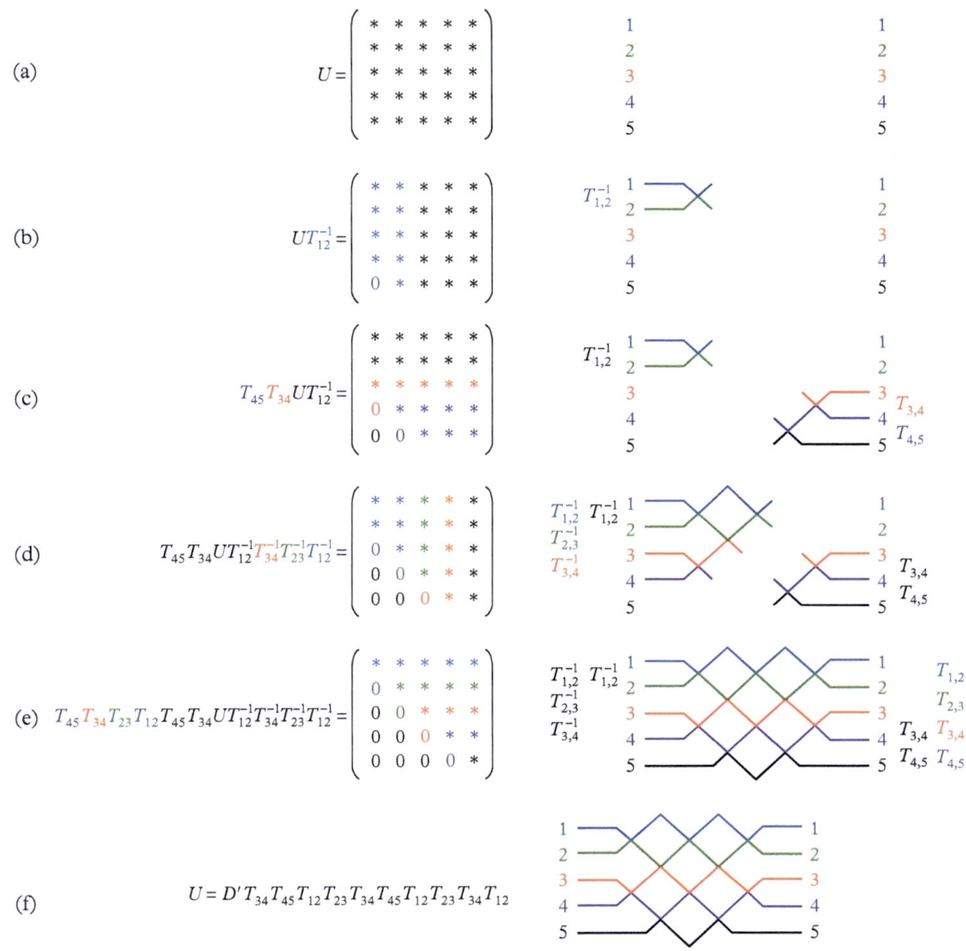

Figure 4.3 *Illustration of the algorithm for programming a universal multiport interferometer for U(5) interferometer (after [2]). The left-hand side presents the decomposition procedure, and the right-hand side shows how the decomposition corresponds to building up the corresponding interferometer. (a) The staring point is any random unitary matrix U(5), and a blank interferometer. (b)-(e) successive stages of the implementation algorithm. (f) Final expression for U(5) in terms of a cascade of U(2) interferometers and a diagonal matrix.*

The decomposition in Eqn. (4.26) corresponds to the multi-port interferometer shown in Figure 4.3(f), and the values of the θ and ϕ of the T_{mn} matrices determine the values of the beamsplitters and phase shifts that must be programmed to implement $U(5)$.

The process can be generalised to any value of N using the following algorithm:

1: **procedure** DECOMPOSE($\hat{U}$)

2: **for** i from 1 to $N-1$ **do**

3: **if** i is odd then

4: **for** $j = 0$ to $i - 1$ **do**

5: Find a $T^{-1}{}_{i-j,i-j+1}$ matrix that nulls element $(N - j, i - j)$ of $\hat{U}$

6: Update $\hat{U} = \hat{U}\; T^{-1}{}_{i-j,i-j+1}$

7: **else**

8: **for** $j = 1$ to i **do**

9: Find a $T_{N+j-i-1,N+j-i}$ matrix that nulls element $(N + j - i, j)$ of $\hat{U}$

10: Update $\hat{U} = T_{N+j-i-1,N+j-i}\hat{U}$

Which renders the following result

$$\left(\prod_{(m,n)\in S_L} T_{mn}\right) U \left(\prod_{(m,n)\in S_R} T_{mn}^{-1}\right) = D \tag{4.27}$$

where D is a diagonal matrix where the non-zero diagonal elements are modulus one (i.e. phase shifts) and S_L, S_R represent the (m, n) index ordering for the T_{mn} and T_{mn}^{-1} matrices respectively.
Equation (4.27) can be transformed into

$$U = \left(\prod_{(m,n)\in S_L} T_{mn}\right)^{-1} D \left(\prod_{(m,n)\in S_R} T_{mn}^{-1}\right)^{-1} = \left(\prod_{\substack{(m,n)\in S_L \\ \text{reverse order}}} T_{mn}^{-1}\right) D \left(\prod_{\substack{(m,n)\in S_R \\ \text{reverse order}}} T_{mn}\right) \tag{4.28}$$

Further, by using property 5

$$U = D' \left(\prod_{\substack{(m,n)\in S_L \\ \text{reverse order}}} T'_{mn}\right)\left(\prod_{\substack{(m,n)\in S_R \\ \text{reverse order}}} T_{mn}\right) = D' \left(\prod_{(m,n)\in S} T_{mn}\right) \tag{4.29}$$

where S is the order of the beamsplitters corresponding to the desired circuit.

The rectangular interferometer requires $N(N-1)/2$ beamsplitters to implement a $U(N)$ transformation, just like the triangular arrangement. However, its depth is only N, which is roughly half required for the triangular design and this leads to lower balanced losses, which are related to the propagation loss of the longest path in the interferometer [2]. In addition, its symmetry leads to improved loss tolerance as related to the unbalanced loss due to the different losses experienced by the signal propagating through the various paths of the interferometer. More information regarding the comparative performance of these and other multi-port interferometer appears later in this chapter.

4.3.2 Integrated Rectangular Interferometers

Rectangular interferometers can also be integrated on a chip following a similar procedure to that of the triangular arrangement using, for instance, the TBU of Figure 2.12. Figures 4.4(a) and 4.4(b) show the layout of a $U(9)$ interferometer and its integrated optics design using 3-dB MZI TBUs.

Due to its recent proposal, there have been few experimental reports of the rectangular interferometer. The first one was reported by Pérez and colleagues [4], who emulated a $U(4)$ interferometer in a silicon hexagonal waveguide mesh. Specific rectangular interferometer integrated designs have been very recently reported [13]. Figure 4.4(c) shows the photograph of a modular set composed of three linear optical circuits. Each module consists of ten tunable MZIs placed side by side (Figure 4.4(d)). Modules are

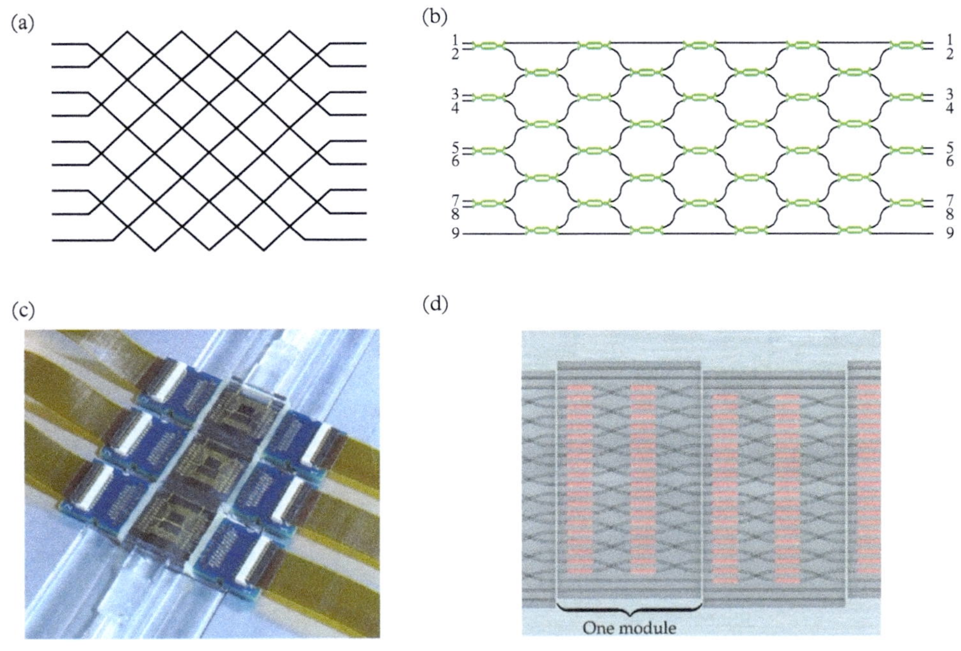

Figure 4.4 *Integrated rectangular interferometers. (a) Design Layout for a U(9) transformer (after [2]) and (b) implementation layout suitable for integrated optics using 3dB-MZIs (after [4]). (c) Photograph of a three module silica on silicon rectangular interferometer (after [13]). (d) Detail of an individual module showing 10 parallel TBUs (after [13]).*

to be tiled to build larger interferometers. For example, N of such modules with at least $N/2$ MZI wide are enough to perform an arbitrary unitary transformation U(N), with optimal circuit depth. In this case, the circuit was employed to implement up to $U(6)$ transformations for different applications, which are discussed in Chapter 7 and Chapter 8.

The chip was implemented in silica-on-silicon technology, where silica glass layers were fabricated in-house by flame hydrolysis deposition (FHD) on a silicon wafer and waveguides were directly written using a 244 nm frequency-doubled argon-ion laser into this planar structure (i.e. a focused UV beam was translated relative to the photosensitive sample on precision air-bearing stages, producing buried channel waveguides). Phase shifters were then patterned through contact lithography and lift-off of a 170 nm e-beam evaporated nichrome layer, while the wiring was a 200 nm copper layer deposited in the same manner. Computer control of the on-chip phase shifters was accomplished using custom drive electronics, producing 8-bit pulse-width-modulated drive signals at up to 20 V. Due to the fact that silica on silicon was employed very low fibre-to-chip coupling losses (0.8 dB) and propagation losses (0.3 5 dB/cm) were reported. Chip-to-chip coupling loss was around 0.2 dB and coupler excess losses where in the range of 2–2.4 dB.

4.4 Specialised Interferometers

Specialised interferometers are multi-port devices configured to perform a fixed task. This means that in general they cannot support arbitrary unitary transformations and in most of the cases, their range of applications and reconfigurability is very restricted. Therefore, from the point of view of programmable photonics they are of less interest. However, there are some designs that either allow some flexible programmability or implement quite useful functionalities. We briefly review some of these examples.

A first structure that has been employed in the context of boson sampling and quantum transport applications is the trapezoidal mesh of reconfigurable beamsplitters (RBSs) [9], [14], [15] shown in Figure 4.5(a)–(c). Figure 4.5(a) shows the circuit layout to be implemented and Figure 4.5(b) the actual physical realisation fabricated using a complementary metal-oxide semiconductor (CMOS)-compatible silicon photonics process, which is composed of eighty-eight MZIs, twenty-six input modes, twenty-six output modes and 176 phase shifters, each of which can impart a phase shift of more than 2π radians. Each RBS is composed of two 50 per cent directional couplers separated by an internal (θ) and an external (ϕ) thermo-optic phase shifter, as shown in Figure 4.5(c) and applies the rotation given by Eqn. (2.111). Although the structure is neither a triangular nor a rectangular interferometer, it can be programmed so part of it can actually operate as one of them, featuring, however, a lower number of input/output port count than its nominal figure, i.e. 26. For example, Shen and colleagues [16] employed part of the chip to implement a programmable $U(4)$ transformation (Figure 4.5(d)), which is required to support the linear stage of a neural network (more details are provided in Chapter 7). Actually, this trapezoidal RBS mesh can be seen as an

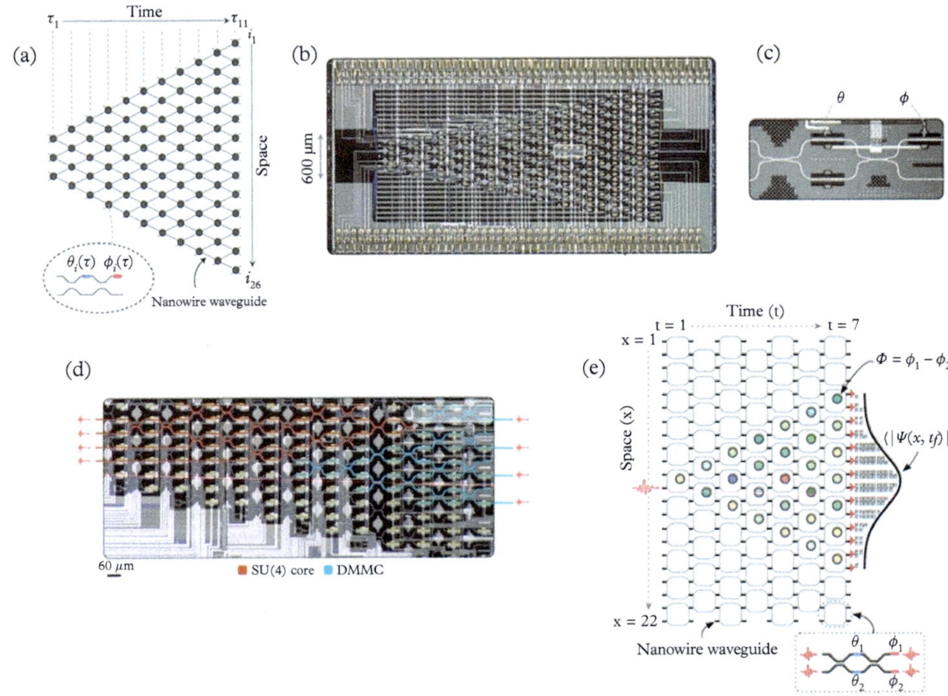

Figure 4.5 *Application specific multiport interferometers (a) Layout and (b) physical realization of a 26 input/26 output trapezoidal RBS mesh interferometer designed for boson sampling and quantum transport applications (after [9]). (c) Detail of the RBS. Emulation of a triangular U(4) transformer using part of the former circuit (after [16]). (e) Embedding of the trapezoidal interferometer in a more general rectangular mesh of RBSs (after [17]).*

intermediate stage between ordinary multi-port universal programmable interferometers considered in Section 4.2 and Section 4.3 and the programmable waveguide meshes, which will be considered in Chapter 5. For instance, Figure 4.5(e) illustrates the embedding of the trapezoidal mesh within a rectangular mesh of RBSs. This rectangular mesh is quite similar to the rectangular universal interferometer shown in Figure 4.4(b).

Another interesting structure due to the huge applications of Fourier transforms are the so-called *fast* interferometers [17]–[19]. The name comes after the fact that these structures are designed to implement the fast Fourier transform (FFT) by hardware implementation of the Cooley–Tukey (CT) algorithm [20] using linear optics and RBSs.

In essence, the fast interferometer seeks to implement the linear matrix transformation corresponding to the discrete Fourier transform (DFT) of a $N = 2^n$ digital vector given by the unitary matrix

$$DFT(n) = \frac{1}{2^{n/2}} \begin{pmatrix} 1 & 1 & 1 & 1 & \cdots & 1 \\ 1 & u & u^2 & u^3 & \cdots & u^{2^n-1} \\ 1 & u^2 & u^4 & u^6 & \cdots & u^{2(2^n-1)} \\ 1 & u^3 & u^6 & u^9 & \cdots & u^{3(2^n-1)} \\ \vdots & \vdots & \vdots & \vdots & \ddots & \vdots \\ 1 & u^{2^n-1} & u^{2(2^n-1)} & u^{3(2^n-1)} & \cdots & u^{(2^n-1)(2^n-1)} \end{pmatrix} \tag{4.30}$$

$$u = e^{j\frac{2\pi}{2^n}}$$

This operation is key in many applications in classical and quantum optics signal processing [21]–[25] and can be implemented using the combinations of beamsplitters and phase shifters described in Chapter 2 and Eqn. (2.111) and Eqn. (2.112). In particular the method outlined in [18] is based on a more complex $U(2)$ unit featuring phase shifters both at the input and output ports of the 3dB MZI tunable coupler as shown in Figure 4.5(a). In this case, following the same procedure described in Chapter 2, it is straightforward to find the scattering matrix for such unit:

$$U(\alpha,\beta,\delta,\gamma) = e^{j\alpha} \begin{pmatrix} e^{j(\gamma+\delta)}\cos(\beta) & e^{j(-\gamma+\delta)}\sin(\beta) \\ -e^{j(\gamma-\delta)}\sin(\beta) & e^{-j(\gamma+\delta)}\cos(\beta) \end{pmatrix} \tag{4.31}$$

The process, for the design of the fast interferometer involves four steps. In the first and once the value of the number of bits n is specified, the required $DFT(\mathrm{n})$ matrix is computed from Eqn. (4.30). In the second step, the CT algorithm is employed to break down the $DFT(\mathrm{n})$ matrix into a product of n matrices $S_k^{(n)}$, where $k = 1, 2, \rightleftharpoons n$ is associated with each stage of the CT algorithm. Each matrix is implemented by means of 2^{n-1} TBUs, described by Eqn. (4.31). In the third step, the CT algorithm is employed to obtain a graphical representation of the optical circuit and finally, in step four the graphical representation and the matrices $S_k^{(n)}$ are employed to find the design parameters (given by Eqn. 4.31) of each $U(2)$ TBU in the circuit.

The case of a two-dimensional ($N = 2$, $n = 1$) DFT is straightforward. Here Eqn. (4.30) transforms to

$$DFT(1) = \frac{1}{\sqrt{2}} \begin{pmatrix} 1 & 1 \\ 1 & -1 \end{pmatrix} \tag{4.32}$$

which corresponds to a Hadamard transformation [25] and, incidentally one 2×2 TBU, hence $DFT(1) = S_1^{(1)}$. The graphical representation corresponds to Figure 4.6(a) where the TBU parameters are given by $TU_1^{(1)} = U(\pi/2, \pi/4, 0, \pi/2)$.

In the case a four-dimensional ($N = 4$, $n = 2$) Eqn. (4.30) transforms to:

$$DFT(2) = \frac{1}{2}\begin{pmatrix} 1 & 1 & 1 & 1 \\ 1 & j & -1 & -j \\ 1 & -1 & 1 & -1 \\ 1 & -j & -1 & 1 \end{pmatrix} \tag{4.33}$$

The second step yields

$$DFT(2) = S_2^{(2)} S_1^{(2)}$$
$$S_1^{(2)} = \frac{1}{\sqrt{2}}\begin{pmatrix} 1 & 0 & 1 & 0 \\ 0 & 1 & 0 & 1 \\ 1 & 0 & -1 & 0 \\ 0 & 1 & 0 & -1 \end{pmatrix}; \tag{4.34}$$
$$S_2^{(2)} = \frac{1}{\sqrt{2}}\begin{pmatrix} 1 & 1 & 0 & 0 \\ 0 & 0 & 1 & j \\ 1 & -1 & 0 & 0 \\ 0 & 0 & 1 & -j \end{pmatrix}$$

This two-stage process corresponds to the graphical representation shown in Figure 4.6(b), where each intersection between lines of the same colour represents a TU. The $S_1^{(2)}$ matrix represents the first stage of the CT algorithm and here two *TUs* combine input ports labelled 0 and 2 and inputs labelled 1 and 3, respectively. The second stage matrix $S_2^{(2)}$ is implemented using two *TUs* that combine respectively the addition of input ports 0 and 2 signals with the addition of input ports 1 and 3 signals and the subtraction of input ports 0 and 2 signals with the subtraction of input ports 1 and 3 signals. The overall $DFT(2)$ circuit is composed of twice the previous $DFT(1)$ circuits, where the output of each circuit is connected to the output of the other through one TU component, similar to that of $DFT(1)$, that is: $TU_1^{(2)} = TU_1^{(1)} = U(\pi/2, \pi/4, 0, \pi/2,)$ and a new one $TU_2^{(2)} = U(\pi/4, \pi/4, 7\pi/4, \pi/2,)$ (Figure 4.6(c)). In total, four *TUs* are required as compared to the six units required by the triangular multi-port interferometer. Figure 4.6(d) shows an implementation of the circuit based on bulk beamsplitters.

The process can be generalised by induction [18] to any $N = 2^n$ system leading to the combination of 2 DFT (n – 1) circuits as shown in Figure 4.6(a). Each of the outputs of these circuits is combined with the outputs of the other using TUs such that they consist of a series of 2^{n-2} components similar to those employed in the DFT (n – 1) circuits and 2^{n-2} components that differ by a phase factor of $\pi/2^n$.

A main limitation for the implementation of fast interferometers in integrated optics is connected to the unavoidable waveguide crossing that is required to implement the architecture of Figure 4.7(a). One way to overcome this limitation is to resort to three-dimensional integrated waveguide circuits [26]–[29], which can be written by exploiting the capabilities of femtosecond laser micromachining [30], [31]. Figure 4.7(b) shows

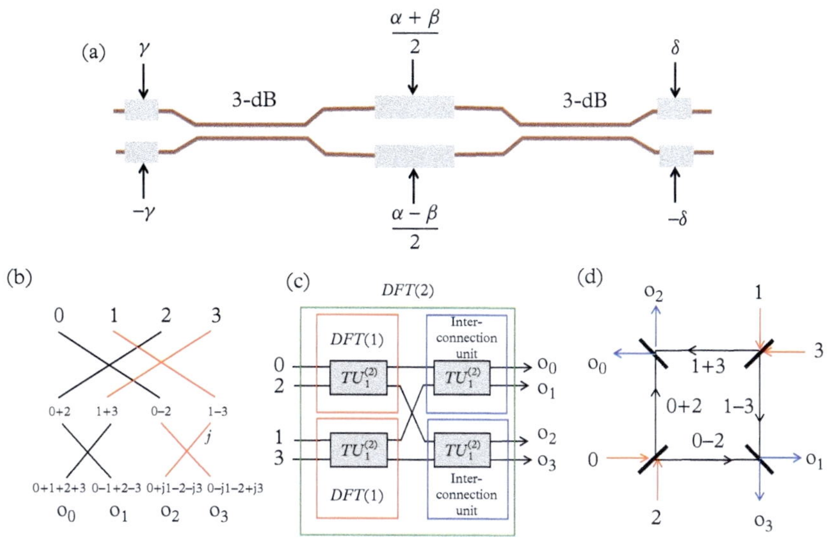

Figure 4.6 *FFT interferometers. (a) Basic Tunable Unit TU layout for a DFT(1) operation (after[19]). (b) Connection diagram for a DFT(2) unit. (c) Block diagram for a DFT(2) unit in terms of DFT(1) and interconnection TUs. (d) Bulk optics implementation of (c).*

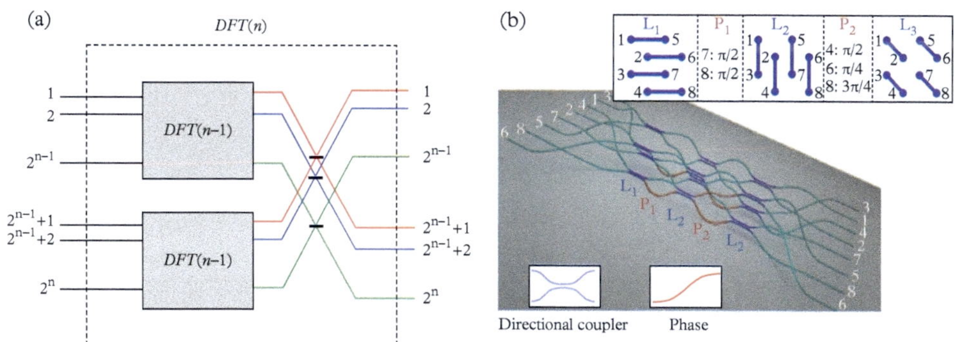

Figure 4.7 *(a) Generalization process to construct a DFT(n) interferometer using two DFT(n-1) transformers and 2^{n-1} interconnection TUs (shown as horizontal black lines). (b) Implementation of a DFT(3) interferometer using 3D integrated waveguides, couplers and phase shifters inscribed in a borosilicate glass chip by means of femtosecond laser micromachining (after [28]).*

the implementation of a $DFT(3)$ circuit fabricated in a borosilicate glass chip [27]. Thus, the connection scheme is quite involved even for a small number of port counts $(2^3 = 8)$ and in many cases it requires a complete port relabelling. Fast interferometers, though not supporting arbitrary unitary evolutions, can implement a number of linear optical networks, most notably Hadamard transformers, with a substantial reduction in the number of necessary optical elements.

4.5 Interferometer Benchmarking

The main types of multi-port interferometers reported here have been compared in the literature [19] against a series of figures of merit and characteristics. Some of the main results are shown in Table 4.1.

The main comparison for arbitrary unitary transformations was performed by means of the *fidelity* parameter defined as

$$F = \left| \frac{\mathbf{Tr}\left(U_{imp}^{\dagger}U\right)}{\sqrt{N}\mathbf{Tr}\left(U_{imp}^{\dagger}U_{imp}\right)} \right| \tag{4.35}$$

where U denotes the ideal $N \times N$ unitary transformation to be implemented by the interferometer, U_{imp} the imperfect version due either to losses or fabrication noise and **Tr** is the trace operator. Comparisons were made by means of random Haar transformations [19]. Table 4.1 shows results that suggest the rectangular design performs better, in terms of fidelity, between ideal and lossy evolutions with respect to the triangular one, even though the latter exhibits a slightly enhanced resilience to fabrication noise. Thus, the rectangular design is preferable in practical applications involving high-dimension circuits where the issue of propagation losses becomes critical. Fast architectures on the other hand represent a specialised design to implement a significant class of unitary evolutions and are highly optimised for the realisation of Fourier transformations.

Table 4.1 *Summary of the main features of the triangular, rectangular and fast multi-port interferometers (after Flamini, et al., [19]).*

	Triangular interferometer	Rectangular interferometer	Fast interferometer
Symmetry	No	Yes	Yes
Number of elements	$\frac{N(N-1)}{2}$	$\frac{N(N-1)}{2}$	$\frac{N}{2}\log_2 N$
Depth	$2N-3$	N	$\log_2 N$
Arbitrary unitary transformations	Yes	Yes	No
Hadamard-type transformations	Yes	Yes	Yes
Fidelity for arbitrary unitary transformations due to losses	Worse	Better	N/A
Resilience to fabrication noise for arbitrary unitary transformations	Better	Slightly worse	

4.6 Linear Universal Components

Linear optic components were addressed in Section 2.6, which provided a general description and a mathematical model based on the singular value decomposition (SVD) theorem. Miller's [7] linear universal component (LUC) is a generalisation of this concept where multi-port interferometers are employed in the implementation. The core of the LUC is designed to operate with spatial modes but the overall component can handle other types of encoding, including polarisation and spectrum [7].

4.6.1 Bulk Optics Implementation

The LUC can implement any linear optic transformation between an input and an output set of modes. This transformation does not need to be square or even unitary, although it has to be represented by a compact operator. In this case the SVD approximation given by Eqn. (2.116) can be employed. This transformation can be implemented over spatial modes by means of two universal self-aligning beam couplers (labelled CI and CO, respectively) as described in Section 2.7 connected back to back through modulators that can set amplitude and phase independently, as shown in Figure 4.8 for a 4×4 device.

The CI device is programmed to implement the unitary transformation $\mathbf{U}^{\dagger}$ in Eqn. (2.116), following a similar procedure to that described in Section 2.7.1, bearing in mind that the CI operation as a mode converter is equivalent to a unitary linear transformation according to [32]. We now briefly describe this procedure specialised to render the values

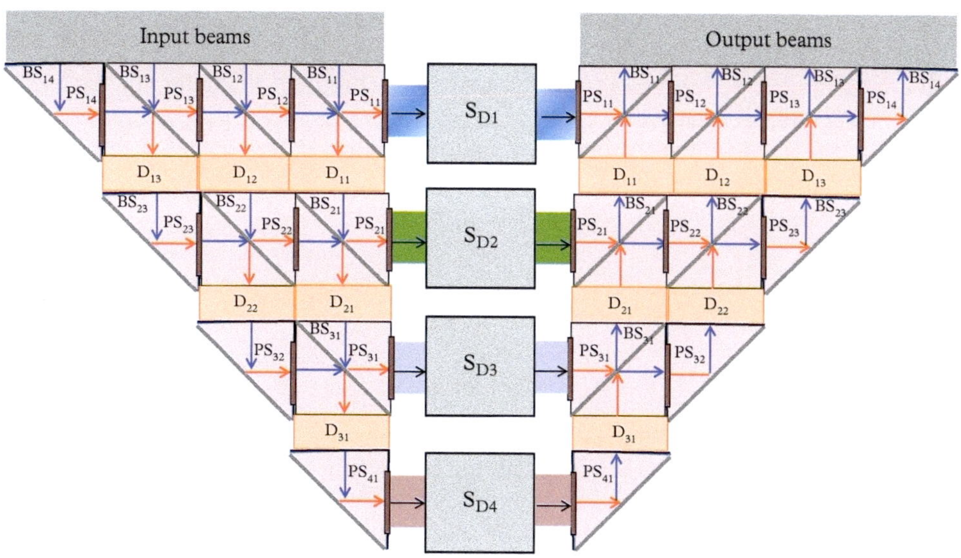

Figure 4.8 *Schematic layout of a 4x4 linear universal component core operating with spatial modes. Diagonal grey rectangles are controllable reflectors. Vertical red rectangles are controllable phase shifters (after [7]).*

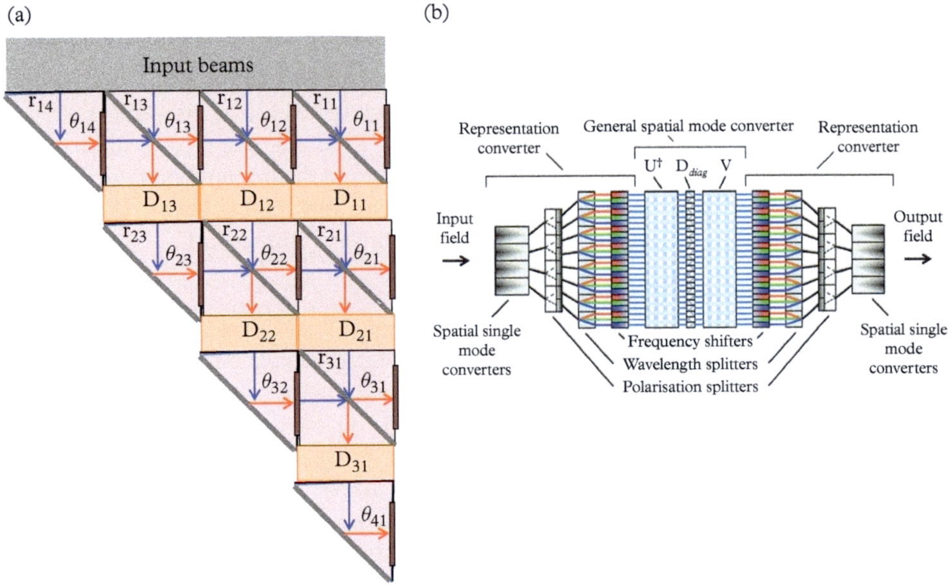

Figure 4.9 *(a) Schematic of the Mode transformer for the operator $U^{\dagger}$ for N=4 with the reflectivities and phase shifts labeled for each beam-splitter block (After [7]). The diagonal mirrors have 100% reflectivity. (b) Full layout of a universal linear component including representation converters to operate with polarization and frequency modes (after [7]).*

of the beamsplitter reflectivities and phase shifters (Figure 4.9(a)) for the case of a 4×4 ($N = 4$) transformation.

Assume that the input space basis functions $|\phi_{\mathbf{D}Ik}\rangle$ are expressed as a linear combination of modes corresponding to the inputs to the individual columns in the device. Thus,

$$|\phi_{\mathbf{D}Ik}\rangle = \sum_{n=1}^{N} a_{kn}|\phi_{1n}\rangle \tag{4.36}$$

where $|\phi_{1n}\rangle$ is the input mode (rectangular spatial function) incident on (and filling only) the top row in the nth column. As in Section 2.7.1, we aim to design the components in the first row so all the power exits out from port 1 at the right. The second step aims to design the components in the first row so all the power exits out from port 2 at the right, and so on, proceeding row by row. We assume that in principle we deal with components fulfilling the requirements for lossless, reciprocal and time reverse operation discussed in Chapter 2. For example, the phase delays and the magnitudes of the reflectivities are the same for forward and backward propagation directions.

We start by considering the operation of the device in the backward propagation direction, i.e. we shine a beam into port 1 on the right and look at the beams coming out of the ports at the top of the device. In this case, and according to Eqn. (4.36), we expect to have the complex conjugate amplitudes a^*_{kn}, in these ports. Then, and according to Eqn. (2.65),

$$r_{11}e^{j\theta_{11}} = a^*_{11} \Rightarrow \begin{cases} r_{11} = |a_{11}| \\ \theta_{11} = -Arg\left[a_{11}\right] \end{cases} \tag{4.37}$$

which automatically gives $t_{11} = \sqrt{1 - |r_{11}|^2}$. From here and Eqn. (2.65) we get,

$$jt_{11}e^{j\theta_{11}}r_{12}e^{j\theta_{12}} = a^*_{12} \Rightarrow r_{12}e^{j\theta_{12}} = a^*_{12}\frac{je^{-j\left(\theta_{11}+\frac{\pi}{2}\right)}}{t_{11}} \tag{4.38}$$

which automatically gives $t_{12} = \sqrt{1 - |r_{12}|^2}$. Continuing with this procedure, we discover

$$r_{1n}e^{j\theta_{1n}} = a^*_{1n}\frac{e^{-j\sum\limits_{p=1}^{n-1}\left(\theta_{1p}+\frac{\pi}{2}\right)}}{\prod\limits_{q=1}^{n-1} t_{1q}} \tag{4.39}$$

Note that, when $n = 1$, it is understood that the summation term will be 0 and the product term will be 1, as well as when $n = N$, $r_{1n} = 1$. This completes the design of the optical components in the first row.

In general, for a field arriving at the top of the rth row of the beamsplitter blocks we have

$$\left|\phi^{(r)}\right\rangle = \sum_{k=1}^{N-r+1} a^{(r)}_k \left|\phi_{rk}\right\rangle \tag{4.40}$$

where $\left|\phi_{rk}\right\rangle$ represents the input (rectangular) spatial mode incident on the rth row in the kth column.

When we shine a second mode $\left|\phi_{\mathbf{D}I2}\right\rangle$ orthogonal to $\left|\phi_{\mathbf{D}I1}\right\rangle$ in row 1, since it is orthogonal, no power from $\left|\phi_{\mathbf{D}I2}\right\rangle$ will be output to port 1 on the right side. However, $\left|\phi_{\mathbf{D}I2}\right\rangle$ will be affected by the setting of the photonic components in row 1 before actually inputting the elements in row 2. This transfer is formalised in terms of a matrix relationship. If $\left|\phi^{(1)}\right\rangle$ represents the field incident on row 1, then the field transmitted to row 2 $\left|\phi^{(1)}\right\rangle$ is given by

$$\left|\phi^{(2)}\right\rangle = C^{(1)}\left|\phi^{(1)}\right\rangle \tag{4.41}$$

The matrix $C^{(1)}$ is the first one of a family of $(N-r)\times(N-r+1)$ matrices given by

$$C^{(r)} = \begin{pmatrix} jt_{r1} & c_{12}^{(r)} & c_{12}^{(r)} & \vdots & c_{1(N-r)}^{(r)} & c_{1(N-r+1)}^{(r)} \\ 0 & jt_{r2} & c_{23}^{(r)} & \ddots & & \\ \vdots & 0 & jt_{r3} & \ddots & \vdots & \vdots \\ & \vdots & \ddots & \ddots & & \\ 0 & 0 & 0 & \dots & jt_{r(N-r)} & c_{(N-r)(N-r+1)}^{(r)} \end{pmatrix} \tag{4.42}$$

where $c_{kl}^{(r)}$ represents the complex multiplying factor of the field incident in column l of row r that contributes to the field incident ion the top of column k of row $r+1$. For the elements in the diagonal it is straightforward to verify from Figure 4.9 that

$$c_{kk}^{(r)} = jt_{rk} \tag{4.43}$$

whereas, for the elements in the right of the diagonal we have

$$c_{kl}^{(r)} = r_{rl}r_{rk}\left[\prod_{p=k+1}^{l-1} jt_{rp}\right]e^{j\sum_{p=k+1}^{l}\theta_{rp}} \tag{4.44}$$

That results from the product of the field reflectivity r_{rl} of the beamsplitter in the extreme left block rl reflecting the signal into row r, the field reflectivity r_{rk} in the downward reflecting beamsplitter of block rk, that reflects down to row $r+1$, the product of the field transmissions of the beamsplitters in the intermediate blocks between rl and rk and the phase factors from all the phase shifters traversed by the signal on its path from block rl to block rk.

Since all relevant parameter values for the components in row 1 are already known, we can compute matrix $C^{(1)}$ by means of specialising Eqns. (4.42)–(4.44), and from it

$$\left|\phi_{\mathbf{D}I2}^{(2)}\right\rangle = \sum_{k=1}^{N-1} a_{2k}^{(2)}\left|\phi_{2k}\right\rangle = C^{(1)}\left|\phi_{\mathbf{D}I2}\right\rangle \tag{4.45}$$

using the coefficients in Eqn. (4.45) we can now proceed in the same way as with row 1. Once the settings of all the phase shifters and beamsplitters in row 2 are found we can calculate matrix $C^{(2)}$ and amplitudes $a_k^{(3)}$ that will appear at the top of the row 3 when illuminating the device from the top surface with the third vector of the basis $\left|\phi_{\mathbf{D}I3}\right\rangle$:

$$\left|\phi_{\mathbf{D}I3}^{(3)}\right\rangle = \sum_{k=1}^{N-1} a_{3k}^{(3)} \left|\phi_{3k}\right\rangle = C^{(3)} \left|\phi_{\mathbf{D}I3}\right\rangle \tag{4.46}$$

Because of the unitary character of $\mathbf{U}^{\dagger}$ orthogonal vectors at the input $\left|\phi_{\mathbf{D}Ik}\right\rangle$ are transformed into orthogonal vectors at the output. Since $\left|\phi_{\mathbf{D}I1}\right\rangle$ is completely routed to output port 1 on the right, this means that no energy from $\left|\phi_{\mathbf{D}I2}\right\rangle$ can leave through this port. As row 2 is configured so $\left|\phi_{\mathbf{D}I2}\right\rangle$ is completely routed to output port 2 on the right then no energy from $\left|\phi_{\mathbf{D}I3}\right\rangle$ can leave through port 2 and port 1. Similar behaviour obtained for subsequent input vectors and output ports is such that, at the end, each input vector $\left|\phi_{\mathbf{D}Ik}\right\rangle$ is directed to a different output port in the right.

In a similar way, the CO is programmed to implement the unitary transformation $\mathbf{V}$ in Eqn. (2.116). Here, the desired output vectors are

$$\left|\phi_{\mathbf{D}Ok}\right\rangle = \sum_{n=1}^{N} b_{kn} \left|\psi_{1n}\right\rangle \tag{4.47}$$

where $\left|\psi_{1n}\right\rangle$ is the output mode (rectangular spatial function), leaving the top of the kth row on the nth column. When the light is shone into a port at the left of the CO we want to create the actual fields for the output basis functions; we take the values of b_{kn} and not the complex conjugates, as in the CI. Finally, the diagonal matrix is implemented with independent modulators connecting the CI outputs and the CO inputs.

The universal linear component can operate with other kind of input/output signals such as polarisation and frequency modes. To this aim, representation converters need to be placed before the input and after the output ports of the spatial mode core converter (Figure 4.9(b)). Several alternatives for these representation converters are possible [7], but practical realisation of such complete devices has not been yet reported.

4.6.2 Integrated Optics Implementation

The spatial core of LUC implementing the SVD transformation given by Eqn. (2.166) lends itself to straightforward integrated optics implementations using the multi-port architectures discussed in Section 4.2 and Section 4.3. Indeed, both the integrated versions of triangular and the rectangular interferometers can be employed to implement the arbitrary unitary transformations $\mathbf{U}^{\dagger}$ and $\mathbf{V}$ required by Eqn. (2.116), while the $\mathbf{D}_{diag}$ matrix can be implemented by inserting TBUs connecting their output and input ports, respectively. The design of each interferometer can be performed following the general guidelines outlined in Section 4.2 and Section 4.3, respectively. Alternatively, we can employ the procedure described in Section 4.6.1, but bearing in mind that the scattering matrix for the integrated 2×2 beamsplitters is now given by Eqn. (2.106) and not Eqn. (2.65), so the final form of Eqns. (4.37)–(4.39), and Eqn. (4.43) and Eqn. (4.44) do

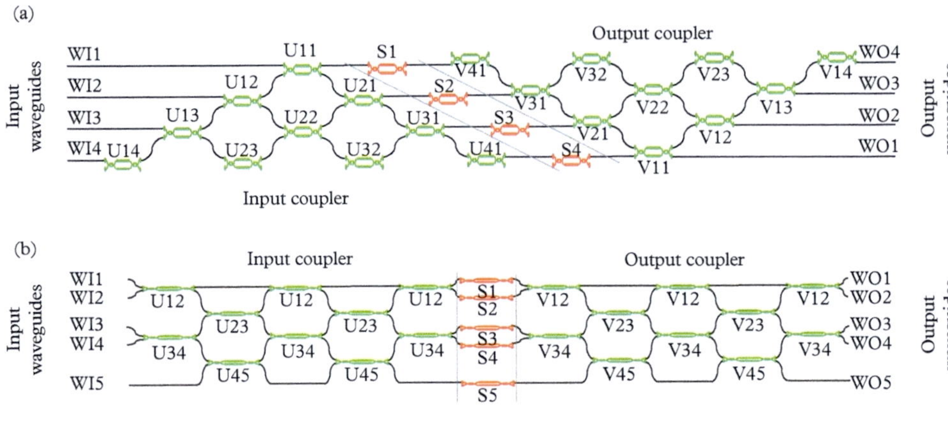

Figure 4.10 *Integrated optics layouts for the linear universal component. (a) 4x4 LUC using the triangular interferometers (after [7]). (b) 5x5 LUC using rectangular interferometers (after [4]).*

change. Figure 4.10 shows two examples of implementation using the triangular and rectangular interferometer configuration.

For instance, Figure 4.10(a) shows the implementation of a 4×4 arbitrary transformation by means of the integrated layout of the triangular interferometer [7]. Note the port number inversion in output waveguides with reference to the port number in the input waveguides. This requirement was already known from our discussion on the integrated optics implementation of the triangular interferometer. The figure shows the TBUs implementing each matrix coefficient for $\mathbf{U}^{\dagger}$,$\mathbf{V}$, and also those (in red) implementing the singular values. Figure 4.10(b) shows the implementation of a 5×5 arbitrary transformation by means of the integrated layout of the rectangular interferometer [4]. Here, as expected, no port number inversion is required.

Since TBUs are fully programmable and provide independent amplitude and phase settings, the configurations in Figure 4.10 and their generalisations for $N \times M$ transformations provide the most complete setups for performing feedforward only fully programmable circuits. These cover many, but not all the possible transformations and circuits required in photonic applications. To implement a programmable circuit capable of featuring all the possible transformations we need structure enabling both feedforward and feedbackward simultaneous propagation. These structures are the *integrated waveguide meshes* (Chapter 5).

4.7 Chapter Summary

Integrated multi-port interferometers, the second evolutionary step in programmable photonics, were initially conceived and designed to implement fixed linear optics unitary transformations between N input and N output ports to support mode transformations in quantum optics. Recent application fields, including quantum information processing and neurophotonics, require devices capable of supporting any arbitrary linear transformation by suitable programming no longer restricted to square matrix transformations and unitary operators.

This chapter has provided a broad coverage of the foundations and design principles of the two main configurations (the triangular and rectangular) proposed so far for the implementation of arbitrary unitary $N \times N$ transformations, including their layouts for implementation in integrated optics. It also addressed some specific interferometer designs, which although they do not provide the possibility of implementing general linear transformations, are still optimised for performing important operations, such as the Fourier transform.

Finally, it presented the linear universal component, which employs integrated multi-port interferometers connected by intermediate TBUs to implement arbitrary and programmable non-unitary $N \times M$ transformations. The LUC is probably the most complete feedforward-only programmable device that can be implemented in practice. Hence, it can cover many, but not all, of the possible transformations and circuits required in photonic applications.

REFERENCES

[1] M. Reck, A. Zeilinger, H.J. Bernstein and P. Bertani. Experimental Realization of Any Discrete Unitary Operator. (1994) *Phys Rev Lett* 73, 58–61.

[2] W.R. Clements, et al., Optimal design for Universal Multiport Interferometers. (2016) *Optica* 3, 1460–5.

[3] N. Harris, et al., Large-scale Quantum Photonic Circuits in Silicon. (2016) *Nanophotonics* 5, 456–68.

[4] D. Pérez, et al., Silicon Photonics Rectangular Universal Interferometer. (2017) *Lasers and Photon Rev* 11, 1700219.

[5] G. Alber, T. Beth, M. Horodecki, P. Horodecki, R. Horodecki, M. Rötteler, H. Weinfurter, R. Werner and A. Zeilinger, *Quantum Information:An Introduction to Basic Theoretical Concepts and Experiments*, vol. 173, Springer, 2003.

[6] D.A.B. Miller. Self-aligning Universal Beam Coupler. (2013) *Opt Express* 21, 6360–70.

[7] D.A.B. Miller. Self-configuring Universal Linear Optical Components. (2013) *Photon Res* 1, 1–15.

[8] J. Carolan, et al., Universal Linear Optics. (2015) *Science* 349, 711.

[9] N.C. Harris, et al., Quantum Transport Simulations in a Programmable Nanophotonic Processor. (2017) *Nat Photonics* 11, 447–52.

[10] N. Spagnolo, et al., Experimental Validation of Photonic Boson Sampling. (2014) *Nat Photonics* 8, 615–20.

[11] C. Taballione, et al., 8 × 8 Programmable Quantum Photonic Processor based on Silicon Nitride Waveguides. (2018) *Frontiers in Optics/Laser Science, OSA Technical Digest* paper JTu3A.58.

[12] A. Annoni, et al., Unscrambling Light—Automatically Undoing Strong Mixing Between Modes. (2017) *Light Sci Appl* 6, e17110.

[13] P.L. Mennea, et al., Modular Linear Optical Circuits. (2018) *Optica* 5, 1087–90.

[14] N.C. Harris, et al., Bosonic transport simulations in a large-scale programmable nanophotonic processor. (2015) *Nat Photonics* 11, 447–52.
[15] G. Steinbrecher, et al., Programmable Nanophotonic Processor for Arbitrary High-fidelity Optical Transformations. In Conference on Lasers and Electro-Optics (CLEO). San Jose, CA, 2015. Paper no. FW4A.2.
[16] Y. Shen, et al., Deep Learning with Coherent Nanophotonic Circuits, *Nat* (2017) *Photonics* 11, 441–6.
[17] P. Törmä, I. Jex and S. Stenholm, Beam Splitter Realizations of Totally Symmetric Mode Couplers. (1996) *J Mod Opt* 43, 245–51.
[18] R. Barak and Y. Ben-Aryeh, Quantum Fast Fourier Transform and Quantum Computation by Linear Optics. (2007) *J Opt Soc Am B* 24, 231–40.
[19] F. Flamini, et al., Benchmarking Integrated Linear-optical Architectures for Quantum Information Processing. (2017) *Scientific Reports* 7, 15133.
[20] J.W. Coole and W. Tukey, An Algorithm for the Machine Calculation of Complex Fourier Series. (1965) *Math Comput* 19, 297–301.
[21] P.W. Shor, Polynomial-time Algorithms for Prime Factorization and Discrete Logarithms on a Quantum Computer. (1997) *SIAM J Comput* 26, 1484–509.
[22] A. Ekert and R. Jozsa, Quantum Computation and Shor's Factoring Algorithm. (1996) *Rev Mod Phys* 68, 733–53.
[23] A. Ekert, P. Hayden, H. Inamori and D.K.L. Oi, What is Quantum Computation? (2001) *Int J Mod Phys A* 16, 3335–63.
[24] R. Cleve, A. Ekert, C. Macchiavello and M. Mosca, Quantum Algorithms Revisited. (1998) *Proc R Soc London Ser A* 454, 339–54.
[25] M.A. Nielsen and I.L. Chuang, *Quantum Computation and Quantum Information*, Cambridge University Press, 2000.
[26] K. Takiguchi, M. Oguma, T. Shibata and H. Takahashi, Demultiplexer for Optical Orthogonal Frequency-division Multiplexing Using an Optical Fast-Fourier-Transform Circuit. (2009) *Opt Lett* 34, 1828–30.
[27] A. Crespi, et al., Suppression Law of Quantum States in a 3D Photonic Fast Fourier Transform Chip. (2016) *Nat Commun* 7, 10469.
[28] F. Flamini, et al., (2016) Observation of Majorization Principle for Quantum Algorithms via 3-D Integrated Photonic Circuits. Available at https://arxiv.org/abs/1608.01141.
[29] N. Viggianiello, et al., (2017), Experimental Generalized Quantum Suppression Law in Sylvester Interferometers. Available at https://arxiv.org/abs/1705.08650.
[30] R.R. Gattass and E. Mazur, Femtosecond Laser Micromachining in Transparent Materials. (2008) *Nat Photon* 2, 219.
[31] N. Spagnolo, et al., Three-photon Bosonic Coalescence in an Integrated Tritter. (2013) *Nat Commun* 4, 1606.
[32] D.A.B. Miller, All Linear Optical Devices Are Mode Converters. (2012) *Opt Express* 20, 23985–93.

5

Integrated Waveguide Meshes

Integrated waveguide meshes are the third evolutionary step in programmable photonics. These circuits rely on the large-scale repetition of interferometric waveguide elements and phase actuators conceived and designed to implement a common hardware platform that enables the programming of arbitrary photonic circuit topologies and design parameters by suitable setting of its control signals. In contrast to linear feed-forward-only programmable circuits (Chapter 4), these circuits enable the synthesis of optical cavities, optical loops and feedbackward paths in general. Their improved versatility entails a paradigm shift with respect to the development of application-specific photonic integrated circuits. Here, the long and costly test cycles can be potentially substituted by off-the-shelf ready-to-use circuits. This allows them to be applied to myriad systems requiring optical linear processing and dynamic configuration. This chapter introduces the basic design principles and the implementations of the most popular designs. It then applies a comparative analysis of alternative waveguide mesh arrangements and their performance evaluation. This component is probably the most complete and versatile programmable device that can be implemented in practice and will be the key building block of the fourth evolutionary step (Chapter 9).

5.1 The Concept of Photonic Integrated Waveguide Mesh

Imagine that you are a photonic designer working in a company or a research centre, and you are asked to design an integrated solution to filter an optical signal at 1.54 μm and to distribute the remaining part of the spectrum to two different output ports with a power ratio of 10–90 per cent. The path of the second output must be 1.5 times larger than the first and it must include a phase shifter. After considering different approaches and making your analytical and statistical simulations, you decide to employ an optical filter based on an optical ring resonator (ORR) for the first stage and a splitter followed by two waveguides of length L_{wg} and $1.5L_{wg}$, respectively. In addition, you consider the integration of a thermo-optic-based phase actuator in the last waveguide (Figure 5.1(a)). Since your centre is fabless you need first to plan the selection of the integration platform alternatives. After analysing the trade-offs reviewed in Chapter 2, you decide to go for an MPW run service of silicon-on-insulator to share the cost among users.

Programmable Integrated Photonics. José Capmany and Daniel Pérez, Oxford University Press (2020).
DOI: 10.1093/oso/9780198844402.001.0001

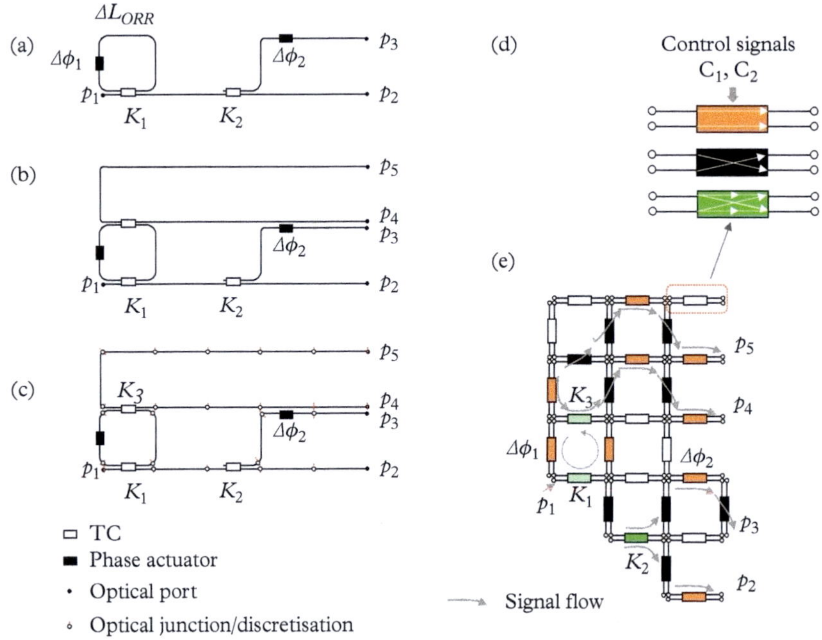

Figure 5.1 *(a) Schematic of a conventional photonic integrated circuit. (b) Schematic extension including two additional ports, (c) discretization of the previous circuit into waveguide elements, couplers and phase shifters, (d) different configurations of a primitive unit and (e) waveguide mesh arrangement.*

After spending a considerable percentage of your project's temporal and financial resources in two design-fab-packaging-test iterations you get the targeted solution working in a lab environment in a typical time of 18–25 months. By that time, the project demands also an additional port (p_5) to monitor the signal filtered at the first stage (Figure 5.1(b)). This small change, or any other design modification, could lead to an extended number of long iterations and a non-negligible increment of the cost.

It is not a coincidence that if we split, break or discretise our design into small building blocks Figure 5.1(c) we find that our design is composed of a well-organised set of fourteen waveguide elements, two phase shifters and two beamsplitters or tunable couplers. Moreover, the combination of these basic building blocks (BBBs) is enough to implement the vast majority of the most simple and complex photonic linear processing architectures. This premise is one of the two fundaments of integrated waveguide meshes.

A second key feature is that a common 4-port tunable basic unit (TBU) can be designed in many different ways to perform as a discrete delay line, an optical switch, a tunable coupler or a phase shifter by suitable programming of its control signals (Figure 5.1(d)).

Under these premises, the combination and interconnection of a large enough number of TBUs describing a waveguide mesh arrangement can be employed as a common hardware photonic platform to program a wide variety of photonic linear processing architectures and design parameters. Figure 5.1(e) shows the independent configuration of each TBU either as a delay line (optical switch in cross or bar state), as a tunable coupler and/or as a phase shifter results in the emulation or synthesis of the circuit in Figure 5.1(c).

At this point, the reader might be finding it difficult to follow the signal flow paths. However, this chapter explains in detail the alternative waveguide mesh topologies, their figures of merit and their programming. The take-away message is that waveguide mesh arrangements make possible the dynamic configuration of a wide variety of conventional photonic integrated circuit architectures by employing the same hardware.

In the previous example, the TBU interconnection resulted in a square mesh topology. Notwithstanding, circuit discretisation, space and power consumption constraints play a critical role in the design of photonic circuits in general, and of optical meshes in particular. The limited area available for growing the optical mesh, as well as the need to reduce to a minimum the number of switching elements required for implementing a set of optical core topologies, call for a careful analysis of possible geometries for the mesh interconnection implementing the optical core.

5.2 Tunable Basic Unit

Before considering the waveguide mesh analysis, it is essential to fully understand the architectures, tasks and fundamentals of their primitive block, the TBU, which, as described in Chapter 2, is a tunable coupler that must provide, independently, a complete splitting ratio tuning and phase response [1]–[3]. By setting its control signals, this switching/tapping/dephasing mechanism can be potentially obtained in many different ways. Figure 5.2 illustrates some of the possible designs for TBUs; most of them are based on the introduction of a phase actuator inside an interferometric structure: balanced and unbalanced 2×2 couplers, dual-drive directional couplers, 3×3 and 3-dB MZIs, and resonant structures. Others rely on the incorporation of micro-mechanical waveguide tuners or in photonic crystal circuits. The phase-change effects susceptible of being exploited must be efficient and must not incorporate excessive additional losses to the unit. Chapter 6 covers the specific behaviour and performance benchmark between most of the reported architectures and phase actuators.

However, this chapter considers 3-dB balanced MZI structures loaded with a thermal tuner on each arm (Figure 5.3(a)). Nevertheless, the treatment is not restricted to this case and can be employed for alternative TBUs structures and tuning mechanisms.

Figure 5.1(d)–(e) shows that, by configuring each TBU placed at each side of the arrangement as a switch (in either its cross or bar state) or as a tunable coupler, we can synthesise a given photonic integrated circuit topology.

The upper part of Figure 5.3(a) shows the TBU composed by the tunable coupler and its access (input/output) waveguides. The geometry of the latter is a function of the bending radius and varies for each mesh topology due to different angle between elements. The basic unit length (BUL) is [2]

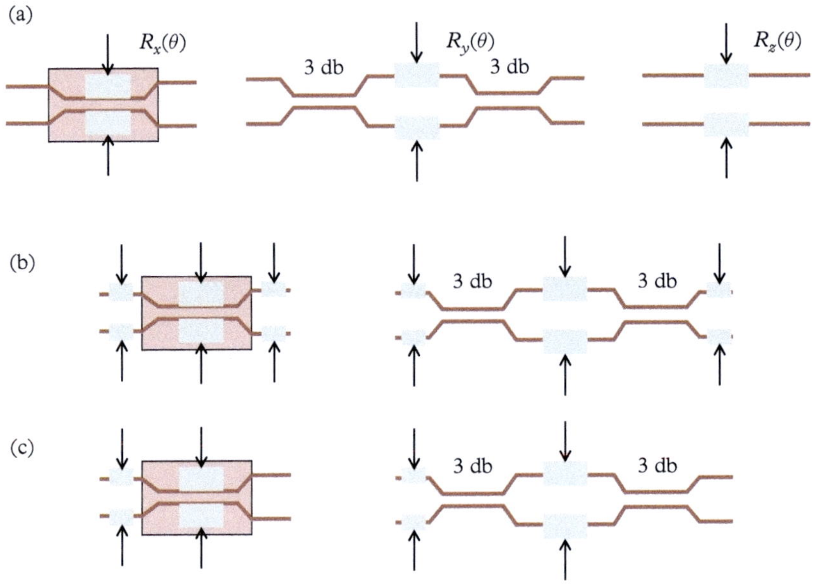

Figure 5.2 *Example of TBU architectures (a) Dual-Drive Directional-Coupler, Mach-Zehnder Interferometer, (b) TBU architectures with four extra phase tuning elements, (c) TBU architectures with additional input phase shifters.*

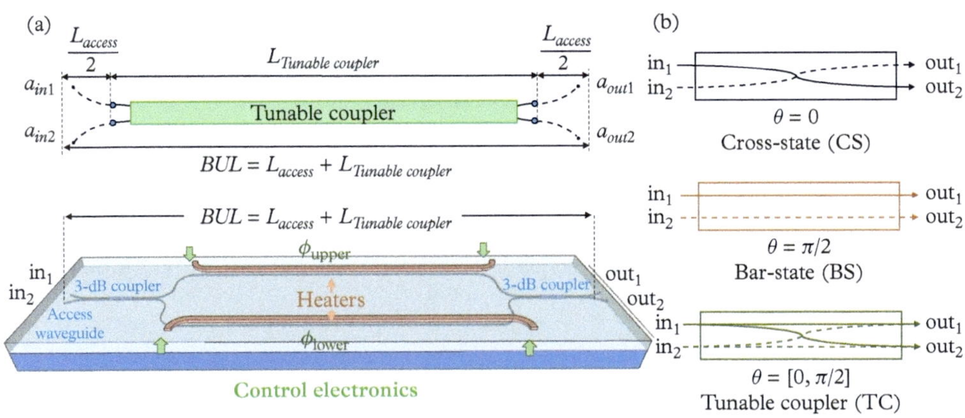

Figure 5.3 *(a) (Upper) Labelled schematic of a general tunable coupler acting as the basic building block of the mesh. The Basic Unit Length (BUL) is illustrated as the sum of the tunable coupler length and the arc length of the access waveguides. (Lower) Particular case of an integrated balanced MZI-based tunable coupler. (b) Signal flow for the different TBU configuration states.*

$$BUL = L_{access} + L_{Tunable-Coupler}, \tag{5.1}$$

where L_{access} is the overall length of the access waveguide segment and $L_{Tunable-Coupler}$ is the length that describes the light-path in the tunable coupler. Most importantly, the time that takes the signal to go through the TBU is called the basic unit delay (BUD) [4]

$$BUD = n_g BUL/c, \tag{5.2}$$

where n_g is the waveguide group index and c is the speed of light in vacuum. Figure 5.3(b) shows how the TBU can implement three different states: cross state switch (light-path connects in_1 to out_2 and in_2 to out_1), bar state switch (light-path connects in_1 to out_1 and in_2 to out_2) and tunable splitter.

For a balanced MZI loaded with heaters on both arms, the splitting ratio is obtained by increasing the effective index due to the Joule effect in the upper or lower arm, which produces a ϕ_{upper} and ϕ_{lower} phase shift, respectively. Once set, a common drive in both heaters will provide a common phase shift, leading to independent control of the amplitude ratio and the phase. The simplified device matrix is defined by

$$h_{TBU} = je^{j\Delta} \begin{pmatrix} \sin\theta & \cos\theta \\ \cos\theta & -\sin\theta \end{pmatrix} \gamma, \tag{5.3}$$

where θ is $\left(\phi_{upper} - \phi_{lower}\right)/2$ and Δ is $\left(\phi_{upper} + \phi_{lower}\right)/2$. The coupling factor K is then defined as $cos^2(\theta)$. A less simplified matrix is presented in Chapter 6.

The prior matrix has a general loss term γ that includes the effect of propagation losses in the access waveguides, the tunable coupler waveguide and the insertion losses for both 3-dB couplers. The device performance can be characterised in terms of the optical power by its insertion losses (*IL*) and optical crosstalk parameters (CT_{Bar} and CT_{Cross})

$$\begin{aligned} IL(dB) &= 20\log_{10}(\gamma) = 10\log_{10}\left(\frac{|a_{out1}|^2 + |a_{out2}|^2}{|a_{in1}|^2 + |a_{in2}|^2}\right), \\ CT_{Bar/Cross}(dB)\Big|_{\substack{|a_{in1}|^2 = 1 \\ |a_{in2}|^2 = 0}} &= 10\log_{10}\left(\frac{|a_{out2,1}|^2}{|a_{out1,2}|^2}\right). \end{aligned} \tag{5.4}$$

These provide valuable information regarding the losses of the tunable units. Since they are connected in cascade configuration to build up the mesh, the overall *IL* of a certain synthetised device will be the sum of the *IL*s corresponding to the units across through which the light has travelled. As an example, if we assume overall device losses of 10 dB and IL of 0.2 dB for each TBU, then the longest path will be limited to fifty units. In the case of the optical crosstalk, the figure determines the leaking of signal that might cause optical reflections and undesired interferometric paths coupled to the desired photonic circuit to be programmed. Chapter 6 is devoted to a deeper analysis in the non-ideal behaviour of the TBU and the waveguide meshes.

In order to reduce the footprint of the synthetised circuits, we allow the possibility of using all the TBU ports independently. For example, in a cross state set both in_1/out_2 and in_2/out_1 connections can be employed to configure a conventional crossing unit.

5.3 Integrated Waveguide Meshes

5.3.1 Square Integrated Waveguide Meshes

Square waveguide mesh arrangements were the first proposed mesh topology. In his pioneer work, Zhuang and colleagues [3] developed the concept where the interconnection of a large number of integrated balanced MZIs with two actuators could lead to the synthesis of a wide variety of PICs. Fig. 5.1(d) demonstrates how this is achieved by discretising conventional circuits into TBU with specific configurations. The proposed interconnection topology enables the routing of the optical signal path to follow orthogonal directions where the repetition of direction is not allowed for two consecutive TBUs. However, the topology is flexible enough to allow the synthesis of discrete waveguides or delay lines, tunable couplers as well as phase shifters, and thus more complex building blocks like optical cavities and unbalanced MZIs. Figure 5.4 illustrates the basic configuration of an optical ring resonator defined by a cavity length of 8 BULs, a delay line of 6 TBUs as well as an unbalanced MZI with arm lengths of 3 and 7 TBUs, respectively.

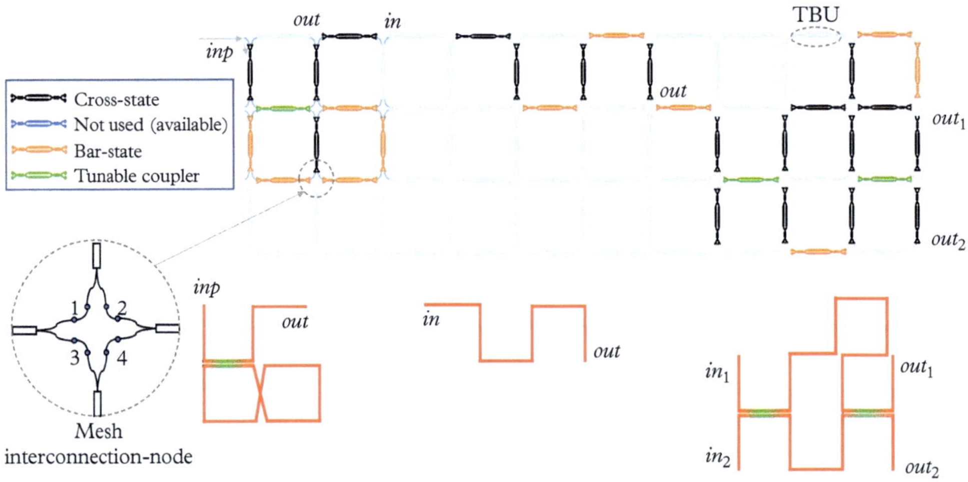

Figure 5.4 *Basic programming of medium-complex circuits in square waveguide mesh topology: delay line, unbalanced MZI and optical ring resonator.*

5.3.2 Triangular Integrated Waveguide Meshes

As shown for the square topology, the interconnection schemes between TBUs lead to regular uniform patterns with different topologies. Another example is the triangular waveguide mesh topology proposed by Pérez and colleagues [2], where three TBUs describing longitudinal orientations of 0° (horizontal plane), 60° and −60° were interconnected through six points resulting in a triangular pattern (Figure 5.5). Notice that the angle described by the longitudinal axis of two connected TBUs is always 60°. Again, we see that the interconnection scheme directly impacts on the degrees of freedom that the light has to flow across the structure and consequently the allowed light-paths and circuits that can be implemented inside the arrangement. These are particularly useful to increase the integration density of TBUs, and to implement optical cavities with reduced cavity lengths. The triangular pattern achieves a better discretisation resolution of unbalanced filter structures as compared to the square topology. The next sections provide an in-depth comparative analysis. We represent in Figure 5.5 the synthesis of a ring resonator with a cavity length of 6 BULs and an unbalanced MZI with arm lengths of 3 and 6 BULs, respectively. In the latter, an internal crossing is implemented.

5.3.3 Hexagonal Integrated Waveguide Meshes

Pérez and colleagues [2] also proposed hexagonal topology. It arises as a more efficient three-point interconnection scheme that resembles the arbitrary linear interferometers described in Chapter 4 with the difference that the proposed pattern also allows the synthesis of optical feedback loops and optical cavities. As illustrated in Figure 5.6,

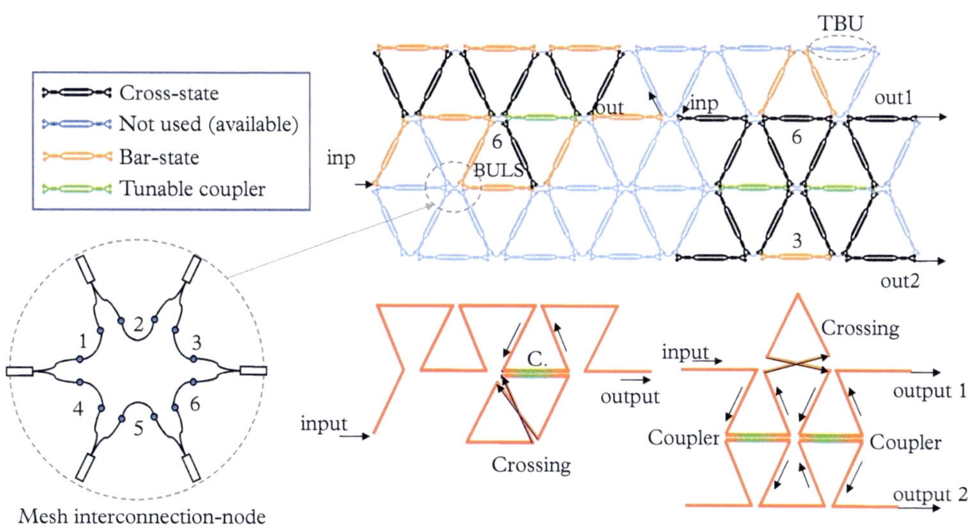

Figure 5.5 *Basic programming of medium-complex circuits in triangular waveguide mesh topology: delay line, unbalanced MZI and optical ring resonator.*

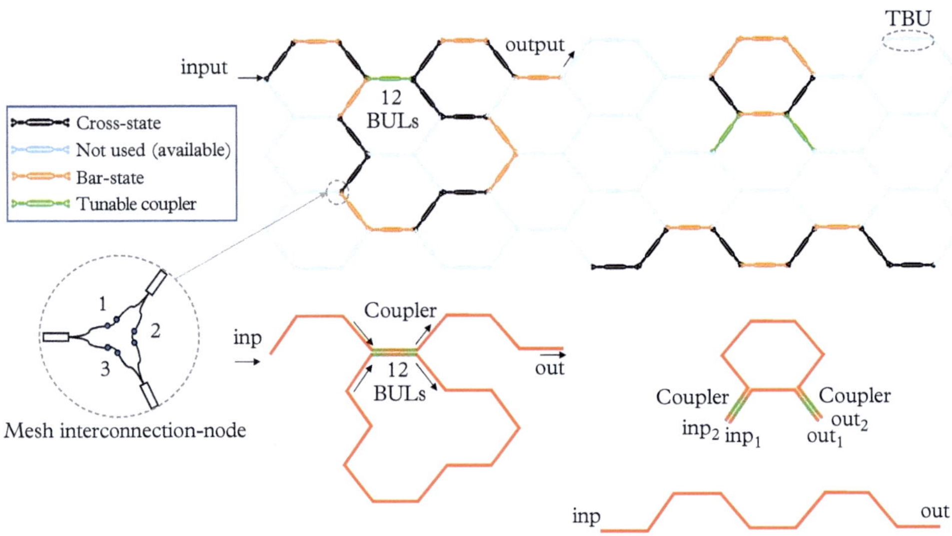

Figure 5.6 *Basic programming of medium-complex circuits in hexagonal waveguide mesh topology: delay line, unbalanced MZI and optical ring resonator.*

the light flow propagates through TBUs with longitudinal axis orientation of ±60° and 0°. However, in contrast to triangular waveguide meshes, the angle described by the longitudinal axis of two connected TBU is always 120°. Figure 5.6 represents a 12-BUL optical cavity, an unbalanced MZI of arm lengths 1 and 5 BUL as well as a discrete delay line of 9 BULs. The main advantages are their versatility and flexibility in fitting conventional circuit programming, their enhanced resolution compared to the two previous topologies and their synthesis efficiency. One of the main drawbacks of waveguide meshes, which is exacerbated in the hexagonal topology, is that the overall mesh footprint can be compromised if the longitudinal TBU dimension is too large.

5.3.4 Flattened Integrated Waveguide Meshes

Previous sections showed that the proposed waveguide mesh circuit topologies arise from the selected optical interconnection node between TBUs following regular and uniform spatial geometries. The resulting pattern is as important as the interconnection scheme. However, here we propose a different design layout for waveguide mesh arrangements and we eliminate the dependencies between interconnection nodes and resulting topologies.

The motivation is threefold. First, the current topologies are geometrically constrained by the patterns originated by their connections. Once we select an option, our arrangement will inherit a series of capabilities, characteristics and performance metrics (analysed in the Section 5.4). As a preliminary example, the square topology will produce a better footprint and TBU density integration, but its interconnection will constrain the versatility/efficiency of the programming.

Secondly, the integration of TBUs with different angular orientations implies the integration of some BBBs with performance metrics that are orientation-sensitive. In particular, 3-dB couplers like directional couplers or multi-mode interferometer designs are more reproducible and robust to fabrication if the longitudinal orientation is kept fixed. In addition, some phase actuators are orientation-sensitive and are more efficient if single or orthogonal orientations are maintained.

Finally, the fabrication costs are directly related to the total footprint of the integrated device. The use of the aforementioned topologies results in an increment of the unused areas in the device, particularly in designs with longitudinally large TBUs. Although it might be convenient to maintain a safety gap for optical and tuning-derived crosstalk, we suggest removing this parameter from the selected waveguide mesh topology.

In the light of these considerations, here we describe the use of parallel waveguide meshes to potentially solve all the aforementioned issues. Since the main performance metrics are related to the optical interconnection schemes, there are multiple possibilities for the definition of their architecture. However, in contrast to the previous patterns, they benefit from maintaining all TBUs with the same longitudinal orientation. Compared to the square, hexagonal and triangular lattices, more complex access paths are required in

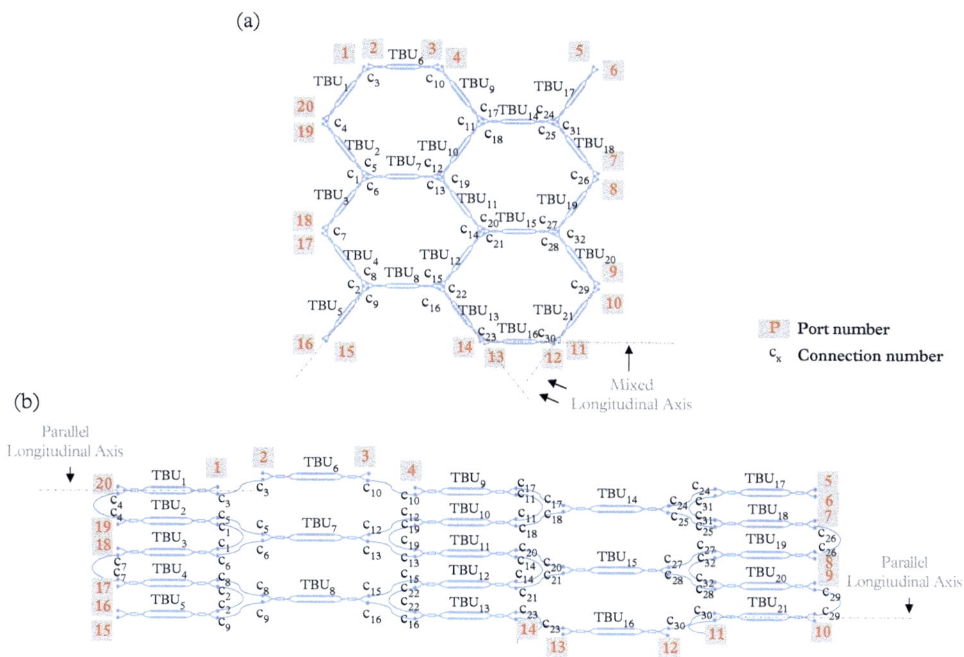

Figure 5.7 *Schematic diagram example of the implementation and interconnection of the (a) hexagonal pattern and (b) its parallel mesh topology homologue.*

the definition of the TBUs for their interconnection, although their impact is mitigated in integration platforms with high-index contrast due to the reduced bend radius.

Figures 5.7–5.9 illustrate the concept by comparing the previous topologies with their homologue parallel configurations. Note that the access waveguides have been exaggerated to enhance the figure readability.

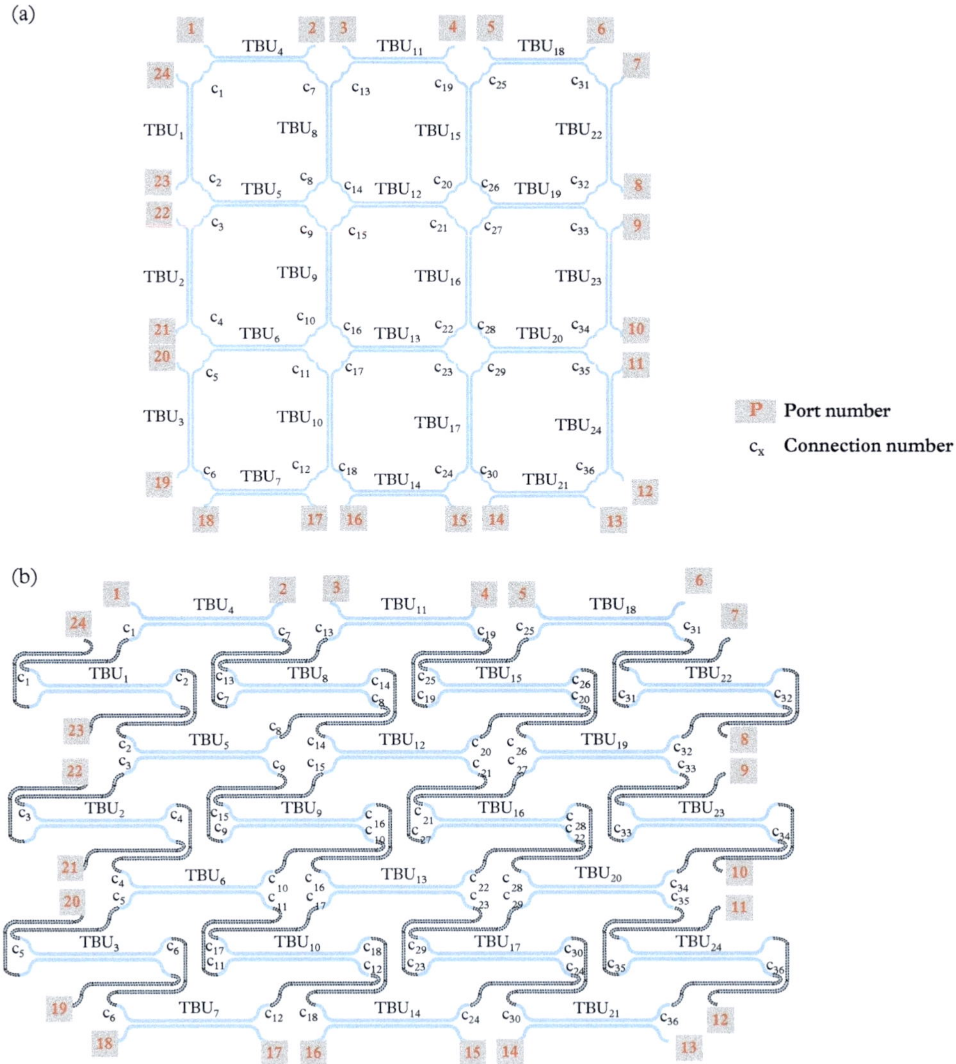

Figure 5.8 *Schematic diagram example of the implementation and interconnection of the (a) square pattern and (b) its parallel mesh topology homologue.*

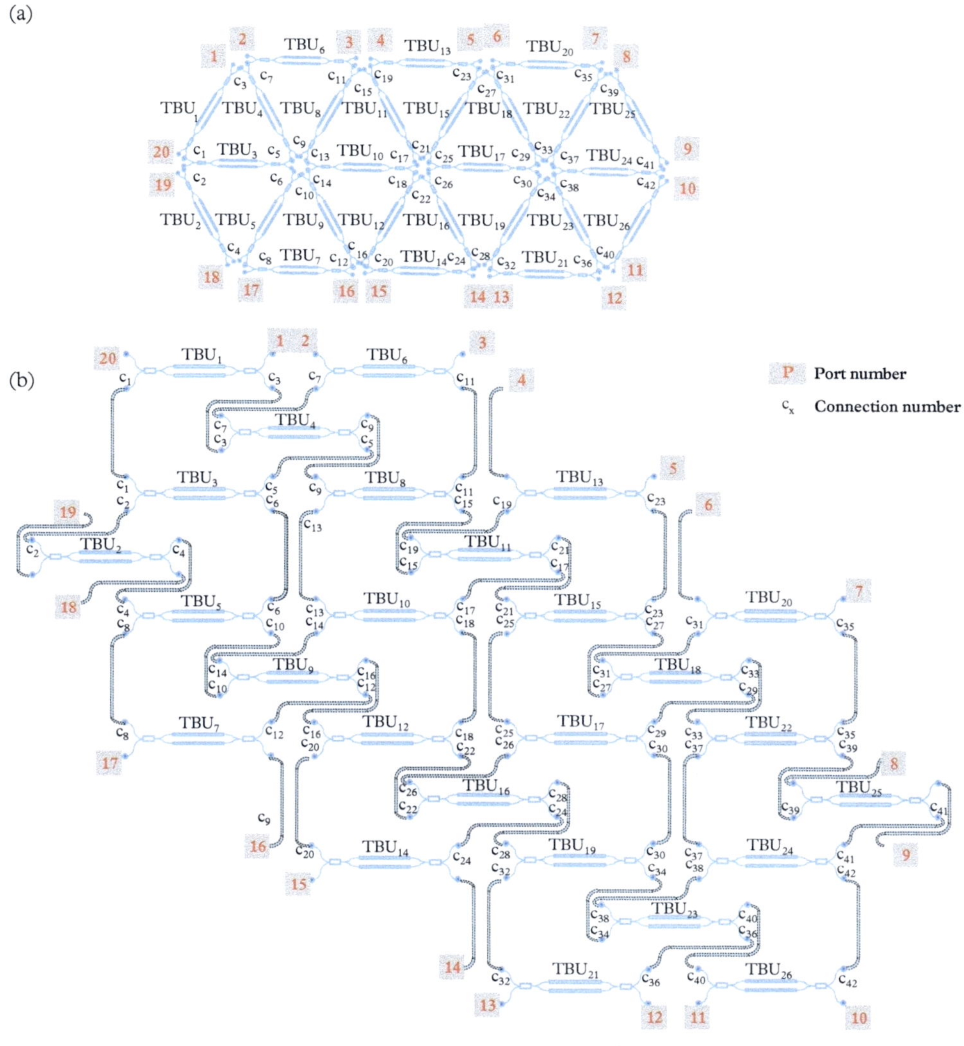

Figure 5.9 *Schematic diagram example of the implementation and interconnection of the (a) triangular pattern and (b) its parallel mesh topology homologue.*

Furthermore, if the bend radius in the access waveguide is small, the designs become narrower and the resulting integration density increases for every interconnection scheme. In addition, we see that during the design of longitudinally parallel meshes, there is more freedom to define the optical interconnection node between TBUs. In the non-parallel implementations, we are constrained to a set of few geometrical patterns. As an example, Figure 5.10 illustrates two alternative parallel waveguide mesh geometries.

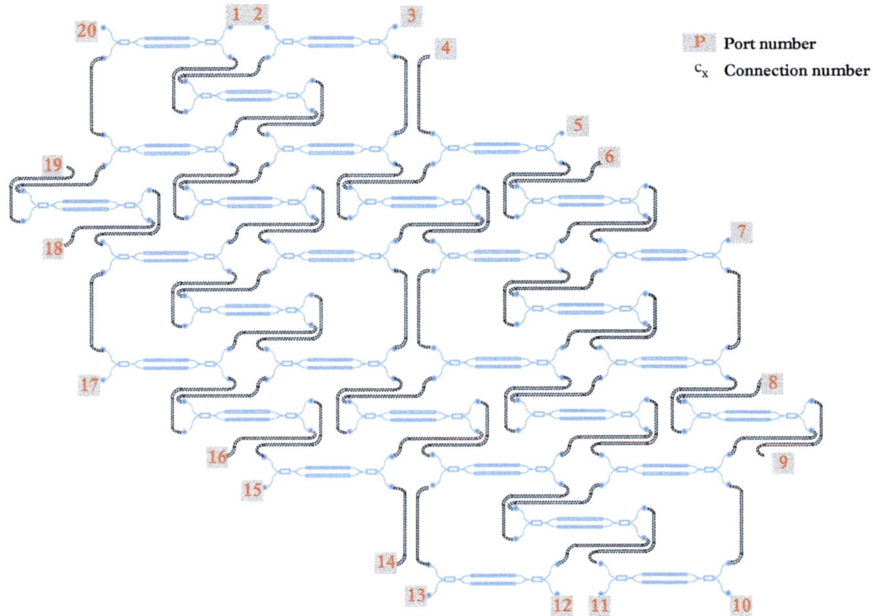

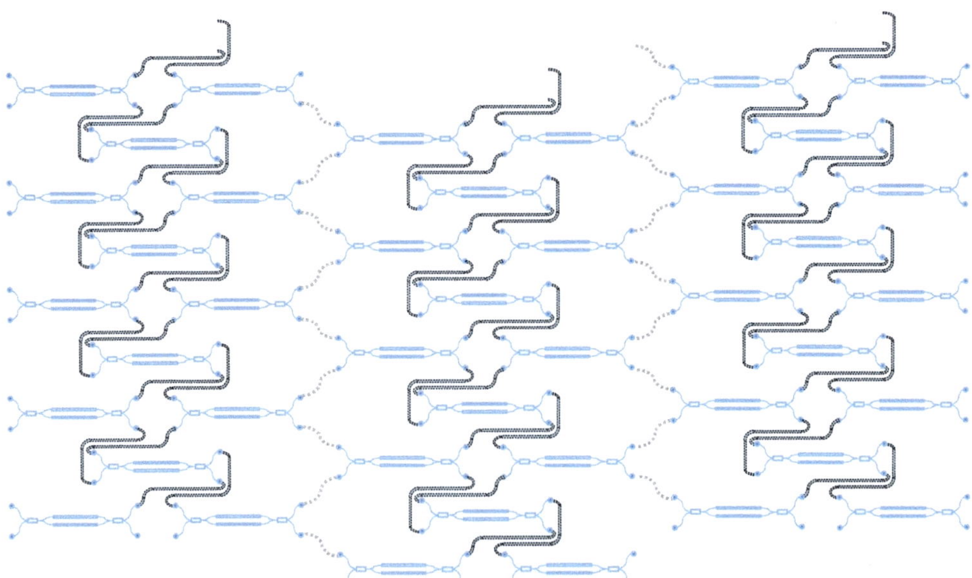

Figure 5.10 *Schematic diagram examples of the implementation and interconnection of two alternative parallel waveguide meshes.*

For most applications, the access waveguides that interconnect the TBUs should be of equal length in order to preserve the same BUL over the waveguide mesh. The design of this access path is straightforward in all the proposed arrangements. However, some applications might find useful a variable non-uniform BUL.

5.4 Comparative Analysis of Waveguide Meshes

Once we complete the introduction of the alternative waveguide topologies, we can foresee they will offer different capabilities and performance. This section analyses and compares the performance of the hexagonal, triangular and square mesh designs as well as the parallel version of the hexagonal interconnection, in terms of a set of figures of merit, which are relevant from a chip integration point of view. For each case, the figure of merit is defined and the quality criterion specified. Finally, it considers the overall performance of each mesh design taking into consideration all the defined figures of merit [2].

5.4.1 Spatial Tuning Resolution Step and Reconfiguration Performance

TBU-based meshes are employed and reconfigured in many occasions to implement either finite impulse response MZIs or infinite impulse response ORR filters. These are spectrally periodic and their free spectral range (FSR) is inversely proportional to the length mismatch between the two arms in the case of MZIs or the cavity length in the case of ORRs. The *spatial tuning resolution step* quantifies which is the minimum step in BUL units by which the arm length mismatch or the cavity length can be increased/decreased. The smaller the value of this figure of merit, the better, as this allows a finer discrete spatial sampling and therefore a wider range of interferometric lengths (L_{ORR} in the case of ORRs) or arm length mismatch values (ΔL_{MZI} in the case of MZIs). The value of $\Delta \upsilon_{\text{FSR}}$, the frequency separation corresponding to the spectral period FSR of the filter is given by

$$\Delta \nu_{FSR} = \frac{c}{n_g \cdot N \cdot BULs}, \tag{5.5}$$

where n_g is the group index of the waveguide and $N{=}L_{ORR}$ or $N{=}\Delta L_{MZI}$ is the representative interferometer length (normalised to the BUL), depending on whether an ORR or an MZI is considered. Note as well that more complex configurations, like coupled resonator optical waveguides (CROWs) [5], side-coupled integrated spaced sequence of resonators (SCISSORs) [6] and ring-loaded MZIs [7, 8] can be configured with any mesh, but in principle, these will be built upon combination of the basic ORR and MZI structures.

For a fixed BUL, each mesh design offers a TBU interconnection freedom that results in a different spatial resolution capacity for discretising the length of the cavities and the

Table 5.1 *Values for the possible cavity lengths (ORR) and arm length mismatch (MZI) in BUL units for the different mesh designs, n ∈ N {0,1,2,...} for MZIs and n ∈ N {1,2,3,...} for ORRs.*

	Square	Triangular	Hexagonal
MZI	$\Delta L_{MZI}^{Sq} = 4n$	$\Delta L_{MZI}^{Tr} = 3n$	$\Delta L_{MZI}^{Hx} = 2n$
ORR	$L_{ORR}^{Sq} = 4n$	$L_{ORR}^{Tr} = 3n$	$L_{ORR}^{Hx} = \{6, 10 + 2n\}$

arm length differences of the interferometric structures. The results of our comparative analysis accounting for the resolution step are shown in Table 5.1.

The maximum FSR value sets the operational frequency limit for some applications. Regarding ORR configurations, the triangular mesh offers the maximum achievable FSR. For cavity lengths above 10 BULs, the hexagonal mesh doubles the resolution of the square mesh.

As far as MZIs configurations are concerned, the hexagonal mesh offers the maximum achievable FSR and, again, doubles the resolution of the square mesh. Both the triangular and hexagonal meshes improve the spatial tuning resolution of the square mesh. The results show that, in general, the finest spatial tuning resolution corresponds to the hexagonal mesh with an incremental step of 2 BULs, although the smallest cavity value in the case of implementing an ORR is 6 BULs followed by a second step of four BUL units.

Given a maximum value available for the interferometric length N, set typically by the maximum allowed insertion losses, we define the *reconfiguration performance* of the mesh as the number of filters with different $\Delta\upsilon_{FSR}$ values that can be implemented. Figure 5.11 plots this figure of merit for the square, triangular and hexagonal meshes as a function of N for the ORR (upper) and MZI (lower) filters. We include here a plot extension to emphasise that the limit in the maximum achievable interferometric length is given by the accumulated losses rather than by the available chip area or the number of cells. In addition note that the waveguide mesh implementation using the parallel version described in the former section would offer a footprint reduction.

As an example, assuming a low-loss 0.3-dB TBU IL, which is below state-of-the-art, a modest hexagonal mesh of seven cells enables interferometric lengths larger than 30 BULs, introducing 9-dB accumulated losses. The same area that allocates seven hexagonal cells allows a larger number of square and triangular cells and, therefore, supports larger interferometric paths. Nevertheless, an interferometric length larger than thirty suffers from more than 9-dB of accumulated losses, so the final design becomes impractical for some applications. Note that the chosen IL per TBU is below the achievable with the current state-of-the-art and that an increment in this figure implies a further reduction in the maximum allowable interferometric length for the synthesised circuits.

Note as well that the hexagonal configuration clearly features the best reconfiguration performance as compared to the triangular and the square meshes for MZI

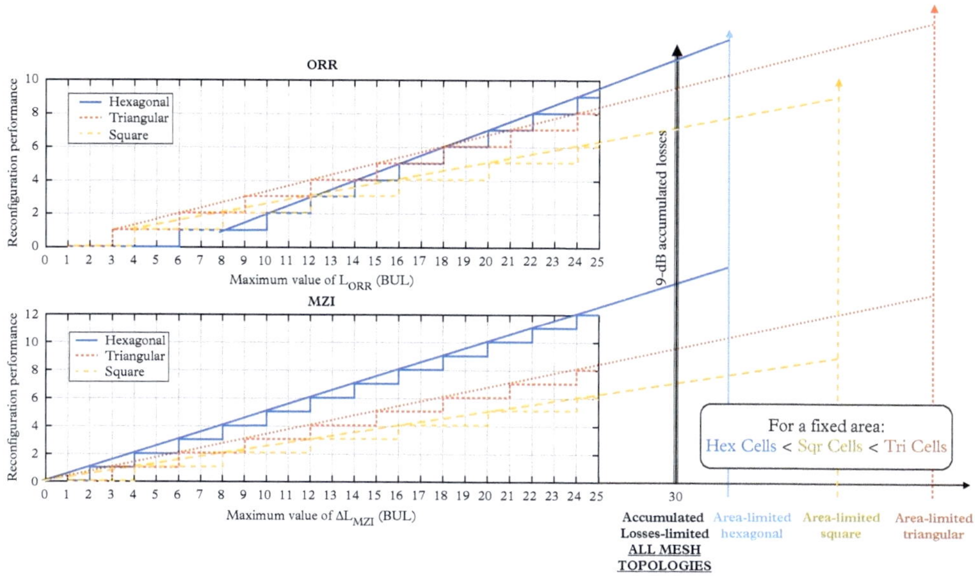

Figure 5.11 *(Left) Reconfiguration performance vs available maximum value of N expressed as an integer number n of the BUL for the square, triangular and hexagonal meshes. ORR (upper) and MZI (lower) designs. (Right) Maximum Interferometric length limitation due to accumulated losses (topology-independent and more restricting) and area constrains for each alternative topology.*

implementation. In the case of ORRs, the hexagonal and triangular meshes provide similar results, although the hexagonal configuration surpasses both the triangular and rectangular configuration for larger values of the cavity length.

5.4.2 Number of Switching Elements per Unit Area

From a standalone point of view of integration density, a mesh having a number of switching elements as high as possible should be preferable, although this is not necessarily so in practice. First of all, the maximum number of TBUs that build up the longest light-path will be limited by their non-ideal propagation and coupling characteristics. In addition, if the tuning mechanism requires metal layers to route the control signals that take up the same layer level, a reduced figure of switching elements per area facilitates this task offering a reduction in the BUL, since there is no need to increase the length of the access waveguides to provide a free way to the metal tracks. If the wire-bonding alternative is preferred, this is also desirable as there is more free space available to prevent the crossing of wires by increasing the separation between metal junctions. Also, in the case of using an MZI as the TBU, a reduced value of this figure of merit offers extra free space to grow further apart the arms of each MZI, reducing the crosstalk due to the tuning mechanism, i.e. thermal crosstalk. Finally, a higher number of switching elements per unit area leads to more power-hungry configurations and does not necessarily mean more flexibility in reconfiguration performance. Thus, for equal *reconfiguration performance*, a mesh design having a number of *switching elements per unit area* as low as possible is preferred.

The approximated footprint of the mesh is obtained first by computing the area of the unitary cell and then multiplying by the number of cells. For the sake of comparison, we normalise the polygons that define each unit cell (square, hexagon and triangle) by their side, i.e. all the meshes have sides with the same BUL. In this way, we neglect the waveguide access difference between meshes, which is the case when the BUL is large as compared to the bending radii of the technology of integration. The unit cell area can be expressed in general as $A_{unit-cell} = k\text{BUL}^2$, where $k = 1$ for the square cell, $k = 3\sqrt{3}/2$ for the hexagonal cell and $k = \sqrt{3}/4$ for the triangular cell. For reference, if we consider a square normalised area that could allocate a modest number of cells, it is straightforward to obtain that the number of switching elements per normalised area and its comparative increase or decrease percentage taking the square mesh as a reference. This figure yields a reduction of the 36.66 per cent for the hexagonal mesh and an increment of sixty-five per cent for the triangular mesh; thus, the hexagonal waveguide mesh provides the best performance in terms of this figure of merit.

Finally, we compare meshes with the same interconnection scheme, like the hexagonal topology and its parallel homologue. Here, they have the same *reconfiguration performance* and the *number of switching elements per area* of the parallel version can be tailored by increasing the necessary distance between actuators when required until the integration density figure is doubled.

5.4.3 Replication Flexibility: Possible Alternative Geometries for Filter Implementation

Another important figure of merit is the *mesh replication flexibility* defined as the number of possible different alternative geometries that a given mesh design topology provides to implement a given ORR (with a fixed cavity length) or MZI configuration (with a fixed arm length imbalance). This metric is a good indicator of the potential for configuring complex devices involving cascaded photonic circuits and also for allocating a given circuit in different locations within the available or unused footprint of a waveguide mesh.

When a new photonic device needs to be added to an already operational mesh, it will occupy part of the unused space available. The more flexible the mesh architecture is to replicate an ORR or a MZI, the easier it will be to allocate it within the spare space. To make a fair comparison between the proposed designs, we have considered the synthesis in an infinite mesh of ORRs with cavity lengths of up to 16 BULs and MZIs with arm imbalances of up to 12 BULs. In order to account for all the MZI designs that provide the same FSR, we have limited the synthesis to MZIs whose minimum arm length is equal or shorter than 3 BULs. Table 5.2 shows the results of our analysis, and the two bottom rows provide a cumulative figure on the total number of replication options and the replication ratio (compared to that of the triangular mesh) for ORRs and MZIs. For ORRs the square mesh features a replication flexibility that is approximately double the hexagonal mesh and approximately a thirty per cent extra replication flexibility as compared to the triangular mesh. For MZIs, the hexagonal mesh benefits from its ability to provide shorter paths (i.e. 1 and 2 BULs) featuring approximately a 236 per cent extra replication flexibility as compared to the triangular mesh and exceeding square mesh metrics by twelve per cent.

In order to illustrate the *replication flexibility* concept, Figure 5.12 plots for each mesh topology the settings to programme fixed-FSR ORRs with the cavity describing different geometries. Here, the nomenclature employed is the following: DEV[NR][ML] where DEV = {ORR, MZI} stands for the device implemented, N is the cavity or arm length mismatch in BUL units, R appears if coupler reutilisation is employed, M is the number of reutilised couplers and $L = \{A, B, C, \ldots\}$ is a letter that designates different designs sharing equal values for the preceding parameters in the code. In particular, we illustrate the case of setting ORRs of the same length in a bus configuration (SCISSORs). The upper left part shows the unique three ways of synthetising ORRs of normalised circumference length equal to 10 BULs using a hexagonal mesh. For the triangular mesh, the two alternatives of synthetising ORRs of normalised circumference length equal to six BULs are shown in the upper right part of Figure 4.9. These results match the ones predicted by Table 4.2. Finally, the lower part of the figure shows the unique five alternatives to programme ORRs of normalised length eight BUL using a square mesh. Notice that the first three designs also re-use one of the TBUs.

It is also worth mentioning that different devices could also share TBUs in order to save space. For example, in the square mesh, between the ORR8A and the ORR8B an ORR8R1C could be synthetised sharing their outer bar-state-configured TBUs.

The results from Table 5.2 confirm that the square mesh is the most flexible when replicating ORRs but, at the same time, they outline that this mesh design is the most restrictive one in terms of *spatial tuning resolution step*. On the other hand, the triangular mesh is the most restrictive in terms of *replication flexibility* due to its mesh complexity and its interconnection scheme (Figure 5.5). Nevertheless, its *spatial tuning resolution step* is better than the square mesh. Finally, the hexagonal mesh presents a good equilibrium between medium-high *flexibility* and the highest *spatial tuning resolution step*.

5.4.4 Losses and Spatial Resolution Associated with TBU Interconnections

Interconnections between the TBUs determine the spatial resolution and the losses due to bending radii and polarisation rotation. To make a fair comparison between the meshes, we consider two benchmarking alternatives. In both cases, the tunable coupler length of the TBU is kept constant: In the first one, the bending radii of the curves are fixed while in the second it is the BUL that is kept constant.

In the first case, the fixed radius is dictated by the integration technology platform. It ranges typically from 5–30 μm in silicon waveguides to 100–250 μm in indium phosphide, 100–1000 μm for silicon nitride platforms and about 1000 μm for silica-on-silicon. In this case, the BUL of each mesh will be different as it depends on a fixed term (the tunable coupler length) and a variable term (the access length) defined by

Table 5.2 *Number of different devices that can be configured for each mesh design.*

ORRs				**MZIs**			
L_{ORR} **(BULs)**	**Square**	**Hexagonal**	**Triangular**	ΔL_{MZI} **(BULs)**	**Square**	**Hexagonal**	**Triangular**
3	–	–	1	0	1	1	1
4	1	–	–	1	–	–	–
5	–	–	–	2	–	2	–
6	–	1	2	3	–	–	1
7	–	–	–	4	3	2	–
8	5	–	–	5	–	–	–
9	–	–	6	6	–	3	7
10	–	3	–	7	–	–	–
11	–	–	–	8	11	8	–
12	20	5	8	9	–	–	3
13	–	–	–	10	–	17	–
14	–	11	–	11	–	–	–
15	–	–	15	12	42	31	8
16	60	22	–	–	–	–	–
Total	86	42	32		57	64	19
Replication ratio	2.68	1.31	1		3	3.36	1

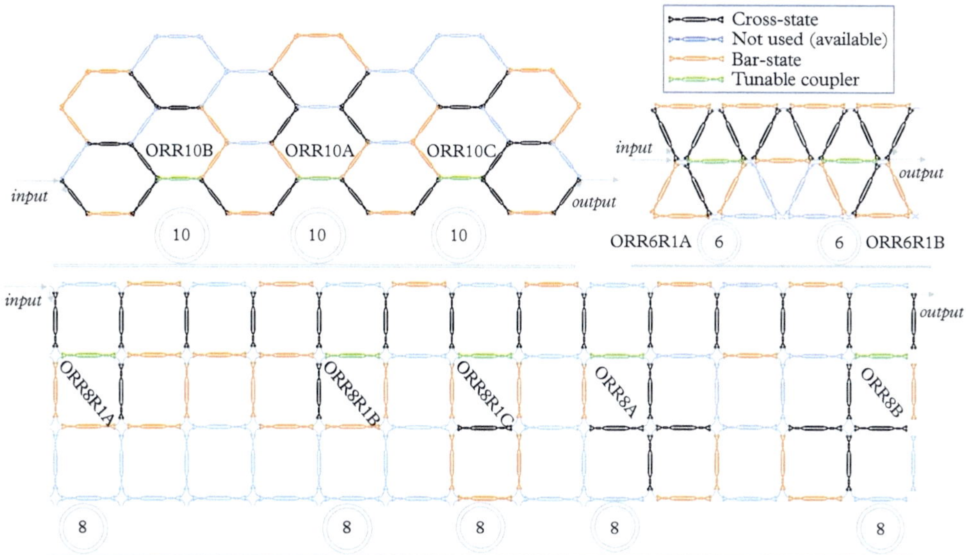

Figure 5.12 *Figure 4.9 Different mesh types configured to synthesise cascaded ORRs of fixed FSR in different geometries. (Up/left) Hexagonal waveguide mesh programming three cascaded ORRs of 10-BUL cavity length, (Up/Right) triangular waveguide mesh programming two OORs of 6-BUL cavity length and (lower) square waveguide mesh programming five ORRs of 8-BUL cavity length.*

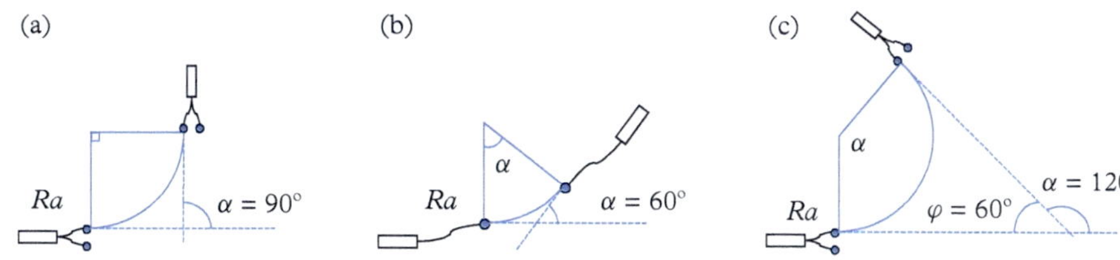

Figure 5.13 *Access length and angle definition for each mesh: (a) square, (b) hexagonal, (c) triangular.*

$$L_{access(both)} = \frac{2\pi Ra}{(360/\alpha)}, \tag{5.6}$$

where α is the angle in degrees of the arc defined by the interconnection, and Ra the radius of curvature (Figure 5.13). Since the angle between TBUs required by the square, hexagonal and triangular meshes are 90°, 60° and 120°, respectively, then the access length of the hexagonal mesh is thirty-three per cent shorter compared to the square mesh, while it is thirty-three per cent longer for the triangular mesh compared to the square mesh.

Therefore, when the radius of curvature is fixed, the BUL, and thus the spatial resolution, will be different for each mesh. Figure 5.14 shows the relation between the

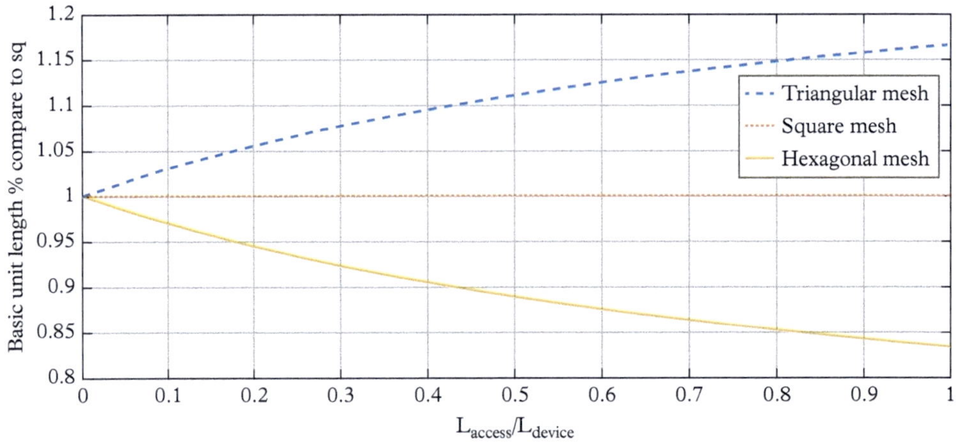

Figure 5.14 *Relation between the BUL of a given mesh design to that of the square mesh as a function of the ratio between the access waveguides and the tunable coupler length.*

BUL of a given mesh design to that of the square mesh as a function of the ratio between the access waveguides and the tunable coupler length. The range of applicability of this curve depends on the material platform used to implement the mesh. For instance, in silicon photonics, the bending radii of 5–30 μm compared with a tunable coupler length of about 500 μm places the region of application in the left side of the graph. In this case, the increase or decrease in the BUL associated to the triangular or the hexagonal meshes will be practically negligible as compared to that of the square mesh. In other platforms featuring higher values for the minimum bending radius, the region of application in Figure 5.14 will move towards the right part of the graph, leading to a higher spatial resolution for the hexagonal mesh.

The second case for comparison consists in fixing the value of the BUL so the product $\alpha \cdot Ra$ shown in the right part of Eqn. (5.6) remains fixed. In this case, the bending radius will be different for each mesh and, more precisely, when compared to the square mesh, the triangular will be a twenty-five per cent smaller whereas the hexagonal will be a fifty per cent higher. Here, since the losses associated to the bending radius and the polarisation rotation are greater for shorter bending radii, the hexagonal mesh again provides the best performance.

Finally, the parallel waveguide meshes will have a larger access length. This fact is more relevant when the bend radius is large when compared to the tunable coupler length.

5.4.5 Summary and Conclusions from the Benchmark Analyses

Table 5.3 summarises the results of the previous subsections carried for the different figures of merit and the three mesh designs. With the exception of the replication ratio of ORRs, the hexagonal mesh design is the most versatile option featuring the best

Table 5.3 *Summary of values for the figures of merit of the different mesh designs.*

Figure of Merit	Triangular	Square	Hexagonal
ORR cavity spatial tuning resolution step in BUL units (the lower the better).	3	4	2* (The first and second step each have a resolution of 6 and 4.)
MZI arm imbalance spatial tuning resolution step in BUL units (the lower the better).	3	4	2
ORR reconfiguration performance (the higher the better) (for $X = 25\ BUL$).	8	6	9
MZI reconfiguration performance (for $X = 25\ BUL$).	8	6	12
Switching elements per unit area compared to *square* % (the lower the better for a fixed value of reconfiguration performance).	+65.00%	0.00%	−36.66%
Replication Ratio for ORR structures up to 16 BUL cavity length (the higher the better).	1	2.68	1.31
Replication Ratio for MZI structures up to 12 BUL interferometric length (the higher the better).	1	3	3.36
$L_{access}/L_{access\ square}$% for a fixed Ra (the lower the better).	+33.33%	+0.00%	−33.33%
Ra/Ra_{square}% for a fixed BUL (the higher the better).	−25.00%	+0.00%	+50.00%

results in all the figures of merit. Its superior performance in terms of spatial tuning resolution step allows for a higher reconfiguration performance, that is, a wider range of spectral periods that can be implemented with complex structures built upon combining ORR- and MZI-based filters. The reduced value in the number of switching elements per unit area allows simpler configurations in terms of fabrication, electrode deposition, control pad interconnections and power consumption. Finally, and equally important, the hexagonal lattice mesh provides shorter curved sections for a given access waveguide bending radius and a fixed BUL value, which, in turn, results in lower propagation losses.

We conclude therefore that the hexagonal mesh is, in general, the most suitable option of the three considered for the implementation of waveguide meshes. In order to reduce the footprint constrains and tailor the integration density, the parallel version featuring the hexagonal interconnection scheme allows a twofold reduction of the footprint while maintaining the outperforming features of the conventional hexagonal pattern.

The analysis and comparative between waveguide mesh arrangements and their figures of merit position the hexagonal mesh configuration and its parallel implementation as the preferable choice for most of the relevant metrics. These results are pertinent for the implementation of a versatile reconfigurable optical platform that allows the setting and configuration of discretised conventional photonic integrated circuits.

5.5 Photonic Circuit Configuration and Programming Using Waveguide Meshes

We have shown that hexagonal-shaped mesh and its parallel implementation feature improved performance in terms of *spatial tuning reconfiguration step*, *reconfiguration performance*, *switching elements per unit area* and *losses per spatial resolution* [2], and enable the simplest implementation of both classic and simple FIR and IIR photonic circuits/components. This section extends the analysis to their combination and configuration to achieve more complex classic filters and arbitrary multiport interferometers described in Chapter 4. Although in the following, we will mainly consider the hexagonal pattern, the reader can find some additional circuit configurations of the remaining topologies in [2].

5.5.1 Finite Impulse Response (FIR) Filters

Unbalanced MZIs (UMZIs) are 2-input/2-output periodic notch filters that constitute the BBBs for lattice and transversal filters [9]. UMZIs find multiple applications [10], including linear phase filters, multi-channel selectors, group delay compensators and biosensors. These filtering structures are all-zero filters in the z-plane.

A multi-stage filter can be realised by cascading single UMZIs structures. In order to programme the optical core to produce a one-stage FIR filter, the first step is the location and configuration of the first TBU that will operate as the input tunable coupler defined by a coupling constant K_1. Secondly, two synthesised waveguide paths have to be configured fulfilling two conditions: they have to maintain the desired differential path

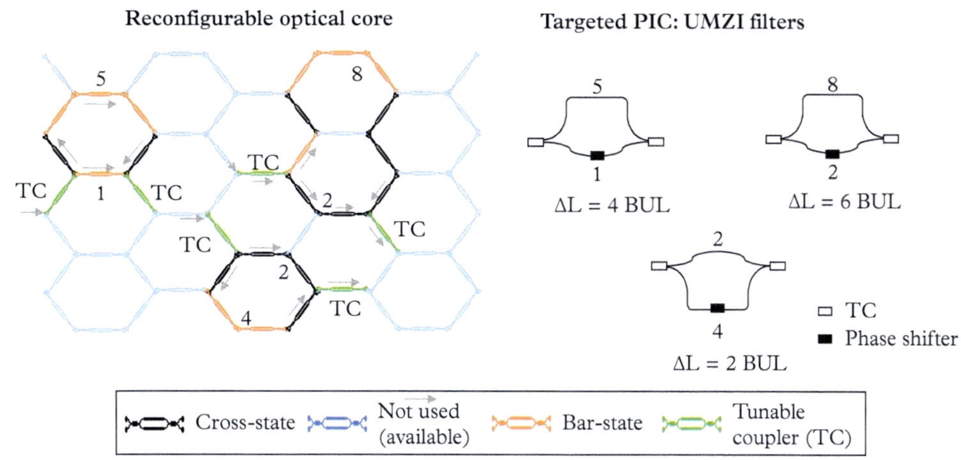

Figure 5.15 *FIR filter implementations. (Left) hexagonal mesh setting for (Right) three different targeted UMZI Filters.*

length ($\Delta L = L_{Longer}$—$L_{shorter}$) or a desired FSR given by Eqn. (5.5) so that at the same time, both final TBUs coincide at each of the input ports of the closing coupler (K_2) of the UMZI. By suitably tuning each TBU within the mesh, we can implement UMZI devices with path unbalances given by $\Delta L = 2n$ BUL, for $n = 0, 1, 2, 3 \ldots$. By taking the minimum possible value for the length of the shorter path, we reduce the insertion loss of the filter and the number of TBUs. Figure 5.15 illustrates the targeted PICs with their corresponding settings of each TBU inside the mesh to obtain three different UMZI filters with different FSRs. Note that the numbers account for the MZI arm lengths in BULs. Similarly, Figure 5.16 illustrates the implementation of two multi-stage filters by cascading single UMZIs, enabling a sixth- and fifth-order filter, respectively.

Different values of the coupling constants K_1 and K_2 change the absolute magnitude of the zero in the UMZI transfer function, which moves it closer to or farther from the unit circle. The closer the zero is to the unit circle, the deeper the notches in the transfer function and the higher the phase shift step in the phase transfer function [9]. A phase shifter can be programmed applying a common phase shift to one of the TBUs while maintaining fixed its splitting ratio. In this way, we can achieve the tuning of the resonance notch position over a complete spectral period by proper phase shifting in one of the UMZI arms. This is equivalent as moving the zero along the z-plane without changing its absolute magnitude.

Cascading this building block is a straightforward way to realise higher-order all-zero responses in the z-plane. A wide variety of algorithms is available in the literature for the synthesis of these so-called lattice filters, which provide the values of the coupling coefficients and phase terms that render the targeted spectral magnitude or phase response [9, 10].

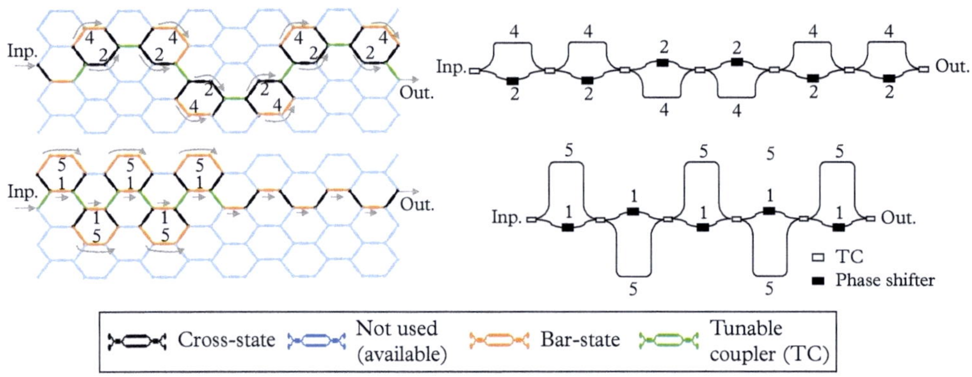

Figure 5.16 *Cascaded FIR filter implementations: (left) hexagonal mesh setting for (right) two different targeted lattice UMZI Filters of 6th and 5th order.*

A general procedure is proposed in the following pseudo-code to programme the waveguide mesh arrangement to synthetise FIR circuits with single or *N*th-order schemes.

FIR 1 stage:

```
procedure placeFIR, (InterLength, PS)
set (TBU_number, TC) % Set input_coupler
for i from 1 to Nshort do
set (TBU_number, DDL) % Set short path
set (TBU_number, TC) % Set output_coupler
for i from 1 to Nshort+ InterLength do
while % closed path between input_coupler and output_coupler
set (TBU_number, DDL) % Set long path
if PS is distinct from 0 then
set (TBU_number, PS) % Set short path/long path phase shift
```

FIR N stages:

```
procedure placeFIR_N, (InterLength, PS,N)
for i from 1 to Ndo
placeFIR (InterLength, PS,N) % Set input_coupler
%last_TBU_output_coupler is the next TBU_number (input coupler)
```

5.5.2 Infinite Impulse Response (IIR) Filters

Ring cavities are either 1-input/1-output or 2-input/2-output periodic filters. In the first case, they implement all-pole IIR notch filters, while in the second they can implement both IIR notch and FIR + IIR bandpass filters [9]. They constitute the BBBs for more complex filter designs such as CROWs and SCISSORs. Ring cavities find multiple applications [7] including integrators, differentiators and Hilbert transformers [11], dispersion compensators [7], as well as tunable radiofrequency phase shifters and true-time delay lines [12].

In order to program such a filter, the first step is the location and configuration of the first TBU that operates as the input coupler defined by a coupling constant K_1. Secondly, a waveguide path has to be configured starting from one output port of the first TBU and ending in one of the input ports of the first TBU—thus providing the desired cavity length (ΔL_{ORR}) or FSR. One of the TBUs within the waveguide path can be set as a second coupler (K_2) to implement a second input/output port and therefore enable a 2-input/2-output filter. By suitably tuning each TBU within the mesh, we can implement single ORRs with cavity lengths given by $\Delta L_{ORR} = 6, 10, 12, 14, \ldots$ BULs. Different values of K_1 and K_2 set the values of the absolute magnitude of the zero and the pole [9]. Since any TBU in the waveguide implement the cavity can be operated as a constant-amplitude phase shifter from 0 to 2π, we can tune the filter resonance position along a full spectral period. Figure 5.17 illustrates the mesh configuration for three IIRs filters with different FSRs with cavity lengths corresponding to 6, 10 and 12 BULs, respectively. Note that, in this example, the shorter cavity has both tunable couplers activated (i.e. it is a 2-input/2-output filter).

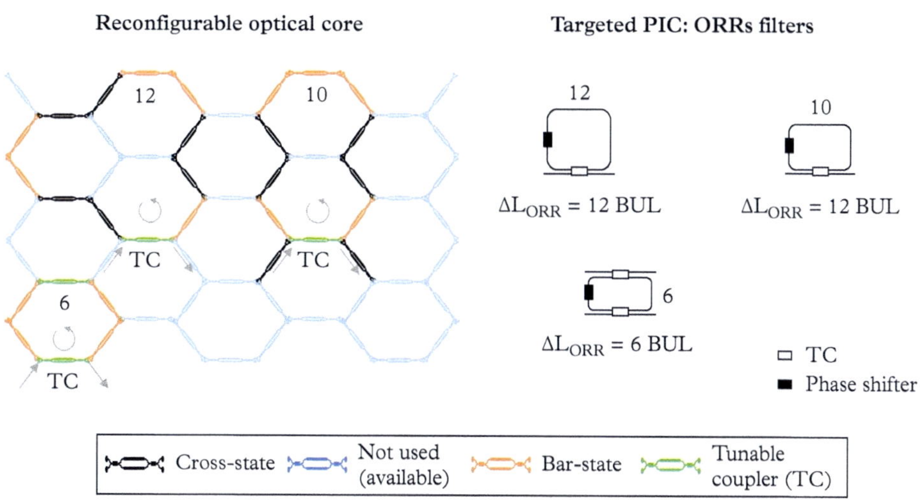

Figure 5.17 *Single-cavity IIR Filter implementations. (Left) hexagonal mesh setting for (right) three different targeted ORR filters.*

5.5.3 Complex Multi-stage Filters

We can build more complex (multi-cavity) signal processing structures like CROWs [13], SCISSORS [14], and ring-loaded MZIs [15] using the former BBBs in the waveguide mesh and activating more TBUs to provide additional propagation paths. These are usually 2-input/2-output filters that are characterised by transfer functions with a higher number of zeros and poles. By suitably tuning the coupling constants and additional phases, we can obtain, for instance, filters with special characteristics in the modulus or the phase response [7].

5.5.3.1 Coupled Resonators Optical Waveguides (CROWs)

An *N*th order CROW consists of a chain of *N* rings where each ring is coupled to one preceding and one succeeding ring (with the exception of the first and last in the chain). The rings in the CROW chain usually have the same length, however, to increase the overall filter FSR, rings with slightly different perimeters in a Vernier configuration can be employed, [9]. Figure 5.18 (a) illustrates the implementation of a 10-stage CROW characterised by 6-BUL cavity lengths.

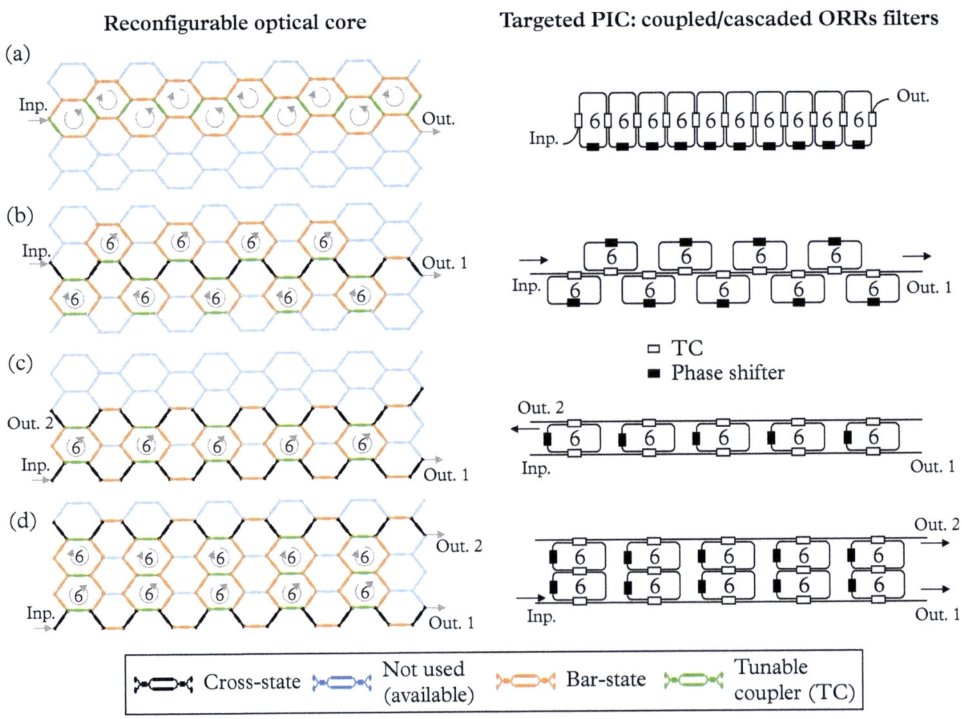

Figure 5.18 *(Left) Settings for CROW & SCISSOR filter implementations in the hexagonal mesh core for (right) (a) 10th-order CROW, (b) 9th-order Single channel SCISSOR, (c) 5th-order Double channel SCISSOR, and (e) twisted double channel SCISSOR.*

5.5.3.2 *Side-coupled Integrated Spaced Sequence of Resonators (SCISSORs)*

Cascading ring cavities in one or two common waveguide buses leads to the implementation of higher-order response SCISSORs filters. Typically, all the ring resonators have the same cavity length. Figure 5.18(b)–(d) illustrates different SCISSOR configurations that can be implemented with the waveguide mesh: (b) is a 9-stage single-channel where all the rings are connected to the same waveguide bus, (c) is a 5-stage doubled-channel, where a second output waveguide bus collects the FIR + IRR bandpass outputs from the ring resonators and (d) is a 5-stage double cavity SCISSOR where two 6-BUL coupled cavities are connected to the collecting waveguide buses per stage. Note that, in this particular case, the signal propagation through the two collecting waveguide buses is in the same direction.

The following pseudo-code proposes a general procedure to programme the waveguide mesh arrangement to synthetise IIR and IIR + FIR circuits with single or *N*th-order schemes.

IIR 1 stage:

```
procedure placeIIR, (InterLength, PS)
set (TBU_number, TC) % Set input_coupler
for i from 1 to InterLength-1 do
set (TBU_number, DDL) % Set cavity
set (TBU_number, TC) % Set output_coupler
if PS is distinct from 0 then
set (TBU_number, PS) % Set short path/long path phase shift
```

IIR N stages (single-bus SCISSOR):

```
procedure placeIIR_N, (InterLength, busLength,PS,N)
for i from 1 to bus_length do
set (TBU_number, DDL) % Set bus channel
for i from 1 to Ndo
placeIIR (InterLength, PS,N) % Set input_coupler
```

IIR N stages (CROW):

```
procedure placeCROW_N, (InterLength, PS,N)
for i from 1 to Ndo
placeIIR (InterLength, PS,N) % Set input_coupler
%last_TBU_output_coupler is the next TBU_number (input coupler)
```

5.5.3.3 Ring-loaded Mach–Zehnder Filters (RL-MZIs)

These optical filtering architectures consist of MZIs with N and M optical ring resonators cascaded on each arm, respectively, [7, 8]. They allow the independent setting of zeros and poles. These structures can be used for implementing compact and footprint-reduced all-pass, and bandpass filters with arbitrary responses. Figure 5.19 shows two examples of single and double-RL–MZI filters where the ring cavity length is 6 BULs.

5.5.4 Programmable Delay Lines

True-time optical delay lines (TTODLs) are key building blocks for a wide range of applications in optical processing when combined with optical couplers and phase shifters. Recent approaches based on cascading tunable couplers and discrete delay lines pursue compact and wideband discrete reconfigurable delays. These implementations are typically suitable for a single specification or application since the delays, even if they are tunable, are interconnected in a fixed configuration. As shown in Figures 5.4–5.6, waveguide mesh arrangements can be employed for the implementation of delay lines with delays equal to $N\,BUD$, where N is the number of TBUs that defines each configured light-path. More complex circuits like arrays of discrete delay lines are the key element in a wide variety of functionalities, e.g. beamforming networks for phased array antennas and multiport processing. These devices are characterised for the selection of a desired delay per output port. Typical configurations provide a sequential delay increment between output/input ports $\Delta\,\tau$ [4].

The TTODL operation can be programmed in waveguide meshes by configuring some of the TBUs as tunable couplers and some as optical crossbar switches, creating adjacent light-paths with an incremental length value ΔL (expressed in discrete values

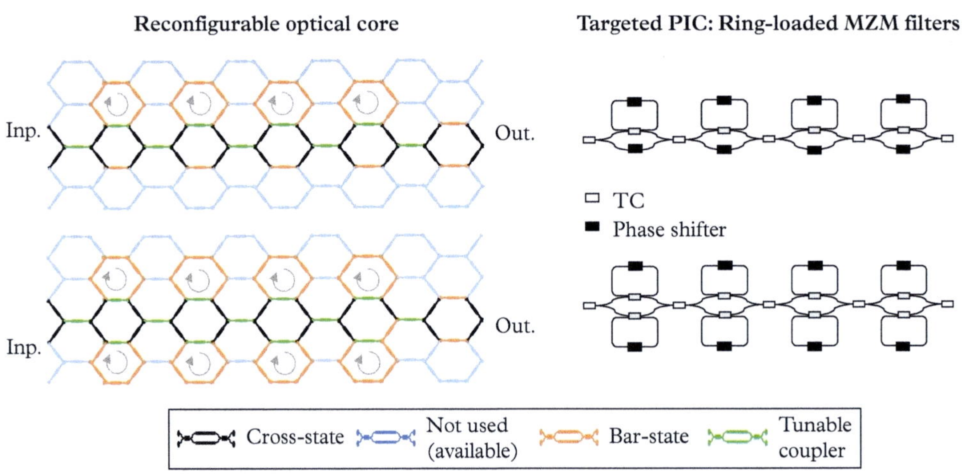

Figure 5.19 *Ring-Loaded MZIs: (Left) Hexagonal mesh settings and (right) targeted single-loaded fourth order and double-loaded eighth order filter, respectively.*

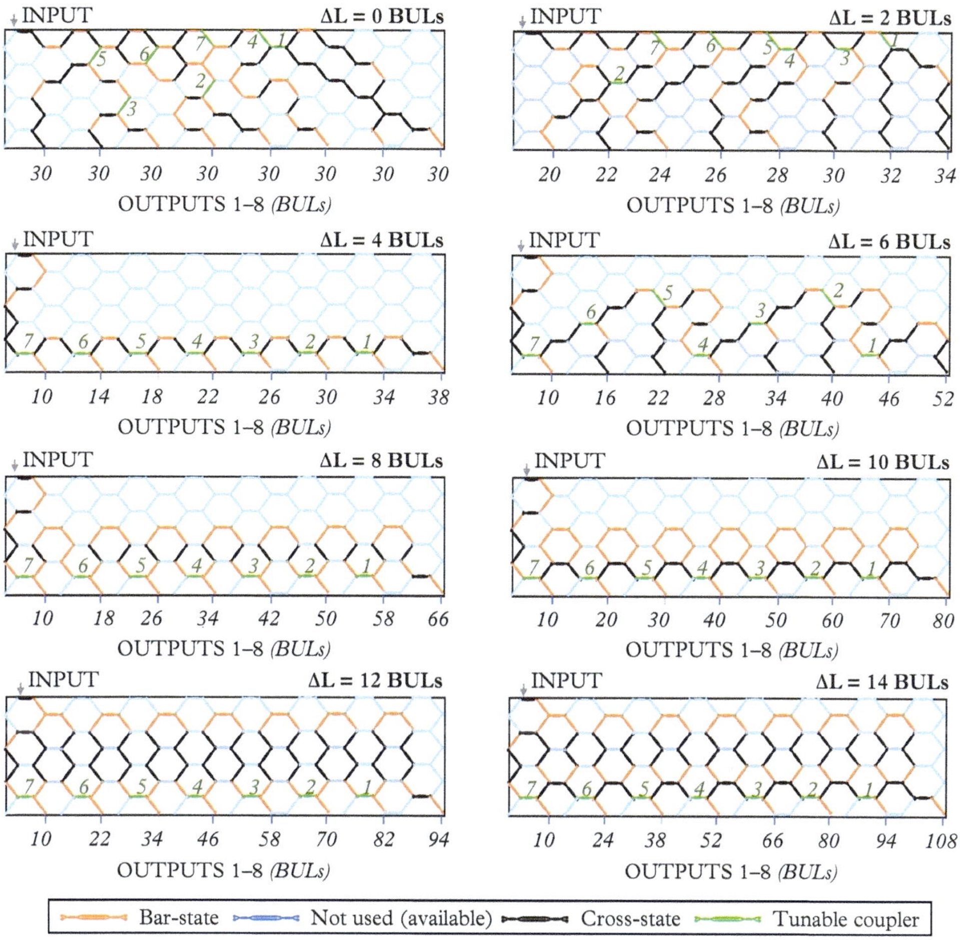

Figure 5.20 *1X8 Delay-line array based on discrete optical delay lines: Configuration examples for 0-, 2-, 4-, 6-, 8-, 10-, 12-, and 14-BUL path difference. Each path length is labeled at each output. The color-coded used for each TBU configuration is: Cross State (black), Bar State (Orange), Tunable Coupler (Green). For each Tunable Coupler mode, a number is given for its reference to the Appendix A and to (5).*

of BULs). This incremental length defines, in turn, the delay that characterises the path associated to each output port. Figure 5.20 illustrates the implementation of a 1×8 beamforming network based on a hexagonal waveguide mesh in a PIC layout.

By suitable configuring each TBU using the defined colour code for crossbar and tunable coupling operation we illustrate the delay-array operation for various differential delays, in particular, for path-length differences of 0, 2, 4, 6, 8, 10, 12 and 14 BULs. For each delay configuration we specify at each output port the corresponding number of TBUs that define the delay. As outlined previously, larger delays suffer from greater losses. These can be compensated by tailoring the coupling constants of the TBUs configured as a tunable couplers. Note that there are seven TBUs in tunable coupler mode for each configuration, labelled and shown in green.

By means of the following design rule, each of these tunable couplers will compensate for the unbalanced subsequent path losses according to

$$K_n = \frac{1}{10^{\frac{L_{n,bar}-L_{n,cross}}{10}}+1}. \tag{5.7}$$

Here $n = 7, 6 \ldots 1$ denotes the tunable coupler number and $L_{n,bar}$ and $L_{n,cross}$ represent the accumulated losses from the output of each coupler to their corresponding beamformer output port. We can differentiate two cases for the definition of the *accumulated loss*. First, if the path from the output of the tunable coupler n is only configured by TBUs in switching states (either cross or bar), then the accumulated loss in this port is

$$L_{n,bar/cross} = N \cdot IL_{TBU}, \tag{5.8}$$

where N is the number of TBUs from the tunable coupler to the corresponding port. As an example, the coupler number two in Figure 5.20 (for the case of $\Delta L = 8$), has accumulated losses in the lower port (*cross*) given by $L_{n,cross} = IL_{TBU}$, since there is only one TBU in bar-switching state in the interconnection path. Secondly, if the path from the output of tunable coupler n and the beamformer port includes at least one tunable coupler TBU, then the nearest one is labelled as coupler q and

$$L_{n,bar/cross} = 10\log_{10}K_q + (r+1)\,IL_{TBU} + L_{q,cross}, \tag{5.9}$$

where r is the number of TBUs between coupler n and coupler q. An example of this case can be observed in Fig. 5.20 (for the case $\Delta L = 8$), where the path connecting the upper (*bar*) port of coupler 2 to beamformer outputs 7 or 8 includes one of the TBUs configured as a tunable coupler.

5.5.5 Universal Linear Optics Transformers

As described in Chapter 4, the function of a multi-port linear optics device is to transform a series of N orthogonal modes ($|\,\phi_I\rangle$) into the corresponding N orthogonal modes at the output ($|\phi_O\rangle$) [16, 17]. This transformation is defined by a unitary matrix $U\left(|\phi_O\rangle\right) = U\left(|\phi_I\rangle\right)$. Linear transformations are the fundamental building block of many applications in quantum information and communication systems, switching and routing, microwave photonics and optical channel management and supervision. The proposed hexagonal waveguide mesh enables the implementation of the two layout versions of the universal linear interferometer proposed in the literature [16]–[26].

5.5.5.1 Reck–Miller/Triangular Arrangement Interferometer

Figure 5.21(a) displays an example of a 4×4 interferometer implemented by means of a triangular arrangement of beamsplitters and Figure 5.21(b) shows the equivalent structure implemented on a hexagonal waveguide mesh. Each beamsplitter can set a certain splitting ratio and a relative phase to the upper output. Reck and colleagues [16] and Miller [25] have developed algorithms to programme and configure the triangular arrangement so it can implement any desired linear unitary transformation [17, 25]. To adapt, for example, the synthesis algorithm developed by Miller to the hexagonal waveguide mesh we must first consider the possible different phase contributions due to the different access paths established between the interferometer inputs and the internal processing elements forming the triangular arrangement of beam splitters and, from these, to the different outputs.

These different phase contributions must be compensated. Then, we need to establish an equivalent configuration—using the available primitive elements in our hexagonal waveguide mesh—to the MZI with a phase shifter in the upper output port employed by Miller (Figure 5.20(c)). In our case (Figure 5.21(d)) the equivalent 'beamsplitter' is implemented using a TBU for the tunable coupler (with a transfer matrix defined by h_{TC} as in Eqn. (5.3)), followed by two TBUs, which are biased in cross state and employed as output connections. In the latter, the upper TBU also implements a phase shifter and is defined by the transfer matrix h_{UPS}. The lower TBU is defined by the transfer matrix h_{LPS}.

Miller's synthesis algorithm is based on writing any of the input basis functions as a linear combination of each input port or rectangular functions ($|\phi_{1n}\rangle$), and configuring sequentially each row of beam couplers for each input mode. These input modes can

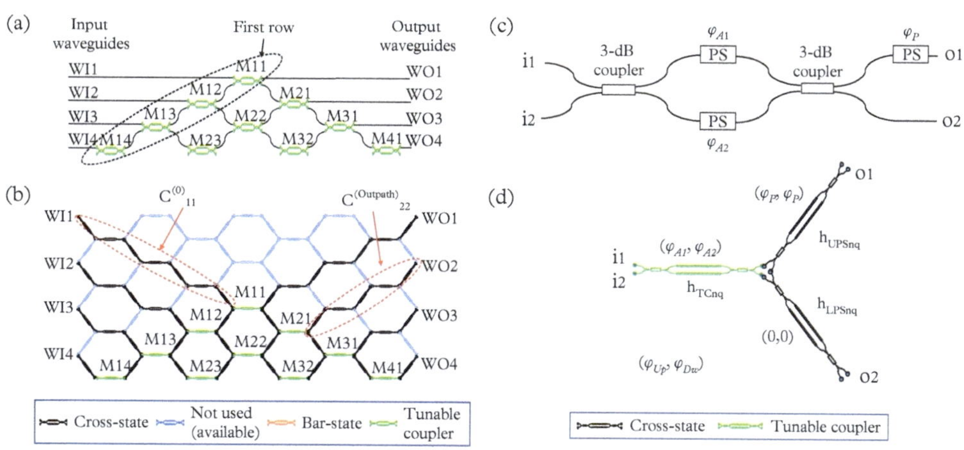

Figure 5.21 *Universal interferometers: (a) Classical triangular arrangement and (b) hexagonal mesh-based implementation of a 4 x4 interferometer. (c) Beamsplitter for the classical approach and (d) corresponding beamspliter implementation with 3 TBUs for the hexagonal waveguide Mesh.*

be obtained from the columns of the Hermitian adjoint of the matrix U. A procedure describing the synthesis algorithm adaptation and the related equations is provided in Pérez [1] and Pérez and colleagues [26].

Figure 5.22 illustrates an application example that includes the corresponding coupling and additional phase values to configure the hexagonal core for a Hadamard transformation of size 4×4, defined as

$$U_{Hadamard} = \frac{1}{\sqrt{2^2}} \begin{pmatrix} 1 & 1 & 1 & 1 \\ 1 & -1 & 1 & -1 \\ 1 & 1 & -1 & -1 \\ 1 & -1 & -1 & 1 \end{pmatrix}. \tag{5.10}$$

The actual implementation of the triangular arrangement for a 4×4 interferometer using a hexagonal mesh with the labelled configuration for each TBU is depicted, for the previous matrix, specifying: *ID* for the TBU identification label, *K* for the coupling constant and ϕ for the additional phase shift. The letter *P* is used to indicate that the TBU remains in passive state (both phase shifters are unbiased). The coefficients and phase terms are found using the configuration algorithm developed in Pérez [1] and Pérez and colleagues [26].

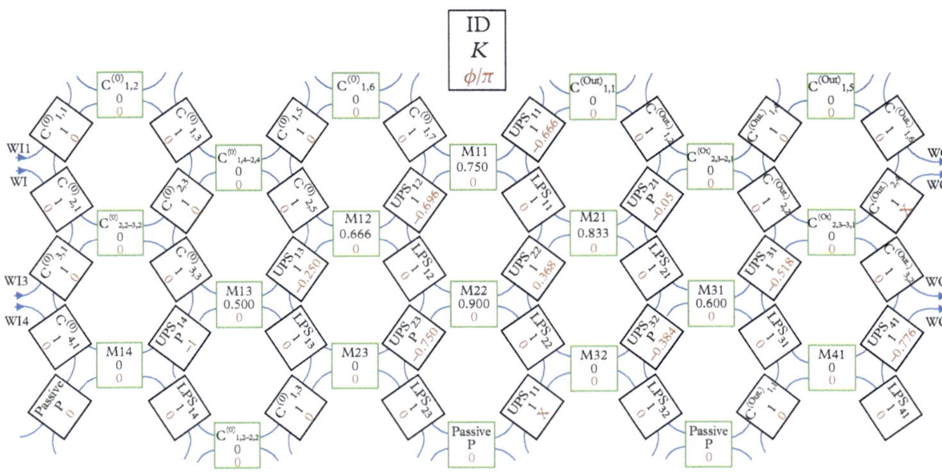

Figure 5.22 *Universal interferometer: TBU settings of the hexagonal mesh for the Triangular arrangement implementation programming a Hadamard 4x4 linear transformation. ID: TBU identification Label, K: coupling constant, ϕ: additional phase shift, P: coupling constant when both phase shifters are unbiased. Green-colour edges for TBUs acting as a tunable coupler and Black-colour for the cross state.*

5.5.5.2 Clements/Rectangular-arrangement Interferometer

Figure 5.23(a) displays an example illustrating the implementation of a more compact 4×4 multi-port interferometer based on a rectangular arrangement [18]. Figure 5.23(b) shows the equivalent structure implemented on a hexagonal waveguide mesh. In the Clements algorithm, each beamsplitter (Figure 5.23(c)) sets a certain splitting ratio and a relative phase sequentially to programme and configure the whole rectangular arrangement so it can implement any desired linear unitary transformation efficiently [18].

To adapt this layout and its synthesis algorithm to the hexagonal waveguide mesh, we need to perform a few modifications. First we must use a different matrix for the beam coupler/TBU structure. Here (Figure 5.23(d)) we employ a TBU for the tunable coupler (coloured in green) defined by a transfer function h_{TC}, and the two precedent TBUs (coloured in black) for the required connections. Here, the upper one operates in cross mode providing an extra phase shifting (upper phase shifter, h_{UPS}), while the lower one operates in cross mode. Note that, incidentally, both the classical and hexagonal approaches of rectangular arrangements need an extra phase shifter at each channel output that is required for certain applications, indicated as (*) in Figure 5.23(b). Finally, some of the outer TBUs that build up the interferometer must be configured to be phase transparent, featuring the phase-shift values indicated in Pérez [1] and Pérez and colleagues [26], where together with a procedure describing the synthesis algorithm adaptation, the reader can find the related equations, incorporating the special case where the element to be null is already equal to 0.

Figure 5.24 illustrates an application example that includes the corresponding coupling and additional phase values to configure the same Hadamard transformation as in

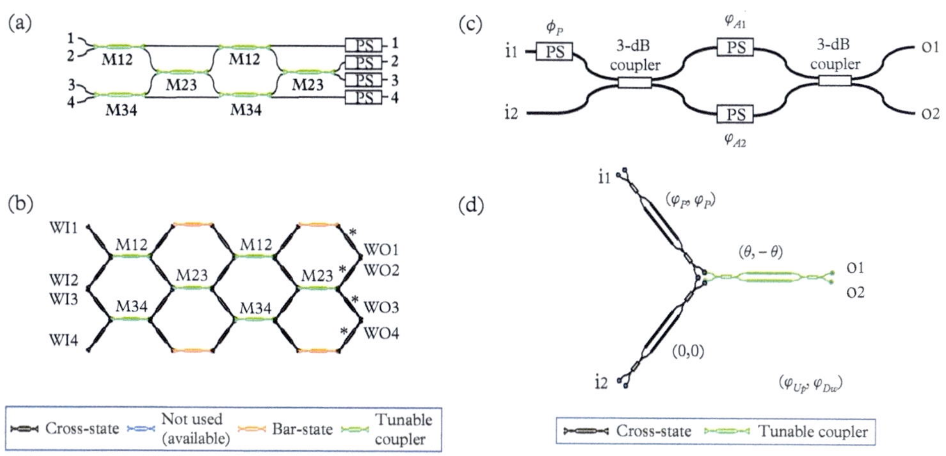

Figure 5.23 *Universal interferometers: (a) Classical rectangular arrangement and (b) hexagonal mesh-based implementation of a 4 x4 interferometer. (c) Beamsplitter for the classical approach and (d) corresponding beamspliter implementation with 3 TBUs for the hexagonal waveguide Mesh.*

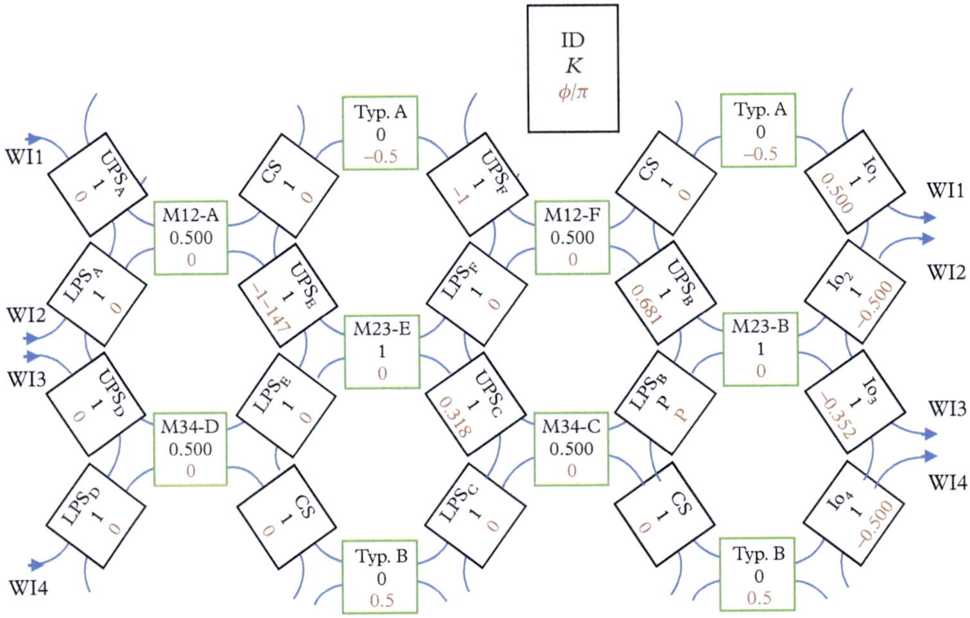

Figure 5.24 *Universal interferometer: TBU settings of the hexagonal mesh for the rectangular arrangement implementation programming a Hadamard 4x4 linear transformation. ID: TBU identification Label, K: coupling constant, ϕ: additional phase shift, P: coupling constant when both phase shifters are unbiased. Green-colour edges for TBUs acting as a tunable coupler, black colour for the cross state and orange colour for the bar state.*

the previous case defined by Eqn. (5.3). Note that the rectangular arrangement is the most compact topology for the synthesis of a certain $N \times N$ linear transformation.

5.5.5.3 *Linear Universal Transformations*

Linear universal components were addressed in Section 4.6 where, after applying the SVD technique described in Chapter 2 to a matrix, it results in the multiplication of two unitary matrices and a diagonal matrix. Waveguide meshes can programme linear universal transformations by programming the unitary matrices with the procedures introduced in previous sections and finding the intermediate paths and configuring the diagonal operations. Figures 5.25 and 5.26 illustrate a 4×4 and 5×5 SVD transformation using the 2D hexagonal integrated waveguide mesh and the triangular and rectangular interferometer configuration, respectively.

For this purpose, the hexagonal interconnection enables a straightforward fitting or discretisation more efficient than the square and the triangular schemes. A key advantage of this hexagonal mesh is the possibility of combining in a flexible way universal interferometers with other classic photonic integrated circuit structures that can be programmed over the mesh as well, e.g. MZIs, interferometric cavities. This comes at the expense of more beamsplitters for a certain $N \times N$ linear transformation,

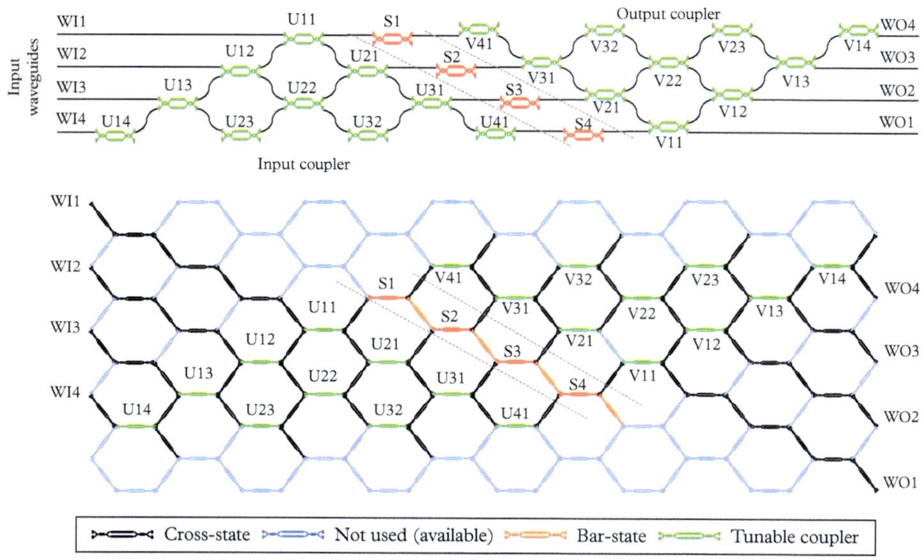

Figure 5.25 *Triangular interferometer layout (Upper) and 2D hexagonal waveguide mesh implementation of a SVD for a 4x4 matrix transformation.*

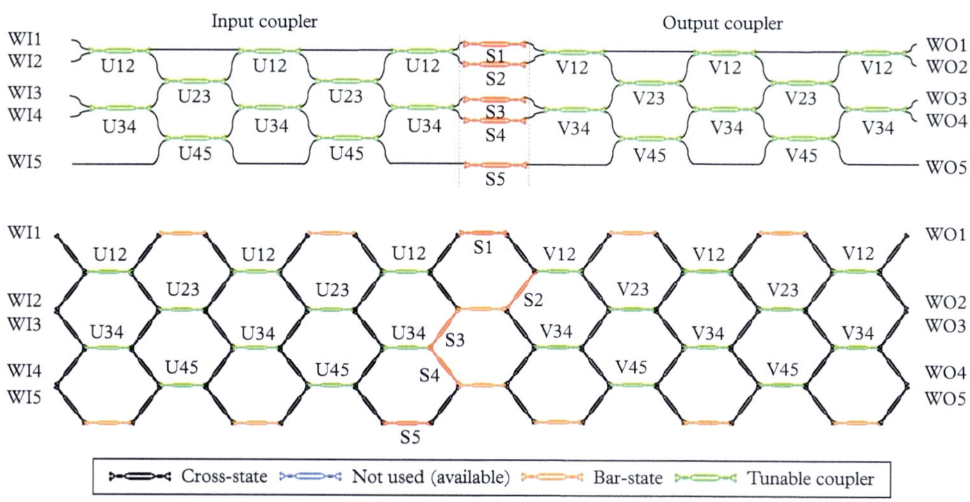

Figure 5.26 *Rectangular interferometer layout (Upper) and 2D hexagonal waveguide mesh implementation of a SVD for a 5x5 matrix transformation.*

which will increase from $N(N-1)/2$ (triangular and rectangular arrangements) to $(3^*N^*(N-1)/2)+2^*N$ for hexagonal applications. As previously outlined, a phase shifter/TBU must be added to each output, increasing the previous numbers by N.

Another advantage of the hexagonal mesh implementation is that, assuming equal losses for all the TBUs, the overall path losses for each configured channel will be equal, as the number of TBUs where the signal goes through is the same.

The implementation and programming of the aforementioned classical PIC in waveguide mesh arrangements can be combined to develop more complex signal processing. Interested readers are referred to the literature for an extended list of applications and mesh configurations [1–3].

5.6 Chapter Summary

This chapter has introduced the theoretical fundamentals of integrated waveguide mesh arrangements. They have been proposed as a PIC layout that enables the programming of complex photonic circuits. This behaviour is achieved by the configuration of each primitive element, and their fixed interconnection schemes. It analysed the resulting mesh topologies and their impact on performance and concluded that the hexagonal interconnection and, in particular, its parallel version outperform the others in terms of figure of merit that account for circuit integration, flexibility and versatility.

In this chapter we have assumed a simplified behaviour of the TBU and the overall circuit. When using non-ideal components, we find more limitations in the insertion loss, the power consumption of the phase actuators, uneven losses due to some fabrication and non-reproducible effects, optical crosstalk that causes reflexion inside the circuit and discretisation limits in real applications. Chapter 6 covers these topics together with the practical implications of a waveguide mesh arrangement design, and the subsystems needed for the configuration, programming and optimization of these components.

REFERENCES

[1] D. Pérez, *Integrated Microwave Photonic Processors Using Waveguide Mesh Cores*. PhD Thesis, Universitat Politècnica de València, 2017.

[2] D. Pérez, I. Gasulla, J. Capmany and R.A. Soref, Reconfigurable Lattice Mesh Designs for Programmable Photonic Processors. (2016) *Opt Express* 24, 12093–106.

[3] L. Zhuang, et al., Programmable Photonic Signal Processor Chip for Radiofrequency Applications. (2015) *Optica* 2, 854–9.

[4] D. Pérez, E. Sánchez and J. Capmany. Programmable True-Time Delay Lines Using Integrated Waveguide Meshes. (2018) *IEEE J Lightwave Tech* 36(19), 4591–4601.

[5] A. Melloni, et al., Tunable Delay Lines in Silicon Photonics: Coupled Resonators and Photonic Crystals: A Comparison. (2010) *IEEE Photonics J* 2(2), 181–94.

[6] J. Khurgin and P.A. Morton, Tunable Wideband Optical Delay Line Based on Balanced Coupled Resonator Structures. (2009) *Opt Lett* 34(17), 2655–7.

[7] K. Jinguji, Synthesis of Coherent Two-port Optical Delay-line Circuit with Ring Waveguides. (1996) *J Light Technol* 14(8), 1882–98.

[8] C.K. Madsen, General IIR Optical Filter Design for WDM Applications Using All-Pass Filters. (2000) *J Light Technol* 18(6), 860.

[9] C.K. Madsen and J.H. Zhao, *Optical Filter Design and Analysis: A Signal Processing Approach*, John Wiley & Sons, Inc., 1999.

[10] K. Jinguji and M. Kawachi, Synthesis of Coherent Two-port Lattice-form Optical Delay-line Circuit. (1995) *J Light Technol* 13(1), 73–82.

[11] W. Liu, et al., A fully reconfigurable photonic integrated signal processor. (2016) *Nat Photonics,* 10, 190–5.

[12] M. Rasras, et al., Integrated Resonance-enhanced Variable Optical Delay Lines. (2005) *IEEE Photonic Tech L* 17(4), 834–6.

[13] S. Mookherjea and A. Yariv, Coupled Resonator Optical Waveguides, (2002) *IEEE J Sel Top Quant Electron* 8(3), 448–56.

[14] J.E. Heebner, P. Chak, S. Pereira, J.E. Sipe and R.W. Boyd, Distributed and Localized Feedback in Microresonator Sequences for Linear and Nonlinear Optics. (2004) *J Opt Soc Am B* 21(10), 1818–32.

[15] J.S. Fandiño, P. Muñoz, D. Doménech and J. Capmany, A Monolithic Integrated Photonic Microwave Filter. (2016) *Nat Photonics* 11, 24–129.

[16] M. Reck, A. Zeilinger, H.J. Bernstein and P. Bertani, Experimental Realization of Any Discrete Unitary Operator. (1994) *Phys Rev Lett* 73(8), 58–61.

[17] D.A.B. Miller, All Linear Optical Devices are Mode Converters. (2012) Opt Express 20(21), 23985–93.

[18] W.R. Clements, et al., Optimal Design for Universal Multiport Interferometers. (2016) *Optica* 3, 1460–5.

[19] A. Ribeiro, A. Ruocco, L. Vanacker and W. Bogaerts, Demonstration of a 4×4-port universal linear circuit. (2016) *Optica* 3(12), 1348–57.

[20] D. A. B. Miller, Self-aligning universal beam coupler. (2013) *Opt Express* 21(5), 6360–70.

[21] D.A.B. Miller, Perfect Optics with Imperfect Components. (2015) *Optica* 2(8), 747–50.

[22] A. Peruzzo, A. Laing, A. Politi, T. Rudolph and J.L. O'Briena, Multimode Quantum Interference of Photons in Multiport Integrated Devices. (2011) *Nat Commun* 2, no. 224.

[23] D. Bonneau, et al., Quantum Interference and Manipulation of Entanglement in Silicon Wire Waveguide Quantum Circuits. (2012) *New J Phys* 14(4), 005043.

[24] B.J. Metcalf, et al., Multiphoton Quantum Interference in a Multiport Integrated Photonic Device. (2013) *Nat Commun* 4, no. 1356.

[25] D.A.B. Miller. Self-configuring universal linear optical component. (2013) *Photonics Res* 1, 1–15

[26] D. Pérez, et al., Silicon Photonics Rectangular Universal Interferometer. (2017) *Lasers and Photon Rev* 11, 1700219.

6

Practical Implementation of Programmable Photonic Circuits

So far, we have covered the fundamentals of integrated multi-port interferometers and waveguide meshes considering their ideal implementation. We saw that in both cases solutions with larger integration density of TBUs enables the synthesis of more complex circuits. However, photonic integrated circuits (PICs) in general suffer from design and fabrication errors, as well as other non-ideal working conditions. This fact compromises performance and imposes considerable scalability limits. In addition, programmable photonic circuits require the development of two additional tiers (electronic hardware and software) to allow their programmability, optimisation and (re)configuration. This chapter introduces the basic practical considerations of programmable PIC design. It then reviews the experimental demonstrations of both multi-port interferometers and waveguide mesh arrangements. Following a general vision, it analyses the main error sources and their impact circuit performance and tries to identify the most challenging evolution obstacles for very large-scale programmable PICs. This chapter also introduces an analytical method for arbitrary waveguide mesh analysis. Finally, it presents a general architecture for the control subsystem, including an introduction to the software framework and the main algorithms.

6.1 Introduction

The implementation of a multi-port interferometer or a feedforward/feedbackward waveguide mesh circuit requires the integration and independent driving of a great number of primitive processing elements, or tunable basic units (TBUs) [1], [2]. The ideal behaviour of the TBU leads to the perfect performance of the programmable PIC. However, in practice, several factors of degradation must be considered: imperfect splitting ratios, phase control, parasitic back-reflections, loss imbalances, fabrication errors (gradients through the circuit in thickness or temperature) and drifts in time [3]–[5]. These constraints, combined with the high cost and large delays in PIC development, challenge the evolution of programmable photonics. Moreover, to achieve a practical use of programmable PICs, they call for a non-trivial assistance from control

Programmable Integrated Photonics. José Capmany and Daniel Pérez, Oxford University Press (2020). © José Capmany and Daniel Pérez.
DOI: 10.1093/oso/9780198844402.001.0001

electronics, the development of a software layer and a demanding package for the electrical and optical interface [1].

As expected, the first experimental demonstrations of programmable photonic circuits based on the replication of TBUs are to be considered to be proof-of-concept examples. However, they are complex enough to highlight their potentiality and help to envision the main scalability issues experienced when more primitive elements are required.

This chapter considers these effects, particularly in the context of waveguide mesh-based PICs. However, most of the practical considerations here are applicable to other programmable circuit classes, e.g. reconfigurable filter banks, reconfigurable lattice filters and topology-constrained programmable photonics in general.

6.2 Experimental Implementation of Multi-port Integrated Interferometers

Chapter 4 introduced the fundamentals of feedforward integrated waveguide mesh arrangements or multi-port interferometers. The first integration of such circuits was mainly driven by quantum information processing community [6]–[8], and was later considered for other applications like mode-processing [9], [10], hardware acceleration for neural networks [11], [12], routing and general-purpose multi-port linear photonic processing [13], [14].

The circuit application has influenced the choice of the integration platform. Silica was initially employed as a solution to exploit its low fibre-chip coupling loss, which is below 0.5 dB per facet [6]. In germanium-doped glass, Carolan and colleagues [6] demonstrated a 6×6 triangular arrangement including 15 TBUs with a single-drive arm and an external phase shifter, resulting in thirty thermal tuning-based phase shifters. The PIC schematised in Figure 6.1(a) was electrically and optically interfaced to an external control circuitry and two fibre arrays, respectively. In addition, it incorporates active temperature management by a Peltier cooling unit. The PIC was configured to perform linear transformations employed in quantum processing, like heralded quantum logic and entangling gates, boson sampling and six-dimensional Hadamard matrices [6].

However, silica-on-silicon waveguides limit the scalability of the circuit. First, if thermo-optic effect is chosen as the phase tuning mechanism, an approximate electrical power of 0.8 W is required for a 2π phase shift per heater. In addition, the large bend radius constrains the miniaturisation of the TBUs and the integration density.

Conversely, silicon photonics is one of the most attractive integration platforms for programmable waveguide meshes as it allows high-volume fabrication as well as high-integration densities because of the low refractive index contrast and moderate propagation losses of 0.5–2.5 dB/cm.

Silicon feedforward waveguide meshes have been reported for the implementation of linear transformations and mode conversion operations [9], [10]. In 2016, Ribeiro and colleagues reported the implementation and demonstration of a 4×4-port universal linear optical circuit based on a triangular multi-port interferometer arrangement [15], where the silicon photonic circuit, combined with electronic control and software

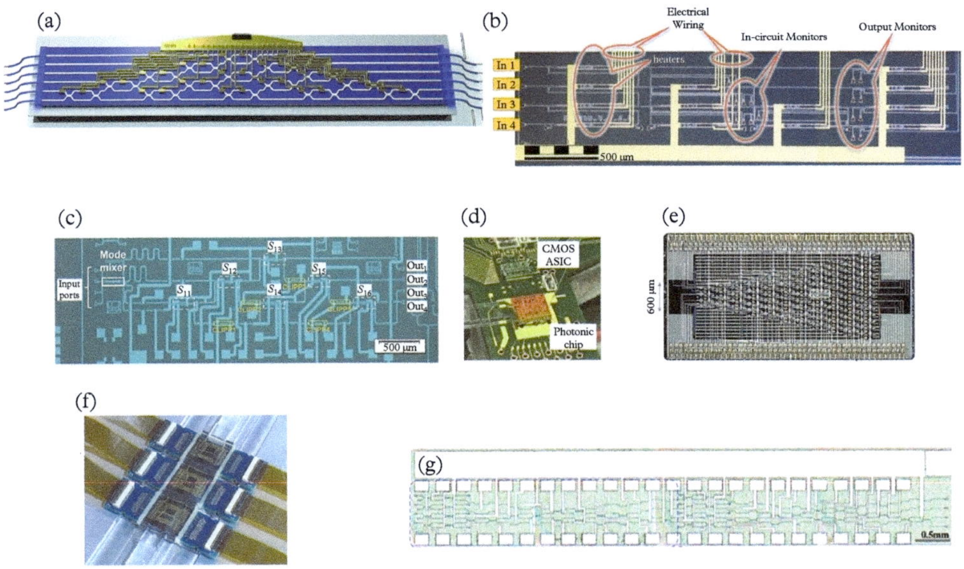

Figure 6.1 *Pictures/schematics of chips with feedforward-only / multiport linear interferometers (a) 6×6 triangular arrangement [6], (b) 4×4 triangular arrangement [9], (c) 4×4 triangular arrangement and (d) its control electronics IC [10], (e) 88-MZIs trapezoid arrangement, (f) modular cascaded of rectangular arrangements [13], (g) 4×4 SVD [12], 5×5 matrix [14].*

feedback, performed any linear operation between its input and output ports. The circuit Figure 6.1(b) consisted of a network of nine thermally tunable symmetric Mach–Zehnder interferometers (MZIs) with phase and amplitude control, in-circuit optical power monitors, and local software-controlled feedback loops. In this case, the optical power monitors were based on the extraction of a small portion of the optical signal by means of a directional coupler and a grating coupler and its monitoring through an infrared vision top-mounted camera.

Also using silicon-on-insulator (SOI) in 2017, Annoni and colleagues [10] reported a 4 × 4 mode unscrambler based also on a triangular Reck multi-port interferometer arrangement (Figure 6.1(c)). The chip was able to automatically unscramble optical beams that had been arbitrarily mixed in a multi-mode waveguide, undoing the scattering and mixing between the spatial modes. The structure incorporated a set of transparent light detectors integrated in a photonic chip and that were used to directly monitor the evolution of each mode along the mesh and allowed sequential tuning and adaptive local feedback control of each beamsplitter. Electronic reconfiguration and set-up were enabled by an assorted complementary metal oxide semiconductor (CMOS) application-specific integrated circuit (ASIC; Figure 6.1(d)).

In 2017, Harris and colleagues [7] reported a reconfigurable chip built upon a waveguide mesh of integrated beamsplitters (Figure 6.1(g)) implemented using integrated MZIs (Figure 6.1(e)). The chip includes eighty-eight MZIs and 176 phase shifters describing a trapezoidal waveguide mesh arrangement and, by suitable programming of the latter, it has been employed, as discussed in Chapter 7, to emulate over 64,000 different quantum particle and phase transport experiments, and as a platform for neural network implementation [11].

In 2018, Mennea and colleagues [13] proposed and demonstrated the implementation of a rectangular interferometer that can be built by cascading modules of TBUs. Each module contains a column of independent phase-controlled MZIs, and several modules can be interfaced to build large reconfigurable interferometers. The team integrated three modules in a silica-on-insulator platform and measured 3 × 3 unitary transformations.

Additionally, in 2018 Taballione reported a triangular arrangement performing an 8 × 8 integrated linear optical network on a silicon nitride platform incorporating 64 TBUs (128 thermally tunable phase actuators) for its application in quantum experiments.

More recently, in 2019 Zhou and colleagues [12] implemented a waveguide mesh arrangement implementing the SVD transformation (Chapter 4) in an SOI platform based on the cascade of two unitary triangular feedforward arrangements, describing 4 × 4 linear transformations. They demonstrated its use for machine-learning applications and basic optical processing. Zheng and colleagues have reported a 5 × 5 rectangular arrangement in silicon nitride. It employs 10 TBUs (eighteen thermo-optic-based phase actuators). However, in this work, only the tunable couplers are driven and no demonstrations beyond cross-bar switching are reported. Table 6.1 summarises the remarkable details from each of the aforementioned contributions.

It is noteworthy that the thermo-optic effect is the preferred tuning option in all these cases. Its rationale relates to its simplicity and robustness in the fabrication process, and the lack of excessive additional loss. Future work will focus on increasing the integration density to obtain more powerful circuits. For this to become a reality, it will require a drop in power consumption per tuning element, a decrement on the tuning-related crosstalk, the minimisation of insertion loss per TBU and a reduction of the TBU footprint.

Finally, for most of them, the electronic and control hardware was based on bulky top-bench equipment. Only Annoni and colleagues [10] integrated the on-chip optical monitors and the control electronics in an ASIC and in the same PCB. The larger the mesh, the more difficult it is to control and manage the whole circuit.

6.3 Experimental Implementation of Integrated Waveguide Meshes

Chapter 5 introduced the fundamentals of feedforward/feedbackward integrated waveguide mesh arrangements, and described them as a PIC based on the replication and interconnection of TBUs. They can be programmed to perform general-purpose photonic processing based on the discretisation of conventional PICs into TBUs. Initially, they were proposed for integrated microwave photonic (MWP) processing [17]–[21], but they rapidly expanded their scope to general purpose photonic processing.

In 2015, Zhuang and colleagues [16] pioneered the field with the experimental demonstration of a square waveguide mesh topology (Figure 6.2(a)). The structure,

Table 6.1 *Experimental demonstrators of feedforward meshes/multi-port interferometers.*

Year	Authors	TBU	PS	OP	dB/ TBU	TE	Foundry	P (mW)/π	RT (us)	Topology	Size (mm $\times$ mm)	PS density (1/mm^2)	CL/facet (dB)	Chip-couplers
2016	Carollan, et al. [6]	15	30	12	TBD	TO	Silica on silicon	400	TBD	Triangular	100×40	0.007	0.5	E
2016	Ribeiro, et al. [9]	9	18	8	1	TO	Silicon	15	250	Triangular	1×3	6.000	TBD	V
2017	Annoni, et al. [10]	6	12	8	1	TO	Silicon	10	10	Triangular	3.7×1.4	2.320	TBD	V
2017	Harris, et al. [7], [11]	88	176	52	TBD	TO	Silicon	10	8	Trapezoidal	2.1×4.3	19.050	3.5	E
2018	Mennea, et al. [13]	30	120	40	4.20	TO	Silica on silicon	10 V (PWM)	TBD	Rectangular	3.05×1.9 ($\times 3$)	20.710	0.8	E
2018	Caterina, et al. [8]	64	128	16	0.27	TO	Silicon nitride	3.35 V	TBD	Triangular (double)	16×16 TBD	0.500	2.9	E
2019	Zhou, et al. [12]	20	48	8	TBD	TO	Silicon	27	TBD	Triangular SVD	1.3×7.5	4.920	3.5	V
2019	Zheng, et al. [14]	10	18	10	0.63	TO	Silicon nitride	330	>1000	Rectangular	11×3	0.550	3.5	E

TBU: Number of TBU, PS: Number of phase shifters, OP: Number of optical ports, P: Power consumption, RT: Response time, CL: Fibre-chip coupling losses, E: Edge-coupling, V: Vertical-coupling

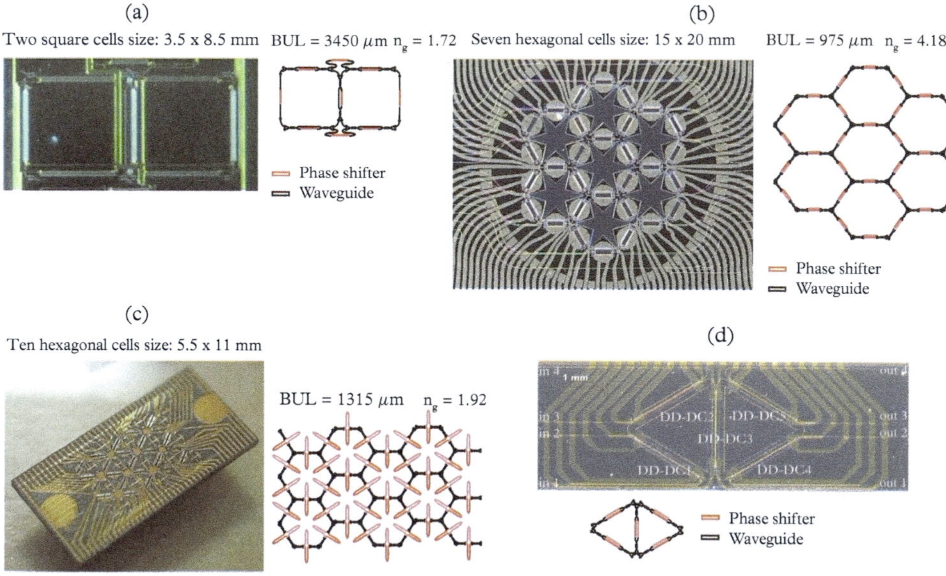

Figure 6.2 *Chip picture and fabricated layout for different feedforward/backward waveguide meshes: (a) Square topology in SiN [16], (b) Hexagonal topology in silicon [17], (c) Hexagonal topology in SiN with modified TBU scheme [20], (d) triangular topology in silicon nitride with directional couplers-based TBU [21].*

fabricated in a silicon nitride platform, comprised two square cells, 5 TBUs and nine phase shifters, was fully programmable and was employed to demonstrate simple FIR and IIR impulse response filters with single and/or double input/output ports of synthetised ORRs.

With a basic unit length (BUL) and basic unit delay (BUD) of 3450 μm and 19.7 ps, respectively, Zhuang and colleagues demonstrated bandpass filters with a tunable centre frequency that spans two octaves (1.6–6 GHz) and a reconfigurable band shape (including flat-top resonance with up to passband–stopband 25 dB extinction). They also demonstrated notch filters with up to 55 dB rejection ratio, Hilbert transformers and tunable delay lines (Figure 6.3). The number of accessible optical ports was four.

In 2017, Pérez and colleagues [17] reported the results of the first hexagonal waveguide mesh composed of seven cells (30 TBUs, sixty thermally tuned phase actuators) fabricated in SOI (Figure 6.2(b)). The device was fabricated at the University of Southampton's Nanofabrication Centre. SOI wafers with a 220-nm thick silicon overlayer and a 3-μm thick buried oxide layer were used. Interested readers are referred to individual studies [1], [17] for more details on fabrication and testing.

Despite the simplicity of the layout depicted in Figure 6.1(b), the seven-cell structure is capable of implementing over 100 different circuits with tunable design parameters for filtering applications (basic MZI, FIR transversal filters, basic tunable ring cavities and IIR filters, as well as compound structures such as CROWs and SCISSORs [1],

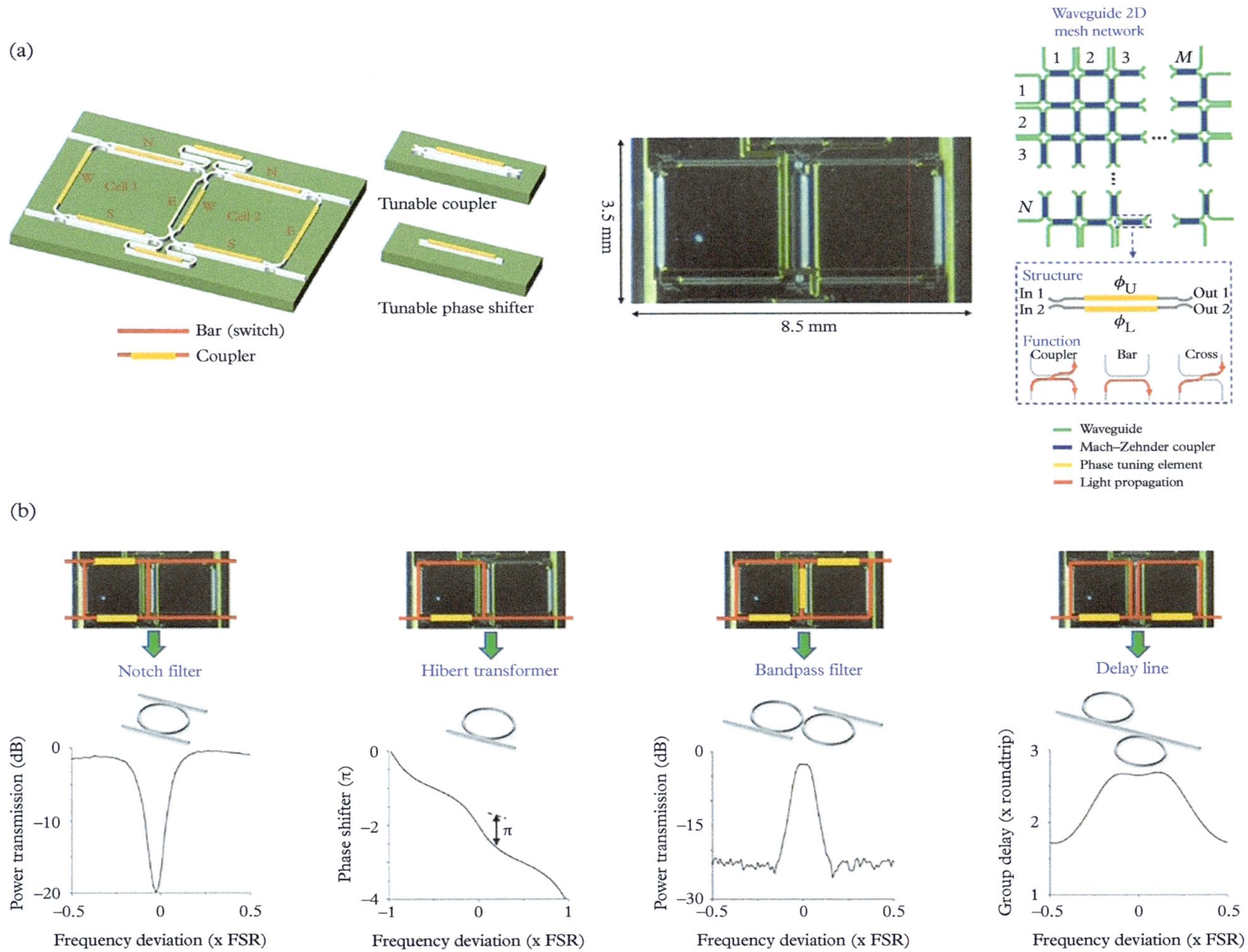

Figure 6.3 *(a) Schematic and a photo of the Si3N4 waveguide technology (TriPleX) chip implementing a 2 square cell waveguide mesh reported in [15]. (b) Experimental results for different programmed circuit configurations obtained by varying phase-tuning elements in the chip and the measurements of their corresponding frequency responses.*

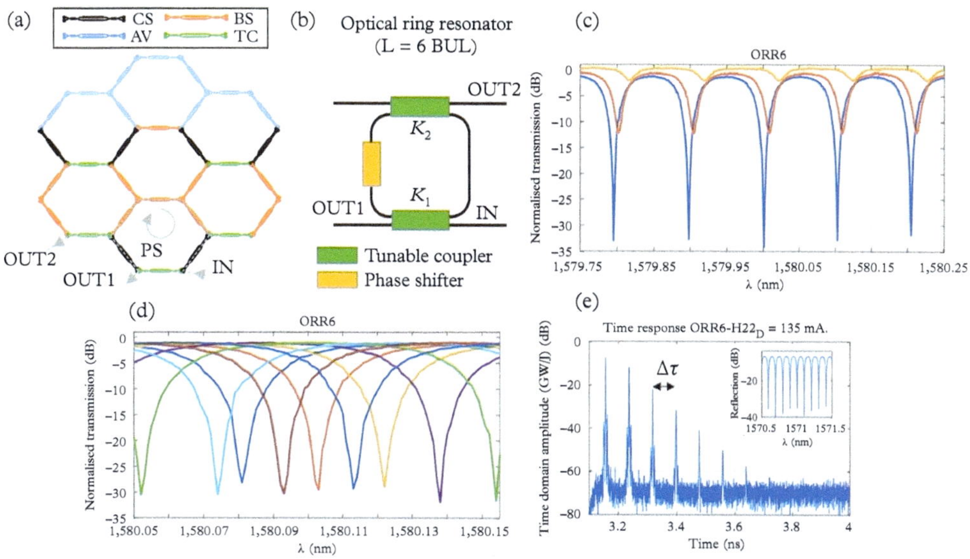

Figure 6.4 *Experimental results for 6-BUL ring resonator IIR and FIR+IIR filters. (a) Waveguide mesh connection diagram, (b) circuit layout and (c) measured modulus transfer function for a IIR filter for different values of the coupling constants K_1 and K_2, (d) IIR filter along a full spectral period for different values of the optical ring resonator round-trip phase shift, (e) time response for critical coupling condition.*

[17]), true-time delay lines, [18] and optical coherent multi-port interferometry [19]. The BUD was 13.5 ps, given by a BUL of 975 μm and a group index of 4.18. The number of accessible optical ports was twenty-four.

For example, Figure 6.4 illustrates a single-cavity optical ring resonator (ORR) with a cavity length given by 6 BULs. The figure shows (a) the waveguide mesh configurations (with the TBU device status according to the colour code previously described), (b) the circuit layout and (c) the modulus response for the OUT_1 port. The measured results correspond to different values of K_1 and K_2, which settle the positions of the zero and the pole. The IIR filter tunability (Figure 6.4(d)) is achieved by exploiting that the coupling constant and the phase shift in any TBU of the mesh can be adjusted independently. Hence, any TBU inside the cavity can be operated as a constant-amplitude phase shifter. Finally, Figure 6.4(e) shows the time response of the ring resonator when the critical coupling is achieved.

More complex filter programming by incorporating additional cavities and delay line paths could also be programmed over the same structure and are reported by Pérez and colleagues [17]. For example, Figure 6.5 shows the results obtained when the mesh is programmed to implement double- and triple-cavity filters. Figure 6.5(a) illustrates a CROW structure where two cascaded ring resonators (input: IN, output: OUTPUT 1) implement a series of reconfigurable filters by arbitrarily moving its zeros and poles.

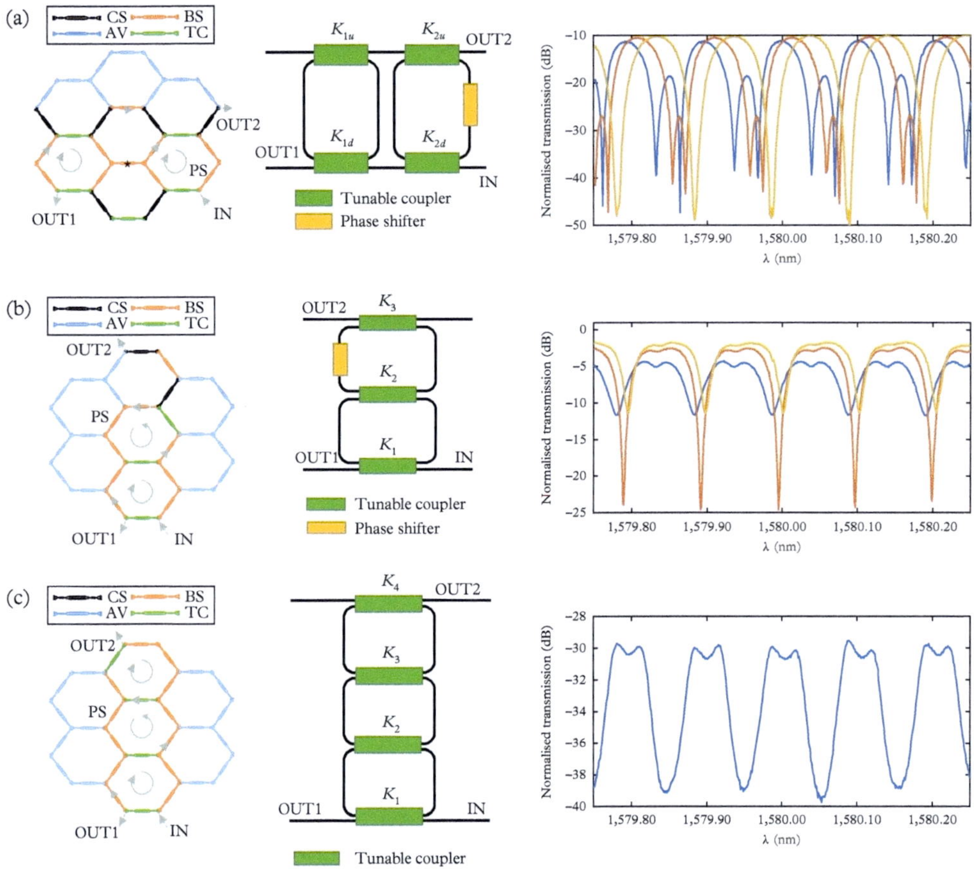

Figure 6.5 *Experimental results for tuneable multicavity ring filters. Waveguide mesh connection diagram (left), circuit layout (middle) and measured modulus (right) for different values of the coupling constants K_1 and K_2 in the case of a), a 6-BUL side-coupled integrated spaced sequences of optical resonator (SCISSOR) filter; b), a 6-BUL Coupled resonator waveguide (CROW) filter; c), a 6-BUL CROW. For each case, the first column shows the 7-cell hexagonal waveguide mesh configuration, where each Mach Zehnder Interferometer (MZI) device is represented by a given colour following the code given in Figure 5. The second column shows the layout of the implemented structure, while the third column shows the measured modulus for the synthesized configuration where the input is in the IN port and the outputs are the OUT1 and OUT2 ports. BUL: basic unit length, PS: phase shifter, CS: cross state, BS: bar state, AV: available, TC: tuneable coupler.*

One of the TBUs identified by an asterisk (*) was kept unbiased. Nevertheless, this is an example of how TBUs can be configured in order to extract non-ideal leaking due to optical crosstalk from the circuit. Figure 6.5(b) shows the reflection response of two coupled ring resonators in a SCISSOR configuration (input: IN, output: OUTPUT 1). This circuit structure is widely used for dispersion compensation and reconfigurable filtering. Figure 6.5(c) shows the transmission response (input: IN, output: OUTPUT 2) of a CROW comprising three ring resonators.

Recently, the same configuration has been employed for the implementation of time-domain true-time delay lines and assorted circuits [18]. The upper part of Figure 6.6 illustrates the programmed configuration of two discrete delay lines enabled by the hexagonal mesh featuring paths of 5 and 11 BULs, respectively. The lower part of the figure displays the measured results obtained by enabling different paths in a recently reported SOI chip where TBUs were implemented using 3-dB MZI-based TBUs.

Note that the amplitudes of the different delayed pulse replicas are different, which is due to the different losses experienced by the signal as it propagates through different paths. In particular, the logarithmic power response vs length (time) was measured to be approximately linear with a decaying rate of 0.59 dB per BUL (per 13.5 ps). Errors in the fitting are mainly related to variations in grating coupler losses (±0.5 dB) and to a lesser extent in TBU losses (< 0.1 dB) [17]. Note that, while different input/output port combinations were employed, these amplitudes nevertheless can be equalised (though

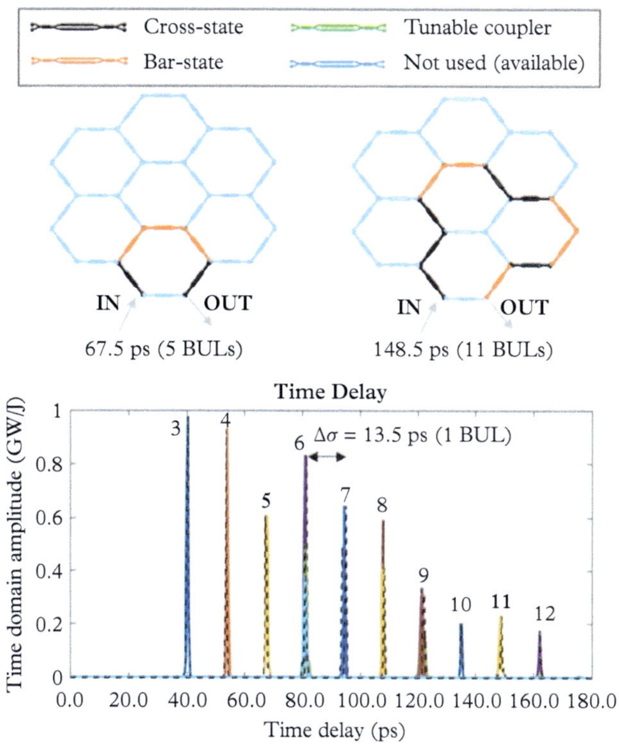

Figure 6.6 *Discrete optical delay lines implemented in a 7-cell layout: (up) settings for two different time delays corresponding to 5 and 11 BULs respectively. (Bottom) Measured delays up to 12 BULs. After [24].*

not shown in the figure) using the ability of TBUs to independently tune their coupling constant values if a splitting tree configuration is required.

Finally, the same hardware structure was employed for the synthesis of several 3 × 3 and 4 × 4 linear unitary transformations as reported by Pérez and colleagues [19]. These are relevant examples of signal processing tasks that are needed in different applications. For example, Figure 6.7(a) illustrates an experimental demonstration of a 3 × 3 linear unitary transformation corresponding to a three-way beamsplitter and a discrete Fourier transform (DFT) configuration. In all cases, the measured results show an excellent agreement with the targeted matrices for the operation wavelength of $\lambda = 1580$ nm with an extinction ratio (ER) > 25 dB between the 1 and 0 coefficients. The required values for the coupling constants and phases of the TBUs used in these implementations were obtained by the synthesis algorithm adaptation explained by Pérez [1] and Pérez and colleagues [19]. The resulting coefficients are translated into the required injected currents to the phase shifters according to the calibration curves obtained for each TBU during the chip characterisation.

More complex 4 × 4 transformations were demonstrated, including that corresponding to the CNOT universal quantum gate (Figure 6.7(b)). Note again that the same waveguide mesh allows for a fast reconfiguration of these N × N transformers to perform different operations according to the synthesised values for their TBU parameters.

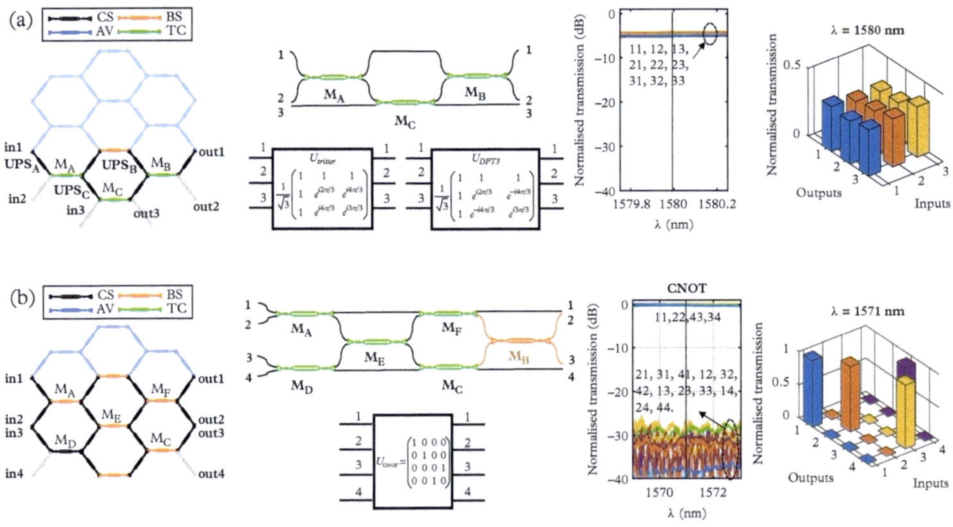

Figure 6.7 *Unitary transformation configuration of (a) 3 x 3 interferometer based on a rectangular arrangement and (b) 4 x 4 interferometer based on a rectangular arrangement to implement a C-NOT transformation. Left Column: 7-cell configuration (CS = TBU in cross state, BS = MZI in bar state, TC = TBU in tunable coupler state, AV = TBU not employed), Central-Left Column: circuit layout of the implemented interferometer, Central-Right Column: spectral measurement of all input/output port connections, Right Column: normalized bar diagram of the resulting unitary matrix (after [15]).*

In 2018, researchers from Spain's Universidad Politècnica de València reported another mesh based on 40 TBUs (eighty thermally tuned phase actuators, see Figure 6.2(c)) [20]. They re-designed the shape of the TBU to achieve a more compact layout and increased the component integration density. This chip was fabricated in a silicon nitride multi-project wafer run at a fabrication platform developed by the spinoff company VLC Photonics and the Centro Nacional de Microelectrónica (CNM). In this design, the BUD is 8.42 ps and the BUL is 1315 μm. The number of accessible optical ports is thirty.

More recently, in 2019, the same group reported the first integration of a triangular topology and the first implementation of the TBU by a dual-drive directional coupler (substituting the balanced MZI) [21]. As illustrated in Figure 6.2(d), the design included 5 TBUs, ten thermal-based actuators and eight optical ports were accessible.

Figure 6.8 shows the single ring resonator implementation. Here, TBU_1 is configured as a tunable coupler and TBU_2 and TBU_3 are configured in their bar state. The reflection response is obtained employing input port 1 and output port 2 (Figure 6.8(b1)). Furthermore, it is possible to exploit the independent phase-shifting capabilities of dual-drive directional couplers (DDDCs) and tune the position of the notch over the full free spectral range (FSR) without seriously compromising the ER.

In a second stage they reprogrammed the mesh to perform as an add–drop filter by modifying the TBU_3 configuration to an arbitrary tunable coupler configuration.

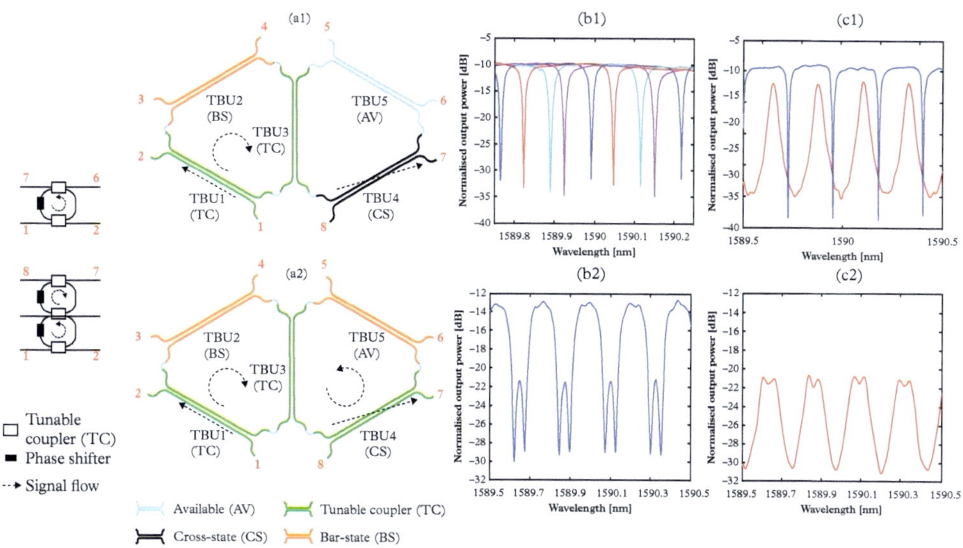

Figure 6.8 *Triangular feedforward/feed-backward waveguide mesh arrangement based on DD-DC units. (a) targeted layout and TBU settings, (b) Reflection response of optical circuits, (c) transmission response of optical circuits, First row: single ring resonator and add-drop, Second row: Coupled Ring Resonator Structure. (After [15]).*

As illustrated in Figure 6.8.(c1), the transmission and reflection responses of the ring exhibit ERs of 24 and 17 dB in the reflection and transmission response, respectively.

A more complex filter formed by two coupled optical cavities was also assembled. Again, the input port is port 1. The reflection response is obtained from output port 2 as seen in Figure 6.8(b2) and the transmission response from output port 7, which is shown in Figure 6.8(c2). In this case, the measured ER was 8 and 9 dB, respectively. The use of thermal-based DDDCs imposes a very large power consumption in the waveguide mesh arrangement.

Table 6.2 summarises the main figures of the aforementioned experimental demonstrations. As in Section 6.2, the dominant tuning effect is again the thermo-optic effect. As explained in Chapter 5, for these applications, the BUD and the interconnection topology set the discrete time resolution of the waveguide mesh. Future work should focus on increasing the integration density to obtain more powerful circuits, reducing the power consumption per tuning element, decreasing the tuning-related crosstalk, minimising the insertion loss per TBU, reducing the TBU footprint and decreasing the BUD. Finally, and similarly to the case of feedforward-only demonstrators, the contributions employed bulky control electronics. The larger the mesh, the greater the challenge to drive and control the whole circuit. The next sections explore the design and analysis of large-scale waveguide meshes, error sources and their impact in performance and the necessary control subsystems and software routines to manage the circuits.

6.4 Design of Large-scale Waveguide Meshes

The study of the whole-mesh performance requires the abstraction of the TBU design. This section performs an explorative exercise with several examples to illustrate the challenges behind large-scale integration. Here, we assume the TBU as a black box defined by its insertion loss, footprint, power consumption, optical crosstalk and bandwidth/uniformity. Under these conditions, we infer the main limits on the mesh scalability for each attribute. Figure 6.9 illustrates few examples of specifications for the mesh and a standalone TBU. Although most of the following considerations apply to both multiport interferometers and general-waveguide feedforward/backward mesh arrangements, some attributes are more applicable to one architecture or the other.

6.4.1 Required Number of TBUs

Ideally, a greater number of TBU results in a more versatile waveguide mesh circuit. In feedforward-only or multi-port interferometers based on waveguide mesh arrangements, the maximum matrix size of the linear operation to be emulated is limited by the number of accessible optical ports and the number of integrated TBUs. Precisely, Table 6.3 includes the required TBUs and phase shifters for a set of given maximum matrix sizes ($N \times N$) for either a Reck [15] or Clements [22] design. We can see that the arrangement scales exponentially with N, resulting in figures that are quite far from the current state of the art.

Table 6.2 *Summary of the main figures of the aforementioned contributions.*

Year	Authors	TBU	PS	OP	dB/TBU	BUL (um)*1	BUD (ps)*2	TE	Foundry	P (mW)/pi	RT (us)	Topology	TBU	Size (mm × mm)	TBU density (1/mm2)	C (CL)
2015	Zhuang, et al. [16]	7	14	4	TBD	3450	19.7	TO	Silicon nitride	300	TBD	Square	MZI	3.5 × 8.5	0.235	E(TBD)
2016	Pérez, et al. [1], [17], [18], [19]	30	60	24	0.59	975	13.5	TO	SOI	110	TBD	Hexagonal	MZI	15 × 15	0.133	V(6.5)
2017	Pérez, et al. [20]	40	80	30	TBD	1315	8.4	TO	Silicon nitride	200	TBD	Hexagonal	MZI	11 × 5.5	0.661	E (3.5)
2018	Pérez, et al. [21]	5	10	8	TBD	1178	TBD	TO	Silicon nitride	300	TBD	Triangular	DD-DC	7 × 2.5	0.285	E (3.5)

TBU: Number of TBU, PS: Number of phase shifters, OP: Number of optical ports, BUL: Basic unit length, BUD: Basic unit delay, TE: Tuning effect, TO: Thermo-optic, P: Power consumption, RT: Response time, C (CL): Fibre-chip coupling method (losses in dB), E: Edge-coupling, V: Vertical-coupling

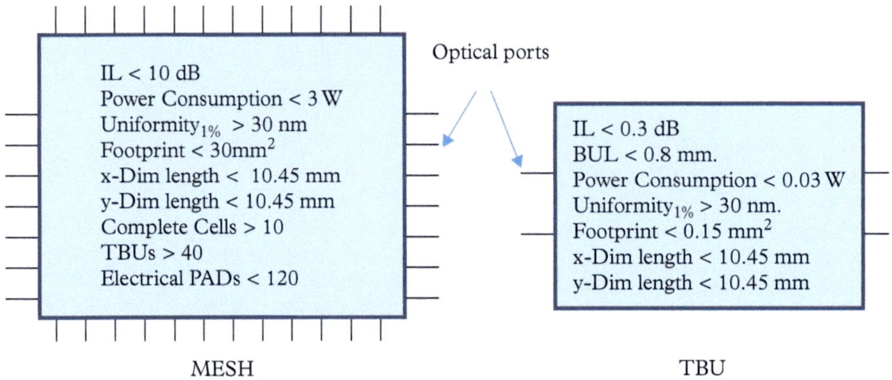

Figure 6.9 *Schematics and example of specifications for a waveguide mesh arrangement and a standalone Tunable Basic Unit (TBU).*

Table 6.3 *Required minimal number of TBU and phase actuators for the emulation of a $N \times N$ interferometer.*

N	**Number of TBUs $N(N-1)/2$**	**Phase shifters $2N(N-1)/2$**
4	6	12
8	28	56
16	120	240
32	496	992
64	2016	4032
128	8128	16256
256	32640	65280

Similarly, ideal feedforward/backward waveguide arrangements offer more versatility and room for the synthesis of more complex and powerful circuits if they integrate a greater number of TBUs. Table 6.4 represents the number of TBUs and phase actuators required for a set of given maximum cell sizes. The current example uses a hexagonal interconnection type, either in the standard or parallel implementation. The number of TBUs per number of cells varies depending on the targeted design and its distribution, although they can be approximated by the trend described by the round number resulting from $3.6\,N_c + 4.54$, where N_c is the number of cells.

Although the numbers in Table 6.3 and Table 6.4 are not comparable and do not target the same applications, the gathered information is useful to establish an order of magnitude or a range of the necessary number of TBUs and phase shifters and their future scalability.

Table 6.4 *Required minimal number of TBU and phase actuators for the integration of N hexagonal or hexagonal-parallel waveguide mesh. (*) represents the approximated number from trend equation.*

N_c	Number of TBUs	Phase shifters
4	19	38
8	34	68
16	62	124
39	140	280
64*	234	468
128*	465	930
256*	926	1852

In practice, both architectures will face scalability constraints that will limit the number of units that can be integrated to get a practical design. The most important are the limited available footprint, the driving and control multi-channel capability and several sources of degradation: accumulated losses, imperfect coupling splitting ratios, optical crosstalk and signal leaking, phase control, parasitic back-reflections, loss imbalances, fabrication errors (gradients through the circuit in thickness or temperature) and drift in time [1], [17].

6.4.2 Maximum Accumulated Loss

The overall mesh performance specifies a maximum number of tolerable accumulated losses depending on the application and required specifications. In standard PICs, accumulated loss results from propagation loss and excess loss of integrated components. In waveguide meshes, the accumulated loss results from the cascaded interconnection scheme between TBUs defining a certain light-path, which sets a limit on the maximum number of TBUs that can define the desired path/circuit. In feedforward-only or multi-port interferometers, this is specified as the depth of the arrangement. For $N \times N$ transformations, it corresponds to N and $2N$-3 for the rectangular and the triangular approaches, respectively [22].

We can compute the maximum number of TBUs that the signal is allowed to go through if we tolerate maximum overall mesh loss of 5, 10 and 15 dB, respectively. From Table 6.5, we can appreciate that for TBU loss of IL_{TBU}0.2 dB, and maximum 10-dB loss for the overall mesh, the maximum light-path is constrained to 50 TBUs. This would limit the maximum size of the matrix to 50 × 50 or 1,225 TBUs in a rectangular waveguide mesh arrangement. In addition, chip-fibre coupling losses will increase the overall loss of the system. The same consideration applies for a general-purpose waveguide mesh like hexagonal, triangular, square and their parallel versions.

Table 6.5 *Maximum number of TBUs defining a synthetised path for a mean insertion loss per TBU and maximum mesh loss.*

	Max. mesh loss = 5 dB	**Max. mesh loss = 10dB**	**Max. mesh loss = 15 dB**
$IL_{TBU} = 0.05$ dB	100	200	300
$IL_{TBU} = 0.10$ dB	50	100	150
$IL_{TBU} = 0.20$ dB	25	50	75
$IL_{TBU} = 0.30$ dB	16	33	50
$IL_{TBU} = 0.40$ dB	12	25	37
$IL_{TBU} = 0.50$ dB	10	20	30

Due to the cascade interconnection, a light increment in the IL_{TBU} dramatically reduces the maximum number of TBUs that can be employed to define a programmed waveguide path. Depending on the TBU design, architecture and material platforms a wide variety of IL and BUL are obtained. Moreover, previous numbers consider perfect dynamic configuration of the TBU. Any drift from the configuration point implies additional excess loss in the light-path. All in all, Table 6.5 reveals that, for a system limited by losses, the maximum number of TBUs in the largest programmed light-path will range between 50 and 25 TBUs for state-of-the-art IL_{TBU}.

6.4.3 Footprint

Space constraints limit the maximum number of TBUs that can be integrated on a chip, and thus, the mesh size. Typical design areas range from 20 to 400 mm^2 in the most commonly employed integration platforms: InP, SOI, silicon nitride and silica on silicon having larger footprints. Within this area, the designer needs to fit the waveguide design, the actuators and metal routing elements to enable the driving of the units as well as to maintain safety gaps for posterior installation constraints, like the inclusion of fibre arrays. For high-density designs it is critical to accommodate more than one metal routing layer (Section 6.4.5). As examples of very large integration, several works reporting the integration of a high-density MZI arrangement for matrix switching operations have succeeded at integrating more than 450 structures similar to a TBU in a single die in an SOI platform, exceeding the Moore law limits [23].

The length and width of the chip are as important as the area because the arrangement topology might exceed from its longitudinal dimensions. The reticle side length available in most silicon photonic foundries is close to 20 mm. Assuming a compact TBU of longitudinal dimension of 300 μm, even the compact version given by the rectangular arrangement would approximately allow a maximum transformation of $N = 64$. Even employing compaction techniques, a value close to N equal to 100 seems hard to achieve, at least with the present state of the art.

For feedforward/backward waveguide meshes, Chapter 5 showed that parallel implementation allows the increment of the integration density. In this case, it is clearer that the main limitation will come from the accumulated loss through the synthetised circuits.

6.4.4 Power Consumption

Equally important is the power consumption of the overall circuit. For a larger number of integrated TBUs, the overall mesh power consumption can be excessive. The electrical power requirements for each TBU are typically referred to the amount of power required to produce a 180° phase shift of the light ($P\pi$). With this power consumption we can tune the switching state of each beamsplitter. The intermediate coupling state requires less power, so the maximum power is not constantly applied for every actuator. Additionally, the setting of an independent phase shift will require a power consumption associated to tuning a phase between 0 and 360°.

The power consumption of the overall mesh is related to the number of TBUs that are active at a certain moment. For example, a multi-port interferometer with an average power consumption of 10 mW per actuator and maximum overall power consumption of 10 W, will be limited to operations of $N = 32$. A decrement up to 1 mW per actuator upgrades this figure into of $N = 100$. The same is experienced by feedforward/feedbackward meshes. However, in this case, a large portion of the TBUs only activates one single phase actuator. As an example, Figure 6.10 considers waveguide meshes of fifty, 100 and 150 active TBUs in single-drive mode. Since the $P\pi$ varies from different tuning effects and different TBU architectures, we list here four typical values. For thermo-optics in SOI platforms, the typical value is around 30 mW [24]. Optimised designs like under-etching allows for values lower than 1 mW [25], [45].

Since the power consumption depends on the configuration of each TBU (not all of them will consume $P\pi$), we have highlighted a range between forty and eighty per cent of the TBU consumption.

Large values of power consumption also impose additional electronic interfacing challenges and call for the availability of high-performance multichannel electronic integrated drivers. Ideally, a close-to-zero power consumption or non-volatile phase shifting mechanism would avoid this scalability limitation. For this purpose, both tunable and field-programmable phase change materials as well as MEMS-based phase shifters are receiving attention from the scientific community [26], [27], [63]. They modify their effective index as a response to an impulse or a combination of impulses by exploiting each effect and maintain the configured phase shift with a static power consumption of five orders of magnitude lower than of traditional thermal phase shifters. However, they must offer a reduced phase actuator length and reduced working voltages to allow very large scale integration.

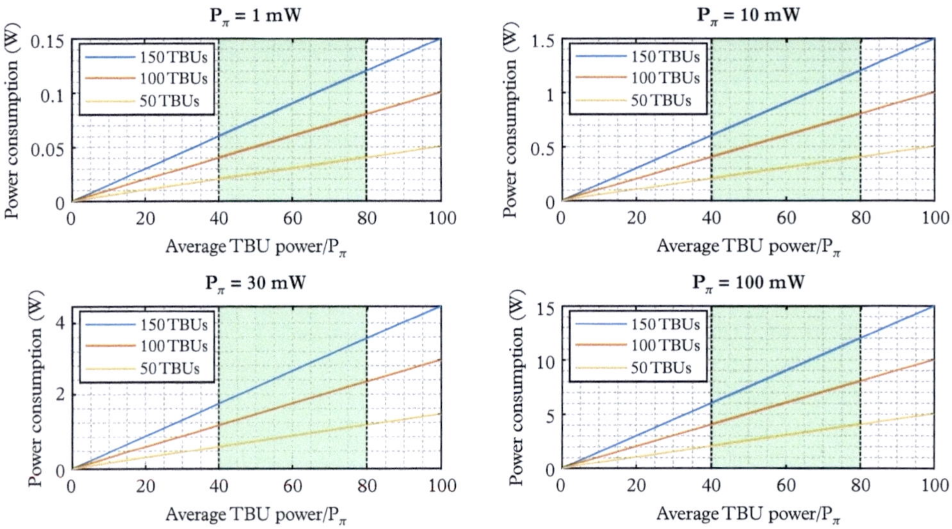

Figure 6.10 *Overall power consumption for waveguide meshes of 50, 100 and 150 activated TBUs for four different TBU power consumption.*

6.4.5 Routing and Metal Layer Design

In Section 6.4.3, we saw that increasing the integration density will be a necessary step in the evolution of programmable PICs. Considering beyond state-of-the-art SOI TBUs with a footprint of 200×100 μm^2, one might think that it is possible to integrate up to 50 TBUs per mm^2, 20,000 TBU (40,000 phase actuators) in a 20×20 mm^2 chip. Nevertheless, the routing of the electrical signals in the chip consumes space and hinders the design process. In order to solve this, it is necessary to employ additional routing layers stacks (isolator, conductive layer, isolator) to allow the routing in different levels and points to join them as desired. This is commonly done in electronic chip integration and two-routing levels plus an actuator level are available in most of the foundries [28].

Interfacing the chip electrically is another major issue of concern and design challenge. If a wire bonding strategy is employed, all the electrical paths are routed to the external perimeter of the die, and to electrical pads at the top layer, to be wire bonded to their corresponding sites at the printed circuit board, [30]. If we consider two electrical pad rows per die-side with 80×80 μm^2, and a pitch of 50 μm, and considering the four sides of the die, this results in an approximate maximum number of pads of 1,230. The number of DC pads is even reduced if we consider that one of the die sides will be employed to make room for the optical interface (a fibre array). On the other hand, if we consider a flip-chip approach with a minimum pitch of 200 μm, we can approximate the number of pads to be around 10,000. Again, we might need to save space for the optical interfacing.

In order to reduce the number of pads required per actuator, we can employ the same solutions as the ones adopted in electronics and in optical switching matrices. These rely on relaxing the number of pads required to drive a large number of actuators. In principle, two pads are required for signal a ground (or larger and lower voltage points). A first solution is sharing the ground pads. This might produce electrical crosstalk due to the possible resistance variation between paths and actuators. In principle, this can be corrected through pre-processing by fully characterising all shared resistances and calculating the voltage differences that compensate the resistance variations, [6]. Other solutions rely on sharing more pads describing a matrix interconnection [32]–[34].

6.4.6 TBU Optimization

During the design process of programmable waveguide meshes, the designer faces important challenges connected to the selection and optimisation of the TBU, while considering the aforementioned limitations.

As covered in Chapter 5, the BUL design is an essential attribute of the TBU. Partially coupled to the footprint concept, it accounts for the tunable coupler length and the access waveguides. It should not be confused with the longitudinal length of the TBU. For feedforward-only arrangements, the TBU miniaturisation helps to increase the integration density. In general-purpose arrangements (feedforward/backward) waveguide meshes, the final BUL and the group index of the waveguide limit the FSR of the interferometric structures and the delay lines to a specific set of discrete values given by the BUD and the mesh architecture [1], [2]. Here, a possible approach is to match one of the frequency grid specifications proposed by the International Telecommunication Union (ITU) to the FSRs of the interferometric structures synthetised in the mesh [1], [64]. This translates to frequency separations of 12.5, 50 and 100 GHz. In this regard, a miniaturisation trade-off appears between the BUD reduction and the accumulated losses per programmed delay. A reduction of the BUL implies that the signal must go through a greater number of TBUs to obtain a desired delay. If the 3-dB couplers limit the overall IL of the TBU, this miniaturisation trade-off must be highly considered. The interested reader is directed to Section 6.4.2 for additional discussion.

For low refractive index difference platforms, the BUL is mainly limited by the tuning mechanism length and the 3-dB coupler lengths. 3-dB couplers in silicon can be reduced to less than 50 μm [35], [36], including the bend sections, while the heaters can be reduced to 62 μm [37]. With the inclusion of bends and straight waveguides sections to increase the distance between both arms of the TBU to decrease tuning-related crosstalk, a total BUL of 240 μm seems potentially achievable. Assuming a typical SOI group index of 4.18, this results in 3.3 ps. With the aforementioned resolution, the maximum FSRs achievable by the hexagonal interconnection topology, or its parallel version, is around 150 and 50 GHz for the synthesis of MZIs and ORRs, respectively.

All in all, the previous features are coupled. The *tuning mechanism* impacts on the final BUL, the tuning crosstalk and the *power consumption*. The use of alternative tuning mechanisms like programmable phase-change materials [26]–[27], MEMS [38], piezoelectrics or electromechanics [39] are promising solutions to reduce the power consumption while enabling a reduction of the distance between the two TBU arms.

In order to further optimise the TBU and the overall mesh, it is essential to know the sources of non-ideal performance.

6.5 Sources of Error and Their Impact over Circuit Performance

At this point it is clear that the use of non-ideal components limits the scalability of the circuit. The circuit propagates and accumulates the performance failures of every component. Many sources of error and parasitic effects must be contemplated in the design of PICs, especially when dealing with large-scale programmable PICs [4], [5].

6.5.1 Waveguide Variations

Loss in waveguides originates from three fundamental types of mechanisms: material absorption loss, scattering and radiation loss. Absorption loss is due to the molecular absorption and band-to-band electronic absorption of electromagnetic energy. It is linked to the imaginary part of the waveguide effective index (material and geometrical properties). Next, scattering loss in integrated planar waveguides occurs from the interaction of the electromagnetic waves with the scattering centres (fluctuation of refractive index and/or surface roughness). Finally, radiation loss due to waveguide bending arises when the electromagnetic signal faces a discontinuity in the bending curvature or when it has a constant bend radius. The first two mechanisms can be reduced by proper material and cross-section design and by improving the fabrication processes. The third can be mitigated if the bend radius is large enough to avoid the region below the critical radius.

Beyond conventional waveguide losses, geometry variations of the waveguides arise due to the non-uniformity of wafers in thickness, and due to small width variations during the fabrication process [39]–[41]. These changes in the nanometre range modify the effective index considered during the design stage. This effect is particularly detrimental in high-index contrast waveguides. Structures like directional couplers are seriously affected and, once fabricated, offer splitting ratio drifts from the one designed. A similar effect is found in PICs including multi-mode interferometers, array waveguide gratings, delay lines and, by extension, in MZIs and ORRs. Considerable efforts are done to achieve designs robust to these variations. In programmable PICs, they mainly affect to the operation of 3-dB couplers and by extinction to the operations carried out by interferometric structures.

6.5.2 Temperature Drifts

Many properties of the materials employed in PIC design are temperature dependent. Similar to the previous point, during the design stage, the designer considers a PIC temperature operation range and employs this dependence to compute the refractive indexes of the core and cladding materials and, subsequently, the effective index of the waveguide modes. The designer must take into account that a later change in the PIC temperature conditions might modify the effective index of guided modes, which could lead to a non-functional behaviour of the circuit. Temperature changes might be due to environmental changes or due to the presence of neighbouring electronic integrated circuits dissipating heat.

In addition, as reviewed in Section 6.2, most of the programmable PICs exploit the localised thermo-optic effect as a tuning mechanism of waveguide properties by dissipating heat through a micro-heater. In this case, a careful design should be implemented that considers the undesired partial heating of waveguides closed to the one thermally actuated. This effect is known as *thermal crosstalk* and imposes a safety distance between actuated waveguides in high-density programmable PICs. Precisely, we can approximate the phase change of a standalone waveguide with a phase actuator of effective length L as

$$\Delta\phi = \Delta\beta \cdot L = \frac{2\pi}{\lambda}\Delta n_{eff} L = \frac{2\pi}{\lambda}\frac{dn_{eff}}{dT}\Delta T_{eff} \cdot L, \tag{6.1}$$

where $\Delta\beta$ is the change in the propagation coefficient, Δn_{eff} is the change in the effective index, given by the multiplication of the thermo-optic coefficient (dn_{eff}/dT) and the effective temperature gradient (ΔT) at the waveguide layer. Since the thermo-optic coefficient is one order of magnitude lower in silicon nitride waveguide than in SOI waveguide, a thermal actuator in silicon nitride requires a $\Delta T \cdot L$ product one order of magnitude greater than that in SOI. This effect translates to the need for considerably longer phase shifters.

Even considering that only one phase shifter is actuated, some tuning effects can produce the spread of a portion of the tuning effect. This leads to an induced and unwanted effective refractive index change in the neighbouring waveguides. In the case of thermal actuators, the heat can flow to the adjacent waveguide and modify its propagation conditions. This effect can be modelled by a thermal crosstalk coefficient (CT). Then, the difference between the phase alteration between an actuated waveguide $\Delta\phi_1$ and a passive waveguide $\Delta\phi_2$ can be modelled as [42]

$$\begin{aligned}\Delta\phi_1 - \Delta\phi_2 &= (\Delta\beta_1 - \Delta\beta_2) \cdot L = \tfrac{2\pi}{\lambda} L \left(\Delta n_{eff,1} - \Delta n_{eff,2}\right) = \tfrac{2\pi}{\lambda}\tfrac{dn_{eff}}{dT} L \left(\Delta T_{eff,1} - \Delta T_{eff,2}\right) \\ &= \tfrac{2\pi}{\lambda}\tfrac{dn_{eff}}{dT} L \cdot \Delta T_{eff,1}\,(1 - CT).\end{aligned} \tag{6.2}$$

where CT is in the range of 0 and 1.

Managing heat flow in a PIC is essential for most of the designs. Solutions can be active, based on the inclusion of a Peltier cell and integrated thermally sensitive resistors in the package, or passive, if during the fabrication process we employ localised temperature-insensitive passive waveguides [43] and/or isolation trenches [44].

6.5.3 Driving Signal Drifts in Programmable PIC Actuators

During the tuning stage of programmable PICs, electrical signals are employed to drive the actuators and configure the PIC as desired. Either from voltage or current source supplies, these driving signals suffer from electrical noise and instability. Moreover, the precision range is limited to their discrete digital resolution. For example, an 8-bit voltage source ranging from 0 to 10 volts, will be limited to 2^8 positions, resulting in steps of 39.1 mV. As another example, we can consider a 12-bit voltage source in the same range but with voltage noise equivalent to 4 mV. In this case, the source offers 212 positions and thus a resolution of 2.4 mV. However, in this case the voltage noise will be dominant over the resolution limited by the digital discretisation.

6.5.4 Impact of Non-ideal Components on the TBU Performance

To fully understand these limits, we now focus our analysis in a standalone vision of the TBU and the implication of these non-ideal effects at a circuit level. This section considers an MZI-based TBU architecture.

The ideal TBU is composed of two 3-dB couplers interconnecting two arms loaded with one phase shifter providing ϕ_1 and ϕ_1 phase shift, respectively (Figure 6.10 (a)). Ideally, by suitable programming each phase actuator, it is possible to get the desired optical power splitting ratio K and impose an additional common phase at the output. However, setting a desired TBU state suffers from several issues that introduce a drift to the targeted working point, both in amplitude and phase. These mainly arise from the imperfect design or fabrication of the waveguides, the 3-dB couplers and the presence of the phase shifters. In addition, the tuning-related crosstalk from neighbouring TBUs and the drift in the phase shifter configuration resulting from the electrical driver resolutions or stability will introduce a dynamic source of configuration errors.

To evaluate the performance of the TBU, we can derive its analytical transfer matrix and simulate the submatrix corresponding to the input/output computational modes given by Eqn. (6.3) [4], [5], [46].

$$
\begin{aligned}
H_{TBU} &= H_{dc\_Coupler,b} H_{2WG} H_{dc\_Coupler,a} = \\
&= \sqrt{1-\gamma_a}\sqrt{1-\gamma_b}e^{-j\omega\tau}. \\
&\begin{bmatrix} c_b & -js_b \\ -js_b & c_b \end{bmatrix}\begin{bmatrix} \sqrt{1-\gamma_c}e^{-j\phi_1} & 0 \\ 0 & \sqrt{1-\gamma_d}e^{-j\phi_2} \end{bmatrix}\begin{bmatrix} c_a & -js_a \\ -js_a & c_a \end{bmatrix} \\
&= \sqrt{1-\gamma_a}\sqrt{1-\gamma_b}e^{-j\omega\tau}. \\
&\begin{bmatrix} c_a c_b\sqrt{1-\gamma_c}e^{-j\phi_1} - s_a s_b\sqrt{1-\gamma_d}e^{-j\phi_2} & -js_a c_b\sqrt{1-\gamma_c}e^{-j\phi_1} - js_b c_a\sqrt{1-\gamma_d}e^{-j\phi_2} \\ -js_b c_a\sqrt{1-\gamma_c}e^{-j\phi_1} - js_a c_b\sqrt{1-\gamma_d}e^{-j\phi_2} & c_a c_b\sqrt{1-\gamma_d}e^{-j\phi_2} - s_a s_b\sqrt{1-\gamma_c}e^{-j\phi_1} \end{bmatrix},
\end{aligned}
\tag{6.3}
$$

where γ_{a-d} describes the loss coefficients of the 3-dB couplers and the interferometric arms, $c_{a,b} = (1-K)^{1/2}$ and $s_{a,b} = -j(K)^{1/2}$ the coupler coefficients and where K is the 3-dB optical power splitting coefficient. We define the input/output splitting ratios K by a Gaussian distribution with mean fifty per cent and a parametrised standard deviation. The excess loss of the couplers is modelled by a Gaussian distribution with a mean corresponding to 0.05 dB /coupler and standard deviation of one per cent. The loss in the arms comes from the propagation losses in the materials and the presence of the phase shifter. Here, we assume values defined by a non-negative Gaussian distribution with mean 5.16 per cent and standard deviation of 2.84 per cent, as discussed by Mower and colleagues [4].

6.5.4.1 *Impact of non-ideal components on the TBU performance. Passive behaviour.*

To evaluate the impact of the static sources of errors (Case A), we perform a Monte Carlo simulation with a number of samples equal to 10^6. Each sample selects the loss and coupling coefficients from the aforementioned distributions and evaluates the TBU model of Eqn. (6.3). Figure 6.11(a) illustrates the parameters. The standard deviation of the couplers is varied from zero to five per cent. The phase shifter values are kept constant at 0 and π, respectively, to set an ideal bar state. As a result, for each sample we obtain the optical power coupling ratio (K) of the TBU, the phase difference of the outputs ($\Delta\phi_{\text{outputs}}$), the cross-port ratio (CPR), through-port ratio (TPR) and the total loss (L), given by

$$K = |H_{TBU,2,1}|^2 = |H_{TBU,1,2}|^2 \tag{6.4}$$

$$\Delta\phi_{outputs} = \sphericalangle H_{TBU,2,1} - \sphericalangle H_{TBU,1,1}, \tag{6.5}$$

$$CPR = 10\log_{10}\left(\frac{|H_{TBU,2,1}|^2}{|H_{TBU,2,1}|^2 + |H_{TBU,1,1}|^2}\right), \tag{6.6}$$

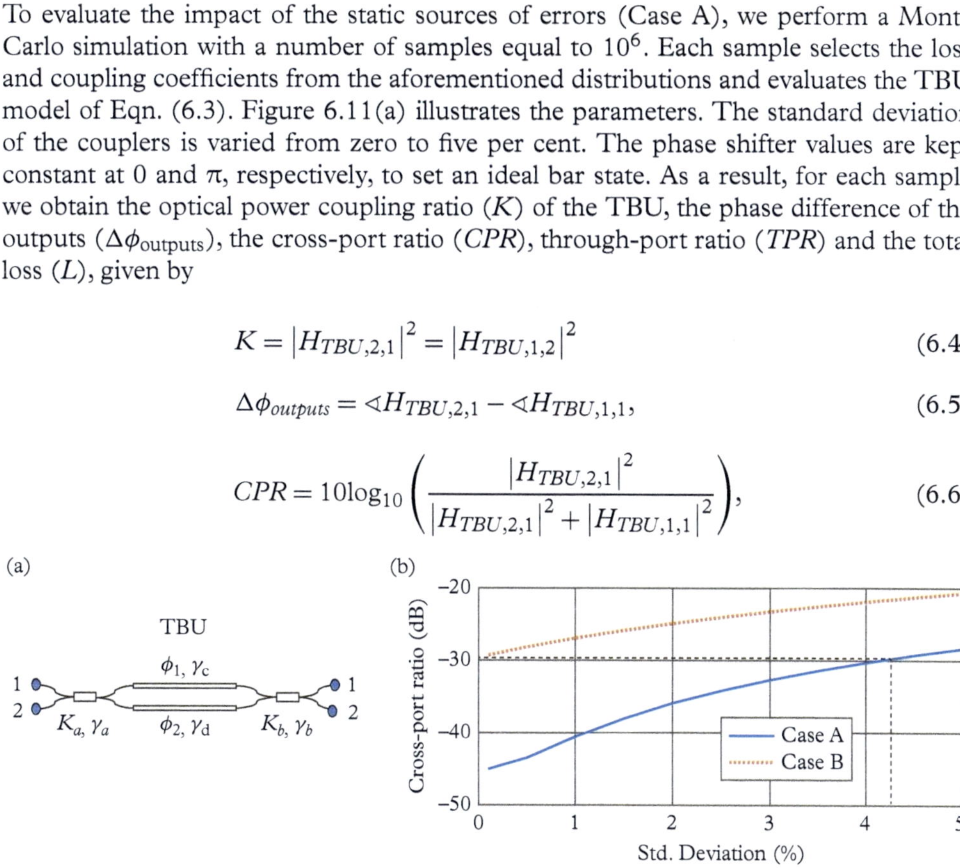

Figure 6.11 *(a) Labeled TBU with coupling, loss and phase parameters following Eq. (6.1), (b) Statistical modeling results for Cross-port Ratio[3]. Case A: No phase-errors, couplers mean $K = 0.5$, σ_K ranging from 0 to 5%. Case B: Fixed $\sigma_K = 4.33\%$. σ_ϕ ranges from 0 to 5%. (Std = standard).*

$$BPR = 10\log_{10}\left(\frac{\left|H_{TBU,1,1}\right|^2}{\left|H_{TBU,2,1}\right|^2 + \left|H_{TBU,1,1}\right|^2}\right), \tag{6.7}$$

$$L = 10\log_{10}\left(\left|H_{TBU,2,1}\right|^2 + \left|H_{TBU,1,1}\right|^2\right). \tag{6.8}$$

The optical coupling ratio deviation from the targeted value will introduce critical performance deviation through the circuit. A good measure for the optical crosstalk is the *CPR* and *TPR*. Ideally, if a bar-state is configured, the *CPR* should be 0 (in linear units) or minus infinite in logarithmic scale, meaning that no-power is measured where no-power was expected. From Case A, we obtain the following results. First, the coupling ratio is shifted from the targeted zero per cent coupling state up to a mean of 0.51 per cent when the standard deviation of the couplers $(\gamma_{a,b})$ is set to five per cent (Figure 6.11 (a)). This fact translates to the addition of additional excess loss to the TBU. According to the overall phase difference between outputs, $\Delta\phi_{\text{outputs}}$ remains equally concentrated at both $-\pi/2$ and $\pi/2$, but with a non-negligible deviation from the targeted value. Following on from Case A, the optical crosstalk can be extracted from the *CPR*, as a measure of the optical leaked power in the cross-port. Since the optical crosstalk is affected by the deviation of the 3-dB couplers and phase errors during the fabrication stage, Figure 6.11 (b) illustrates the resulting *CPR* versus the standard deviation of the 50:50 couplers employed in the TBU $(\gamma_{a,b})$. From the blue trace Case A, we can see that as we vary $\gamma_{a,b}$, the optical crosstalk of the TBU is worsened from -45 dB to -28.41 dB.

6.5.4.2 *Impact of non-ideal components on the TBU performance: dynamic behaviour*

For the evaluation of the dynamic sources of error, we fix the standard deviation value of the 3-dB couplers based on reported wafer-scale analysis to 4.33 per cent [47], [48]. Then, for Case B, we model the phase shifters as Gaussian distributions of mean equal to the set value (as in Case A) but varying their standard deviation from zero to five per cent. As mentioned, this variability will appear due to, for example, waveguide fabrication defects, resolution in the subsystem driving the actuators, tuning-related crosstalk, etc. The remaining parameters continue in the same conditions as in Case A. In this occasion the resulting splitting ratio is shifted up to 1.59 per cent. This is translated to a mean cross-port ratio decrement of 8.61 dB (Figure 6.11). Note that the reported values are relative to the mean values. Figure 6.12 provides an example of the variable distributions at the input and the output of the simulations for Case B where the standard deviation of the phase shifters is five per cent. Output phase difference becomes randomly distributed with more concentration at 0 and $\pm\pi$.

Finally, a third simulation considering Case B distributions with phase shifters means equal to 0 and $\pi/2$, respectively, and phase standard variation of five per cent result in a standard variation of one per cent for the total K variance.

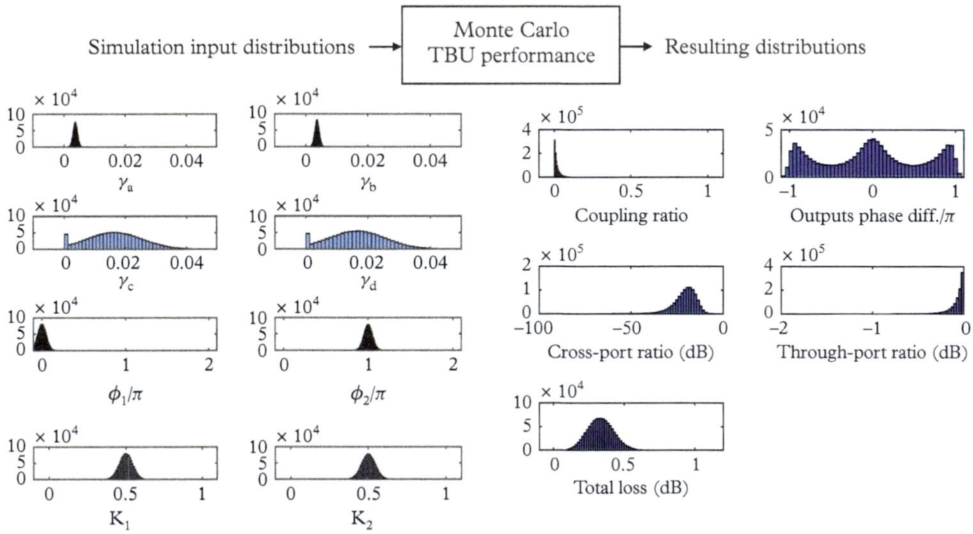

Figure 6.12 *Statistical modeling Case B. Example of input distributions for the TBU variables of Eq. (6.3) and resulting distributions for the overall coupling ratio, phase differences, cross-port ratio, through-port ratio and total loss of a TBU.*

In summary, the design and fabrication defects considered result in state-of-the-art MZI-based TBU with losses ranging from 0.2 to 0.5 dB, optical crosstalk in the range of -35 to -28 dB that with additional dynamic source of errors becomes inside the -30 to -18 dB range. The behaviour of the TBU as a standalone non-ideal component is essential for understanding the implications on a larger-scale circuit. Section 6.5.5 shows that current deviations are translated into accumulated losses and signal leaking.

First, we show a simpler initial example where only one TBU is considered. Here, we employ a TBU as the tunable coupler of a ring resonator filter that requires a driving circuitry to tune the phase shifters of the TBU and the phase inside the ring to tune the resonance to the desired wavelength and the ER [49]. The precision of the control subsystem will affect the performance and the drifts from the ideal point of operation. In this example, we have considered an ORR integrated on an SOI wafer. The group index is 4.18 and the propagation losses are 2.5 dB/cm.

Next, we configure the phase shifters to maximise the ER while keeping the central wavelength fixed. If we assume a drift in the phase shifter driving the MZI-like tunable coupler due to the noise originated at the electrical power supply source or tuning-related crosstalk with a standard deviation of $\sigma_\phi = 0.1$ per cent rads, we can compute through a 1000-instance Monte-Carlo simulation that fifty per cent of the samples obtain a reduction of the ER greater than 15 dB.

This fact is exacerbated in circuits with a large number of dynamically tunable components like waveguide mesh arrangements [5].

6.5.5 Impact of Non-ideal Components on Feedforward Waveguide Meshes

By using the TBU model of Eqn. (6.3) for each TBU, and the multi-port interferometer equations in Chapter 4, it is possible to compare and analyse the performance of the transformation, considering state-of-the-art circuit deviations. The procedure is illustrated in Figure 6.13. First, we start with the desired linear matrix U_0 of size $N \times N$. Then, we employ the equations in Chapter 4 to obtain the phase shifting values of every TBU (coupling and phase shifting). Next, we apply the desired TBU deviations (3-dB coupling ratios, loss and dynamic phase errors) to our model. The whole arrangement of non-ideal TBUs will result in the operation described by matrix U. Now, we can compare both matrices and measure the performance error. Typically, it uses the fidelity figure that employs a normalised Hilbert–Schmidt inner product [4], [22], [50]:

$$F(U, U_0) = \left| \frac{tr\left(U_0^{\dagger} U\right)}{\sqrt{N \cdot tr\left(U^{\dagger} U\right)}} \right|^2, \tag{6.9}$$

The normalised fidelity makes sure that $F(U, U_0) \leq 1$. For this reason, it does not distinguish between matrices that differ by only a constant multiplicative factor and concentrates instead on unbalanced loss.

Clements and colleagues [22] applied the procedure to compare Zheng and colleagues' triangular [14] arrangement to their own arrangement. In the analysis, a set of fixed TBU insertion loss from 0 to 1 dB is employed in simulations with matrices of different sizes (from N equal to 0 to 50). It is concluded that the method reported by Clements and colleagues [22] for decomposing unitary matrices leads to a rectangular and compact configuration that outperforms the design proposed by Reck [14] and is more loss-tolerant.

Compared to fixed implementations, programmable networks can compensate at a certain degree the passive deviations, like the splitting ratios and some phase errors [4]. For example, whereas a fixed circuit find an insuperable splitting deviation, the TBU will be affected by a reduction of the achievable transmission states. However, it will allow its tuning over a remarkable range. This can be overcome if additional tuning structures are

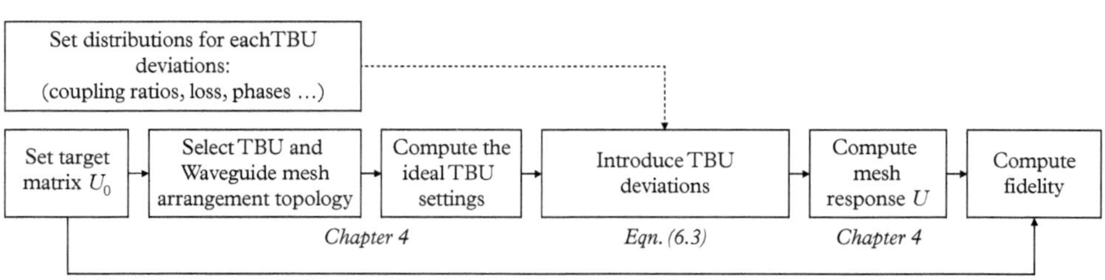

Figure 6.13 *Procedure for the performance evaluation of a feedforward waveguide mesh arrangement.*

employed in the construction of the TBU, as proposed by Thomas-Peter and colleagues [51]. However, this imposes additional overhead per TBU, which affects the scalability of the circuit.

6.5.6 Impact of Non-ideal Components on Feedforward/Feedbackward Waveguide Meshes

In a similar way, feedforward/feedbackward meshes can be analysed from a non-ideal TBU operation point of view. The analysis on accumulated losses can be approximated to the one performed in Section 6.4.2. However, dealing with optical crosstalk is a problem that, in principle, is more difficult to analyse. The most important sources of impairment in the operation of waveguide mesh circuits derive from the fact that either imperfect components are obtained as a result of the fabrication process, or that the setting values for the structure phase shifters depart from the ideal values required by design. The result is close to that experienced in multi-port interferometers (Section 6.5.5) with the difference being that the signal leakage travels forward and backward inside the mesh [1], [5].

Figure 6.14 shows that, according to the PIC point of view, this effect is similar to the addition of interferometric structures through different points in the programmed circuit with reduced coupling coefficients and lossy leaking paths. Figure 6.14(a) shows the synthesis of a finite impulse response filter representing the light-path in black and the signal back-reflections and leakages in red and grey. The signal leakage guided through the non-used TBUs is susceptible to add parasitic signals compromising the signal performance. Undesired paths containing reflected/leaked signals will suffer from additional leaking and propagation losses. We can define first-grade leakage (FGL) as the grouping of reflected and non-ideally coupled signals that go through a number of TBUs lower than six before going into the ideal light-path again. They are represented in red.

Similarly, we can define the second-grade leakage (SGL) as the grouping of reflected and parasitic signals that go through a number of TBUs greater than five before going into the ideal light-path again. They are represented in grey. By definition, SGL signals are less likely to occur, and will probably show a more attenuated contribution. The rationale here is that the larger and the further the parasitic path travels, the less probable that it will be compared to one that travels through fewer TBUs. Note that the FGL effects will add a ripple with a FSR larger than the one produced by the SGL. Both effects will typically manifest inside the FSR of interest. Note that SGL signals travel through paths shared with FGL signals.

Zero-grade leakage (ZGL) is one produced inside the TBU. Reflections inside this structure will generate a perturbation with an FSR that is typically greater than the frequencies achieved by the interferometric structures that can be programmed in the mesh. However, their contribution should be minimised as well. Figure 6.14(c)–(d) represents the case for a programmed delay line.

In practice, we can tune the non-used TBUs to guide the reflected and leakage signals as far as possible from the defined circuit(s) or to a defined set of drain optical ports, as

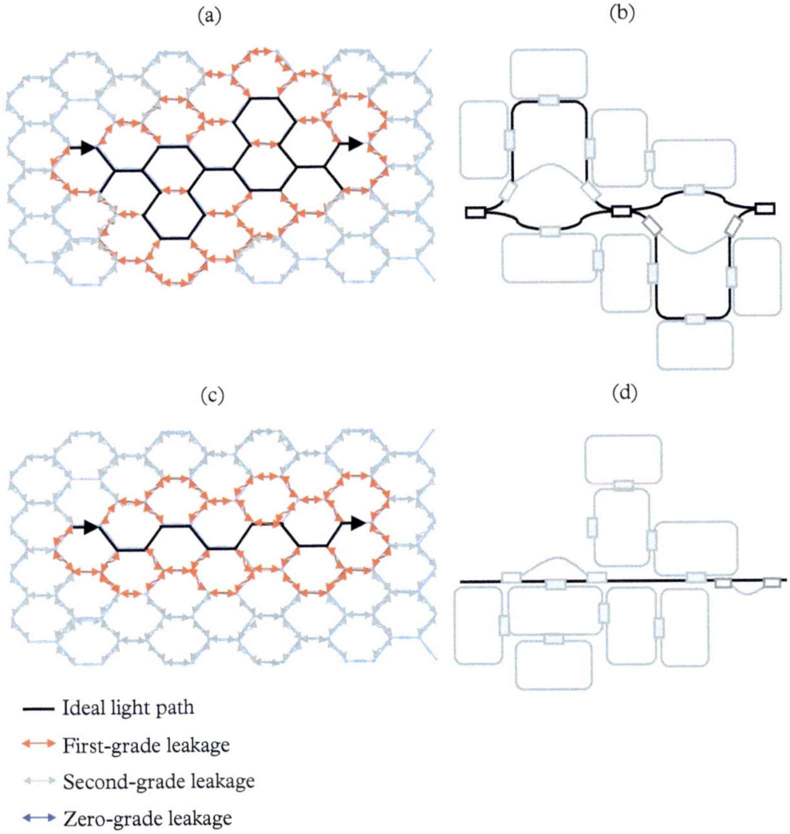

Figure 6.14 *(a) Layout and mesh-configuration for two cascaded imbalanced MZI, (b) Schematic illustrating the associated undesired interferometric structure, (c) Layout and mesh configuration for a programmed delay line, (d) Schematic illustrating the associated undesired interferometric structure.*

shown by Pérez and Capmany [5]. In order to extend the study and be able to apply the same methodology as in Figure 6.13, we must analytically compute an arbitrary waveguide mesh arrangement given the status and performance of each individual TBU. The analytical model developed by Pérez and Capmany [5] performs that computation. It is based on a mathematical induction method and gets the spectral response of the full scattering matrix: amplitude and phase for each port, as well as for the desired wavelength. As an example, we illustrate the process on two different waveguide mesh circuit configurations: a hybrid multi-port interferometer and an optical filter. For both cases, we follow the procedure in Figure 6.13, although for the sake of simplicity, we change the TBU model to a simplified one

$$TBU = \alpha \begin{pmatrix} ie^{i\phi}\sqrt{1-K}\cdot e^{iw_n} & ie^{i\phi}\sqrt{K}\cdot e^{iw_n} \\ ie^{i\phi}\sqrt{K}\cdot e^{iw_n} & -ie^{i\phi}\sqrt{1-K}\cdot e^{iw_n} \end{pmatrix}, \tag{6.10}$$

where K is the coupling factor in power, ϕ is the overall phase term, ω_n is the normalised frequency (to the BUL) and α is the loss term. However, any model could be straightforwardly employed, including experimental data, or the one in Eqn. (6.3).

First, we start with the programming of two 4×4 linear transformations on the same waveguide mesh, working in parallel. Figure 6.15(a) illustrates the implementation example of a linear transformation of a controlled-NOT (CNOT) gate and simultaneously, a discrete Fourier transform (DFT) operator, both described by unitary 4×4 matrices (Eqn. (6.11)).

The CNOT gate is a 2-bit gate where one bit acts as a control bit while the other is the information bit. If the control bit is '0' then the logical value of the information bit at the input port is left unchanged at the output port. If the control bit is '1', the logical value of the information bit at the input port is swapped (from 0 to 1, or from 1 to 0) at the output port. The CNOT gate is the fundamental building block of quantum information circuits. In photonics the CNOT gate is implemented using a dual rail approach. This means that the logical bit is implemented using two waveguides and the information bit is implemented using another two waveguides. The presence of light in one waveguide of the pair indicates a logical zero, while its presence in the other indicates a logical 1. In this way, a photonic implementation of the CNOT gate requires four input and four output ports and its operation is a 4×4 unitary matrix.

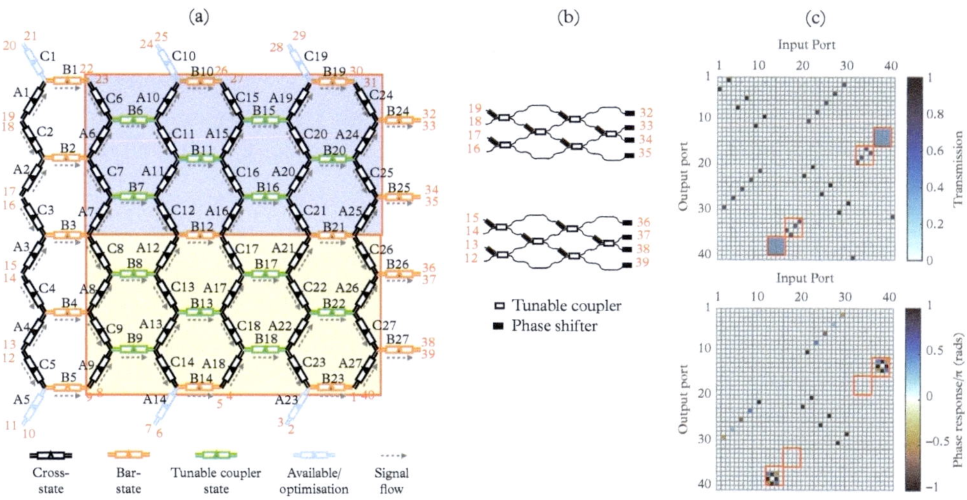

Figure 6.15 *Scalable Analysis Method application to single wavelength operation of waveguide mesh configuration for universal linear interferometers. (a) Mesh architecture and configuration for simultaneously implementing two 4x4 linear transformations. (b) Equivalent circuit layouts with indication of the input and output ports in red ink. (c) Moduli and phases of all the 40x40 matrix coefficients when the 2x2 transformations are programmed to implement a CNOT (Blue background) and Discrete Fourier Transform operator (Ochre background), after [5].*

The coupling and phase coefficients of each TBU are computed following the algorithm routine from Pérez [19] and specified in Pérez and Capmany [5]. We can program the mesh to implement these gates in a very compact layout. The results illustrated in Figure 6.15 match with the targeted values of amplitude and phase and the 40 × 40 waveguide mesh scattering matrix provides information related to the additional undesired input/output port relationships that are enabled by the waveguide mesh programming:

$$H_{DFT4} = \frac{1}{2}\begin{pmatrix} 1 & 1 & 1 & 1 \\ 1 & -i & -1 & i \\ 1 & -1 & 1 & -1 \\ 1 & i & -1 & -i \end{pmatrix}, \quad H_{CNOT} = \begin{pmatrix} 1 & 0 & 0 & 0 \\ 0 & 1 & 0 & 0 \\ 0 & 0 & 0 & 1 \\ 0 & 0 & 1 & 0 \end{pmatrix}. \tag{6.11}$$

As in feedforward-only meshes, the non-ideal behaviour of the TBUs leads to additional losses, scattering and errors when setting the values for the phases and coupling factors. In the particular case of the implementation and analysis of linear transformations, we use a simple model that assumes equal insertion loss for every TBU and normal coupling factor and phase drift distribution with standard deviations corresponding to σ_K andσ_ϕ, respectively for Eqn. (6.10). We can quantify the fidelity of the transformation $NxNU$ compared to the ideal/targeted transformation U_0 by using the same figure of merit in Eqn. (6.9).

If σ_K and σ_ϕ are equal to 0, the fidelity of the hexagonal waveguide mesh remains equal to 1 (100 per cent) independently of the circuit size and the loss per beamsplitter. This is not the case for the triangular interferometer design [14] and the rectangular design [22]. The reason is that, for the synthesis of linear transformations in the hexagonal waveguide mesh configuration, the number of TBUs where the signal goes through is always equal for any possible light-path. However, this comes at the cost of larger insertion loss and a greater number of TBUs.

Figure 16.16 illustrates, as an example, the non-ideal component fidelity analysis for the two transformations (CNOT and DFT4). Here, we illustrate the test for two different σ_K equal to 0.5 per cent and one per cent, whereas σ_ϕ is varied between zero and 0.5 per cent.

For the second example, we illustrate the performance deviation due to errors in the coupling ratio and the overall phase at each TBU when the waveguide mesh is programmed to synthetise an optical filter. For the statistical analysis, we again resort to a Monte Carlo iterative process, where the operation of each TBU is modelled by two Gaussian random variables centred at its ideal setting and featuring a standard deviation σ_K and σ_ϕ accounting for random fluctuations around the mean of the coupling coefficient (K) and phase term (ϕ), respectively. Notice that, again, we could have employed Eqn. (6.3) instead. The mesh is programmed to implement a double ring-loaded MZI filter employed to implement maximally flat passband Butterworth- and Bessel-type filters [52], [53]. Figure 6.17(a) depicts the waveguide mesh programming

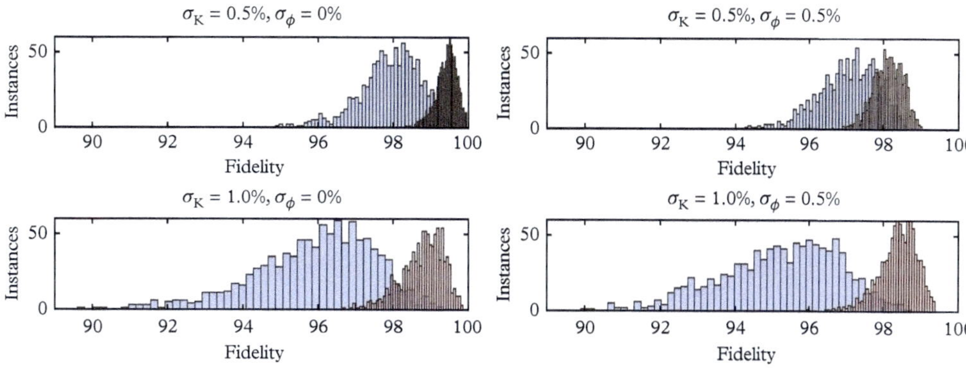

Figure 6.16 *MonteCarlo test to obtain the fidelity variations to non-ideal effects of the TBUs. (Blue instances are related to CNOT transformation and red instances are related to DFT4 of Figure 6.14.*

to implement the circuit layout shown in Figure 6.17(b). Phase shifter coefficients for ideal operation are discussed by Pérez and Capmany [5] and the results of the Monte Carlo analysis after 1,000 realisations are shown in Figures 6.17(c)–(e). Each TBU coupling constant is modelled by means of a Gaussian random variable centred at a mean value corresponding to its ideal setting and featuring a typical [54], [55] value for the standard deviation $\sigma_K = 1$ per cent accounting for random fluctuations around the mean. It should be noted here that the Monte Carlo analysis considers both the spatial and spectral behaviour of the mesh although here only the transfer function $h_{36,15}$ is displayed so the input vector to the waveguide mesh is, in this case, $I = (i_1, i_2, \ldots i_{40})$ where $i_k = 0$ unless $k = 16$ and $i_{16} = 1$. Again, each realisation takes only a few seconds as it is based on an analytic/statistical method. From the resulting spectra we can select or build up the figures of merit we want to analyse. Here, we select relevant performance parameters such as the filter's ER, insertion losses and passband ripple, which can be displayed in terms of histograms from which averages and standard deviation values can be extracted.

All in all, we cover the parasitic effects in both feedforward and feedforward/backward waveguide mesh arrangements. In the latter, we appreciate that the signal leaking in the circuit will deteriorate the circuit performance. However, a smarter configuration of the circuits and a *smart routing* of these parasitic signals can enhance dramatically the final circuit performance [5]. This optimisation deals with the configuration of both used and non-used TBUs. For this to become practical, we can employ analytical models and their combination with optical power monitors at strategic points in the circuit.

In both waveguide mesh arrangement approaches, a considerable electronic circuitry and software framework is required to enable the configuration of the circuits as well as to fully exploit the benefits and advantages of programmable PICs. Both control electronics and software are a fundamental part of the system and are introduced in Section 6.6.

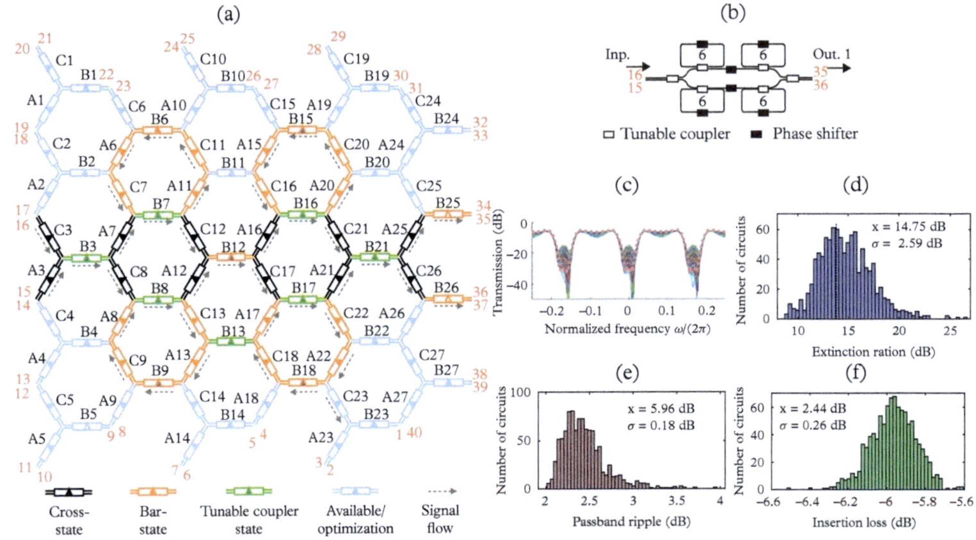

Figure 6.17 *Scalable Analysis Method application to multiparameter error analysis of a waveguide mesh implementing a feedforward/feedbackward double ring loaded MZI with cavity lengths (6 BULs). (a) Mesh architecture and configuration for implementing the double ring loaded MZI. (b) Equivalent circuit layout with indication of the input and output ports in red ink. (c)-(f) Results of the Monte Carlo analysis of the spectral transfer function $h_{35,16}$ after 1000 realizations, where Each TBU coupling constant is modelled by means of a Gaussian random variable centered at a mean value corresponding to its ideal setting and featuring a standard deviation $\sigma_K = 0.1$ accounting for random fluctuations around the mean. (c) Spectral transfer function realizations, (d) Filter Extinction ratio statistics, (e) Filter passband ripple statistics, (e) Filter insertion loss statistics*

6.6 Monitoring and Control of Programmable Photonic Integrated Circuits

The dynamic (re)configuration of programmable PICs requires, by definition, the assistance of a software framework as well as electrical control and driving circuitry. The previous sections showed that the performance and scalability of large-scale programmable PICs requires the precise driving of a set of localised actuators. The control subsystem is completed with the inclusion of optical power monitors to assist during the configuration stages and to perform optimisation tasks that mitigate the non-ideal behaviour of photonic components.

In order to consider the dynamic behaviour of large-scale programmable PICs, we describe here a disaggregated closed-loop control system (CCS) architecture [49]. As illustrated in Figure 6.18, the CCS consist of an electronic processing unit (PU) or microprocessor, a multi-channel electronic driver array (MEDA) and a multi-channel electronic monitor array (MEMA).

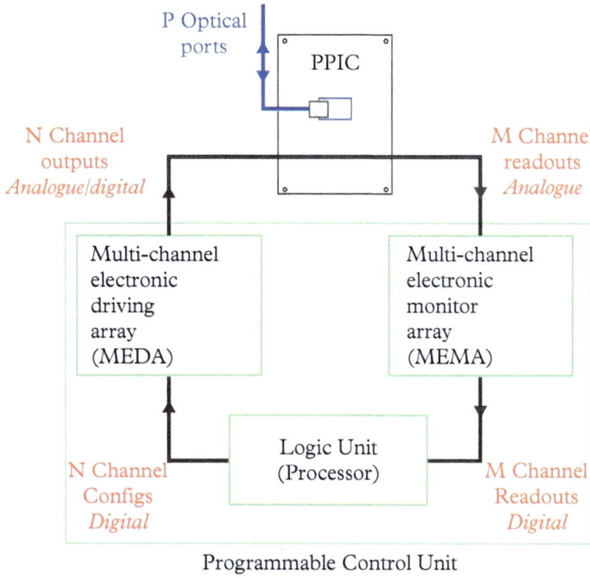

Figure 6.18 *Schematic proposed Programmable Control Unit for programmable photonic integrated circuits and waveguide mesh-based circuits.*

6.6.1 Processing Unit

The PU configures and sets the signals sent to the driving system circuitry. The configuration is chosen based on the desired application to be performed using the photonic circuit. The prospect of using such algorithms in programmable PICs is a relatively unexplored area of research. Only few concrete algorithms have been proposed for specific waveguide mesh circuits and they have been limited to feedforward networks only [3], [4]. Additionally, these algorithms might require a feedback signal coming from the output ports of the photonic circuit, or the local power monitoring at every tunable primitive unit output [9], [10]. This is supplied by the MEMA. Note that the electronic PU receives and sends digital electronic signals.

6.6.2 Multi-channel Electronic Driving Array (MEDA)

The MEDA is the electronic system that receives a digital signal from the PU. The said signal contains the configuration of every actuator in the system and translates it into the required driving signals. The format of these signals varies depending on the type of integrated actuator in the PIC. Typically, reconfigurable multi-channel current or voltage sources are employed for this purpose. The precision, linearity, stability and low-noise features are essential to ensure the correct performance of the PIC. The format of the signal varies from constant direct current driving [11]–[17], [56] to digital pulsed formats [32]–[34].

6.6.3 Multi-channel Electronic Monitor Array (MEMA)

This subsystem obtains electronic readouts from the programmable PIC. Most of them carry signals that are proportional to the optical power (i.e. signals from opto-electronic transducers) in order to monitor the PIC behaviour at specific points in the circuit. Others monitor different parameters like the thermistor in a temperature-controlling closed loop. The signal is then converted into a digitised format as required by the PU after which they are aggregated into an information array.

With regard to the optical power monitors, their number and location vary for each design. On-chip monitors enable a compact and small form-factor system. These can be classified in two approaches. First, the most conventional is the use of photo-diodes like those offered by MWP platforms such as Jeppix [57] and EuroPractice [28]. Typically, such devices exhibit a responsivity in excess of 0.7 A/W along with dark currents less than 50 nA especially when used as low-speed, monitor elements. These can be placed either at the optical termination of the chip or elsewhere in the circuit preceded by an optical tapping element that extracts a portion of the signal of a waveguide and delivers it to the photo-diode. Depending on the power budget and the specifications, this might be a viable option, but it is definitely not ideal. Secondly, photonics elements such as contactless integrated photonic probes (CLIPPs) [10] can provide a non-invasive solution, making them a more ideal option.

Both MEDA and MEMA subsystems imply the consumption of resources (area, electrical power, driving components, processing elements, memory, etc). This overhead can be represented by the following ratios

$$
\begin{aligned}
OR_{MEMA\_ACTUATORS} &= \frac{M}{N_{act}}, \\
OR_{MEMA\_O-PORTS} &= \frac{M}{N_{opt.ports}}, \\
OR_{MEMA\_TBU} &= \frac{M}{N_{TBUs}},
\end{aligned}
\tag{6.12}
$$

which represent the overhead ratio of the MEMA channels (mainly number of optical power monitor points (M)) versus the number of actuators (N_{act}), the ratio between M and the optical ports$\left(N_{opt-ports}\right)$ and between M and the number of TBUs in waveguide mesh arrangements, respectively.

6.7 Software Layer and Smart Operations

As fundamental as the control electronics is the software layer. The software-defined tag has been used in MWPs [58] and the software-radio [59] communities for the last few years. It implies that a particular architecture or system can be programmed/reconfigured after fabrication and thus provide a sort of adaptability degree.

In programmable photonics, the software framework calls for the aggregation of a wide variety of algorithms, methods or routines that are run by the PU. They are mainly employed for setting certain functionality in the PIC or to perform necessary backplane optimisations. In general, they employ input arguments that come from the user settings, the data acquired by the MEMA and/or data from analytical models [5], [22]. Then they decide on the configuration for the MEDA parameters. In some cases, optimisation tasks and other dynamic operations are required in order to compensate the operational drifts due to any tuning-related crosstalk of the programmed components and environmental changes. In these cases, the *PU* can launch recurrent or iterative procedures.

A programmable PIC's architecture can be reconfigured by means of its electronic control signals. These signals can be grouped together into the processor state vector $\mathbf{C} = (c_1, c_2, c_3, \ldots, c_N)$, where each c_i identifies a particular parameter of a given photonic actuator. For example, in the context of optical modulators, $c_i = \{0, V_\pi/2, V_\pi\}$ could define the voltage values for null, quadrature and maximum bias point of an internal modulator; $c_i = \{V_X, V_\|, V_K\}$ the voltage values for cross and bar operation of a 2×2 optical switch or TBU defined by a K coupling factor; $c_i = \left\{V_0, V_{\pi/2}, V_\pi, V_{3\pi/2}\right\}$ the voltage values required to achieve a given phase within an optical phase-shifting element in a waveguide mesh arrangement. The common hardware platform would then implement different functionalities by changing the values of the components in its state vector [1], [60]. Depending on the software capabilities, for a targeted functionality the programmable PIC might be able to configure itself to optimally perform the application.

Figure 6.19 illustrates an example of a simplified workflow of a programmable PIC based on a waveguide-mesh arrangement. Generally, as an initial step, the pre-evaluation or characterisation of the subsystems is necessary for correct performance or to provide a first evaluation of the circuit. After this task, which can be performed by the processor during its initialisation, the device is ready to programme diverse functionalities. First, the subsystems implied in the operation must be enabled with the correct control signals. Next, the circuit topology and/or optical interconnections to be synthetised in the reconfigurable optical core must be chosen together with the high-level design parameters. In the case of an optical filter synthesis based on cascaded ring resonators, the number of rings, coupling constants and differential phases applied to each ring must be computed. Once obtained, these results have to be programmed by properly feeding

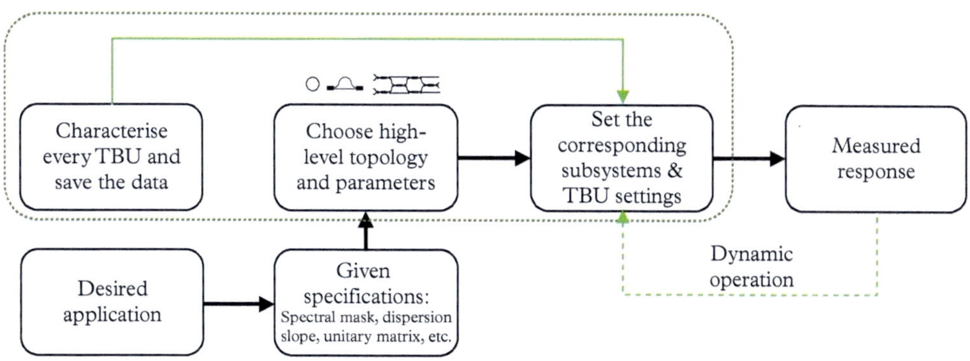

Figure 6.19 *Programmable PIC simplified workflow including optional dynamic operation.*

this data to the MEDA. Dynamic operation based on signal monitoring can be employed to optimise and improve the overall performance in real time by reading from the MEMA and making any adjustments based on software protocols in the PU.

The software capabilities of the processor determine the complexity level of the tasks that it can perform independently, i.e. without the user's assistance. For our current waveguide mesh example, we can define three tiers of software-definition [1].

Tier one implies the user ability to change the lower-layer settings such as TBU phase-shifter currents that are directly related to the TBU coupling coefficient and the additional phase response. In this case, the circuit topology and design parameters are chosen by the user, who must enable and configure each TBU to obtain the desired light-path. The user oversees enabling and configuring all the subsystems required for the desired operation.

Tier two is defined as that which includes the capacity of the electronic PU to choose the best alternative to synthetise/emulate a certain configuration. For example, if the user decides to programme an ORR of a certain FSR and ER, the processor will internally configure itself the circuit topology and design parameters in an efficient way. In this case, the processor performs a system pre-evaluation that allows knowing the status of each subsystem. For the waveguide mesh arrangement, the processor performs an algorithm to optimally allocate the synthesised circuit and to set the suitable high-level design parameters.

In contrast, in tier three the user specifies a desired application and the specifications, while the processor chooses the best circuit topology to accomplish the previous requirements, allowing the user to become totally disregarded of the internal configuration. The last two levels could optionally provide self-healing attributes. This characteristic can be achieved if the processor automatically detects a defective phase shifter, TBU, waveguide or port and prevents the signal from going through this element. These software-defined tiers can actually co-exist Figure 6.20. Several examples of software procedures are included in Chapter 5.

As in any software framework, the combination of simpler functions can be employed to solve complex operations. In waveguide mesh arrangements, examples of basic operations are the programming of *all-cross* and *all-bar* TBUs, *pathfinding* or finding the path that connects to points in the arrangement with a certain length, dynamic *self-characterisation* and *scattering matrix programming*[5]. The completion of these algorithms and their combination with the software procedures described in Chapter 4 and Chapter 5, result in a well-suited stack of algorithms to be employed for the construction of more complex routines. Fortunately, *path-finding* routines have been already demonstrated for their use in myriad applications and constrained and partially modified adapted versions can be applied to waveguide mesh arrangements [60]. Similarly, software routines that exploit graph theory can find application in photonic programming, leading to completely new fields of research [61]. Following graph theory terminology, each TBU port is transformed to a vertex in a non-planar graph that is shared with the adjacent interconnected TBU. The presented parallel version of waveguide mesh arrangements in Chapter 5 is just an isomorphic version of the non-flattened one. A brief introduction to *auto-routing* algorithms is covered in Section 6.7.1.

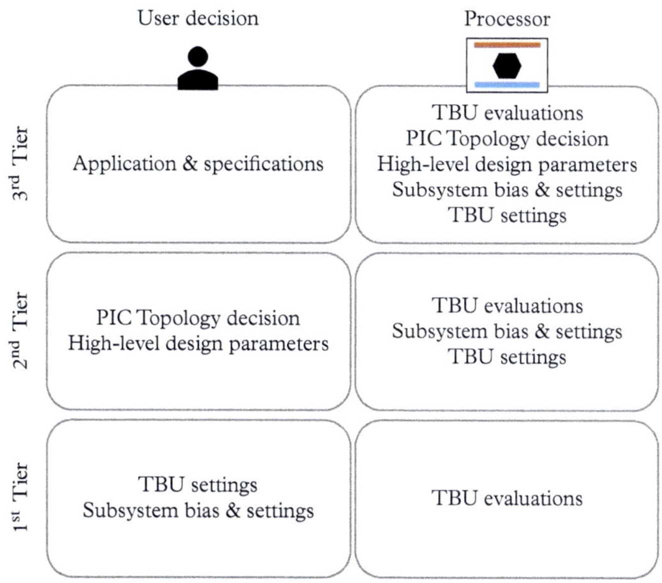

Figure 6.20 *Software-definition levels depending on the capabilities of the signal processor to perform configuration and evaluation tasks.*

As an application example, we propose and present the analytical results of a *self-characterisation* routine. Due to design and fabrication phase errors, the TBUs often present a random coupling factor in their passive state. The presented algorithm and control scheme are employed to compute the phase shifts necessary at each phase actuator to achieve the cross- and bar-states. The architecture requires four optical monitoring points and two optical source points, located as specified in Figure 6.21. At each slot of time only one monitor and one optical source pair is employed. The waveguide mesh arrangement architecture is represented in Figure 6.21. It is composed by 36 TBUs (seventy-two phase actuators) and twenty-four optical ports. The OR ratio defined in Eqn. (6.6) of optical monitoring points per optical ports, TBUs and phase actuators is 0.166, 0.111 and 0.055, respectively. These figures show the reduced overhead of the proposed monitoring system location compared to the mesh size and the actuation system.

For the analytical validation of the scheme and the routine we simulate the waveguide mesh arrangement behaviour with the analytical model developed by Pérez and Capmany [5]. In this case, to increase the difficulty of the scenario, we assume a mean of 0.1 dB/TBU. For the random-coupling passive nature of the TBU we choose for each an initial phase deviation from a uniform distribution between 0 and 2π radians. The *self-characterisation* algorithm does not know and does not have access to the passive phases, which are fixed during fabrication. However, it has non-simultaneous access to the readouts of source-monitor activation pairs. At each combination, it performs optical

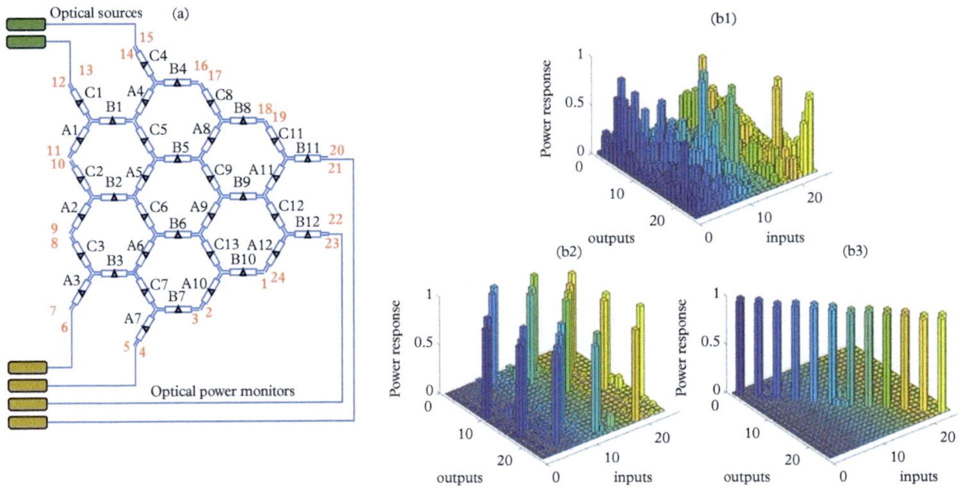

Figure 6.21 *(a) Waveguide mesh arrangement of 36 TBUs and the location of two optical sources and four optical power monitors for the self-characterization routine. (b1) absolute power of the scattering matrix of the waveguide mesh arrangement at its passive random-phase state., (b2) absolute power of the scattering matrix of the waveguide mesh arrangement at its learned cross and (b3) bar states, first at [49].*

power minimisation through the action of the phase actuators in the vicinity but not in the targeted TBU paths and power maximisation through the actuation in the targeted paths. We performed the test 100 times, imposing a fixed step in the phase sweeping of 0.05 rads. Under these conditions, the mean phase error for the cross- and bar-state was 0.021π and 0.015π, respectively. We repeated the test imposing a fixed step in the phase sweeping of 0.03 rads and obtained a mean phase error for cross- and bar-state of 0.011π and 0.013π, respectively. Although these figures are expected to improve with a finer tuning of the phases, the added noise from photodetectors, electrical and optical source variations and tuning-based crosstalk between phase actuators impose additional limitations. Each test performs the characterisation of both phase actuators per TBU. Figure 6.21 illustrates one of the resulting scattering matrices at *all-cross* and *all-bar* states at *b*2 and *b*3, respectively, compared to the passive state of the mesh at *b*1.

Once the characterisation is performed by either a semi- or fully automated process, the PU saves a look-up table with the functions mapping the driving values for setting the TBU in cross or bar. Only then can it start the configuration stage. Figure 6.22 shows how to configure the seven-cell hexagonal arrangement to programme four different circuits by using pre-set configurations.

In this example, the software framework and the control subsystem are based on non-integrated MEDA subsystem based on a table-top multi-channel current source, and an optical spectrum analyser (OSA) as the optical monitor. As for the processor, the work reports the use of a standard personal computer and custom Python-based routines. The communication interface is based on general purpose interface bus (GPIB) protocol.

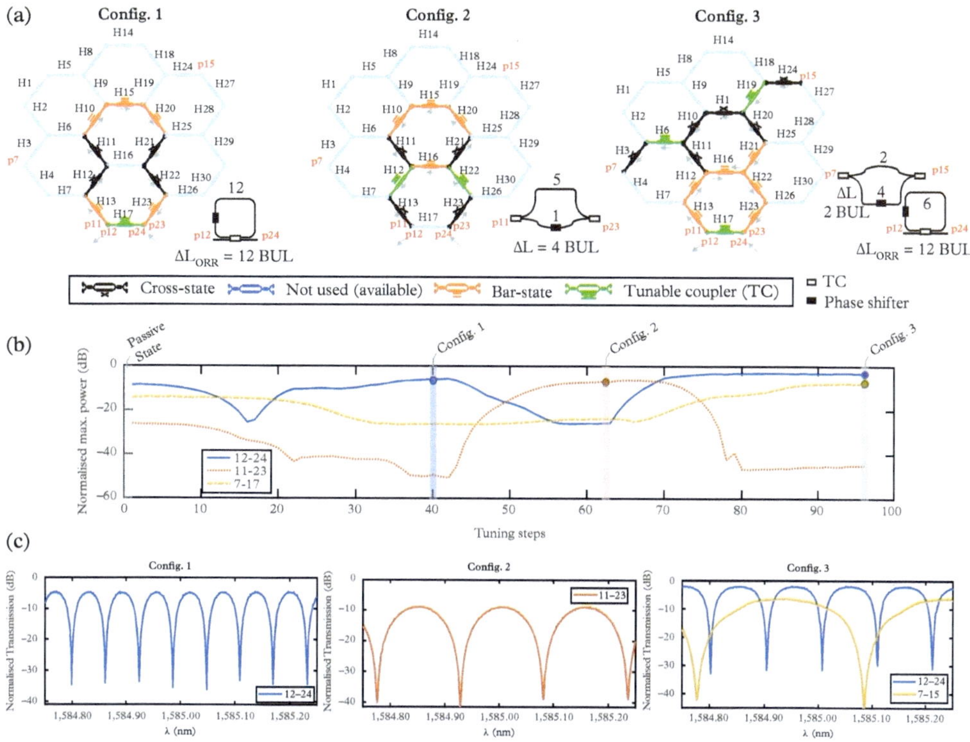

Figure 6.22 *(a) Waveguide Schematic and configuration of the fabricated seven-cell feedforward/backward hexagonal waveguide mesh. Config1: Optical Ring Resonator of 12 primitive units, Config. 2: Unbalanced Mach-Zehnder Interferometer of 4 primitive units, Config3: Two simultaneous circuits ORR and MZI working in parallel., (b)Measured normalized maximum optical power at the channels 12-24, 11-23 and 7-15 specified in (a). Versus tuning steps of phase actuators., Normalized spectral response measured for the different configurations: Configuration 1, Configuration 2 and Configuration 3. Optical channel is specified for each case following the nomenclature of Fig. 1.*

To carry out the experiment, we first employed the control scheme in Figure 6.18 and a semi-automated basic routine to characterise the response of every phase actuator and TBU [62]. The characterisation was saved in a look-up table for its use in the following steps. We saved the required configuration of three different circuit programming scenarios, referred to in Figure 6.22(a) as Config. 1, Config. 2 and Config. 3, respectively: a 10-TBU ORR, a 4-TBU imbalanced MZI, and a simultaneous combination of a 6-TBU ORR working in parallel with a 2-TBU imbalanced MZI. Each configuration is translated to an array with the required electrical current value for each phase actuator in the waveguide mesh arrangement. While the cross-bar values were selected from the look-up table containing the pre-characterisation, the coupling factors of the TBU performing as tunable couplers were selected and saved after measuring the critical coupling conditions.

In order to show the dynamic (re)configuration of the waveguide mesh arrangement, we send to the MEDA the pre-set values of the targeted Config. 1. During the tuning, and for illustrative purposes, we divided the configuration in steps or slots where a maximum change of 5 mA was allowed per channel. Figure 22(b) illustrates the monitored maximum optical value measured between ports 12–24, 11–23 and 7–15, respectively. In the first iterations, we can appreciate the variation of the optical power of the three optical channels when the circuit is configured from its passive state to Config. 1. The number of steps is forty and is given by the configuration of at least one channel from 0 to 200 mA. Here, the optical power has been normalised with respect to a straight waveguide length of 14 mm and 6.5 dB per vertical fibre-to-chip coupler.

As illustrated in Figure 22(a), the 12–24 channel set matches with the optical ports of the programmed ring resonator of 10 TBUs. Once the MEDA reaches the targeted values to set Config. 1, the obtained measured spectrum is illustrated in Figure 22(c). To obtain the spectrum, we performed a sweep with a tuneable laser and an OSA. We see that the targeted optical cavity ring resonator with ER > 30 dB is achieved.

Next, the processor configures the MEDA to evolve from Config. 1 to Config. 2. In this case (Figure 22(a)), the targeted optical circuit is established by setting channels 11–23. The programmed circuit is an imbalanced MZI of 1- (short) and 5-TBU (long) arms, respectively. Again, Figure 6.21(b) illustrates how the maximum optical power in the targeted channel increases and achieves its optimum configuration. The number of steps is dependent to the proximity of the targeted MEDA array and the one from the previous configuration. The resulting spectrum is illustrated in Figure 22(c). The critical coupling is achieved and results in an ER > 30 dB.

Finally, the processor configures the MEDA to evolve from Config. 2 to Config. 3. As depicted in Figure 22(a), the third configuration is a special case were two circuits are programmed to work simultaneously in parallel. In this case, the 11–24 channel set is employed to implement an ORR of 6 TBU and the 7–15 channel set implements an unbalanced MZI with arm lengths of 2 and 4 TBUs, respectively. The achieved spectra are illustrated in Figure 21(c). An important measure in this case is the optical crosstalk between the two circuits working in parallel. Here, we measured the crosstalk of channel sets 7–24 and 12–15 and obtained figures below 40 dB. This is quite remarkable as the circuits share common TBUs inside the circuit footprint.

For the three configurations, we observe how the insertion loss of each circuit varies with the number of TBUs employed, as the loss associated with each TBU is in the order of 0.6 dB. In addition, chip-couplers have loss of 6.5 ± 1 dB.

With respect to the reconfiguration time, thermo-optic-based actuators require hundreds of milliseconds. In our experiments, the electronic circuitry required 400 ms per step, which includes the tuning of thirteen sources, sequentially. Although possible, we avoided the one-step configuration of the phase actuators to reduce risks in the micro-heating structures coming from abrupt differential voltage changes.

Finally, the next two sections provide examples of auto-routing and advanced configuration methods in general-purpose waveguide mesh arrangements.

6.7.1 Auto-routing Algorithm in Waveguide Mesh Arrangements

As covered in Chapter 5 and Chapter 6, most of the circuits programmed in waveguide meshes rely on the configuration of optical delay lines and optical interconnections between optical nodes. In addition, as presented in Chapter 9, the integration of a general-purpose waveguide mesh arrangements with specific high-performance building blocks gives rise to the optical homologue to the field programmable gate array, the field programable photonic gate array (FPPGA) [65]. In both cases, the configuration of optical connections and optical delay lines requires the selection and configuration of the programmable unit cells in the arrangement that will define the path followed by the optical signal. For a given specification, (for example, a delay of 10 ns between two specific optical ports), there will be multiple solutions. In such a scenario, the implementation of a software routine capable of finding the drive power and automatically controlling the optimum path in terms of loss, power consumption and other non-ideal effects is highly desired. Consider the example in Figure 6.23 where an FPPGA is configured to programme simultaneously two optical signal processing circuits.

Following with the example, the first circuit (in red) requires the connection and auto-routing between points AB, CD and EF. The second one requires the connection between GH, IJ and KQ. In addition, the first circuit incorporates the formation of an optical resonator (in blue), whose optical cavity is defined by programmed waveguide or optical delay line and a tunable coupler configuration. This scenario requires an automated routine to build up the interconnection and the delay lines for the structures

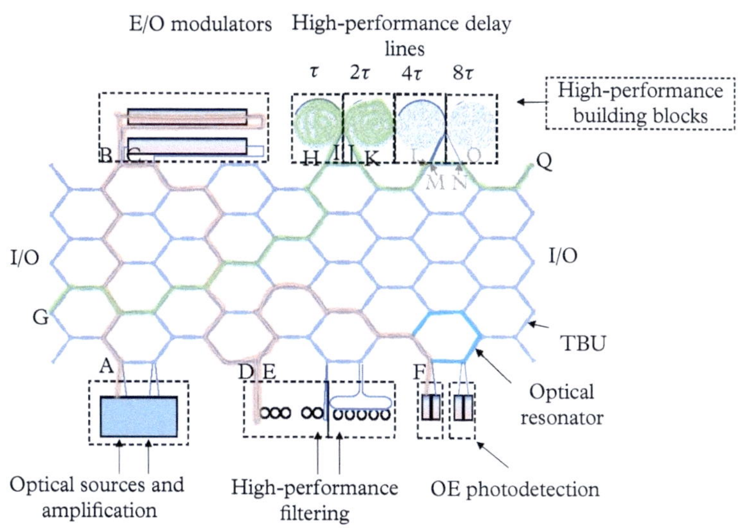

Figure 6.23 *Optical FPPGA with two circuits configured simultaneously. Circuit 1 is defined by the interconnection of labelled nodes A-B, C-D, E-F. Circuit 2 is defined by the interconnection of labelled nodes G-H, I-J, K-Q. E/O: Electro-optical, I/O: optical inputs outputs, TBU: Tunable Basic Unit.*

programmed in the FPPGA core. To address this problem, a recent solution explores the use of *pathfinding* algorithms to find the optimum configuration of a waveguide mesh arrangement. The reported solution is inspired by the modification of a pathfinding algorithm like Dijkstra [66, 67] and search trees, widely used in many different research fields, e.g. IP traffic engineering, artificial intelligence. The auto-routing algorithm is employed to determine automatically any optical path and/or interferometric structure, given a waveguide mesh topology and a set of specifications. In addition, the routine leads to unprecedented self-healing and fault-tolerant capabilities of the PIC, where, given a set of damaged areas of the circuit, the algorithm is able to find alternative sub-optimal paths through the photonic arrangement.

6.7.2 Advanced Optimisation Methods to Configure Waveguide Mesh Arrangements

Computational optimisation and machine-learning techniques have become a new toolbox in different applications fields [68]. In photonics, machine-learning and data driving methods have been employed in the identification of labels in multi-mode optical fibre transmission in highly scattering media [69], in optical mode classification [70], in the prediction of bit error rates of WDM systems [71], in the optimisation of Raman spectroscopy amplification [72], in optical filter optimisation [66] and also to assist in the design of PIC components [73], [74]. In addition, computational optimisation methods and more simplified routines have been applied for the configuration of optical filters [75] and multi-port interferometers [76]. In each, a single application is targeted. These methods are, however, particularly advantageous for the management and control of complex systems, where many variables and parameters are involved that need to be configured, continuously monitored and controlled. Since medium- and high-density programmable integrated photonic circuits fall under the latter category, it is only natural to expect a very high gain in their operation, flexibility and reliability by developing and tailoring machine-learning and optimisation techniques especially suited for them. Indeed, the concept has been demonstrated to several configuration examples, including a waveguide mesh arrangement programming a 1x8 beamsplitter and for optical filters [77].

Mathematical and computational optimisation methods deal with the problem of finding minima/maxima of a cost function. In this case, we can define application-specific cost-functions that are computed based on the specifications of the targeted functionality. As the proper algorithm, per se, the design of the cost function is essential for the success and faster convergence of the algorithm. The cost function can combine several features f and being multiplied by a correction factor (w) to specify its weight in the final metric:

$$CF(\mathbf{v}) = \sum_{n} w_n f_n(\mathbf{v}), \tag{6.13}$$

where n is the number of features considered and v is the vector, including the variables that are tuned in the optimisation process. The vector v includes a set size equal to the

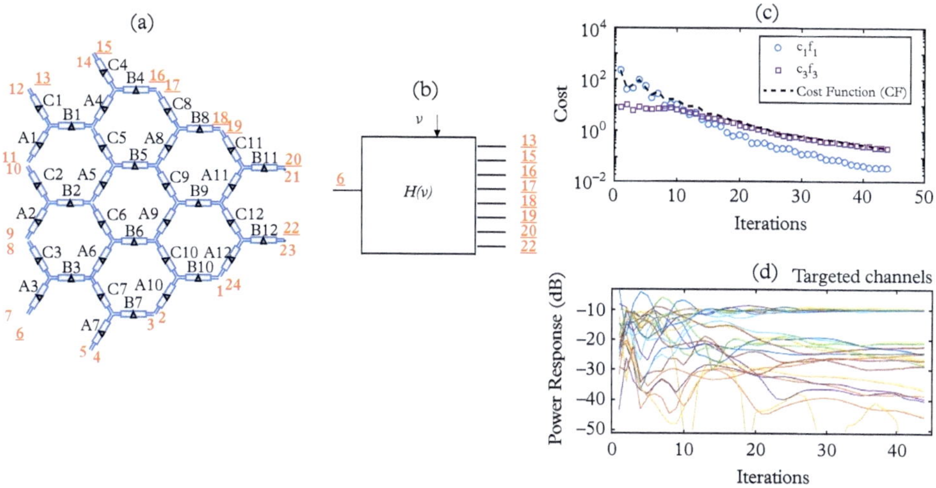

Figure 6.24 *Application of self-configuration algorithm for the programming of waveguide mesh arrangement. (a) Labelled mesh layout, (b) black-box schematic to be programmed, (c) results of the cost function evaluation during the optimization process, (d) results of the power response at every output channel.*

double of the number of TBUs enabled during the optimisation process ($2N_{\mathrm{TBU}}$). The first half of elements deals with the differential drive and actuates only the upper phase shifter of each TBU. The second half deals with the common drive and actuates over the sum of both phase shifters in the TBU. The features are obtained by measuring the optical spectra at each iteration and evaluating and comparing the performance with the targeted value. These features are obtained by optical readout monitors at the external perimeter of the mesh. For example, in Pérez [79], it is demonstrated the configuration process of a waveguide arrangement implementing a 1×8 beamsplitter through an iterative optimization process that measure how far the frequency response is from the targeted optical power and spectral flatness. Once selected, different algorithms can be employed to iteratively search the optimum values to drive the phase actuators. The work demonstrates the use of derivative (gradient-based) and non-derivative computational methods. Figure 6.24(a–b) illustrates the application example of a 36-TBU hexagonal arrangement configured as a 1×8 beamsplitter. Figure 6.24(b) illustrates the problem from the algorithm perspective, where the routine is charged with a black box representing the optical system defined by the targeted input and output ports. Figure 6.24(c–d) shows the iterative process of a gradient descent application, and demonstrates how the minimisation of the cost function leads to the targeted performance where all the optical signals at input port 6 are divided into eight optical channels.

In short, the combination of a versatile and powerful software allows not only the practical driving of complex programmable PICs, but also an alternative to overcome fabrication and non-ideal PIC operation.

6.8 Chapter Summary

This chapter covered the practical issues arising in the development of programmable PICs and the architectures necessary to enable the software-assistance and the programmability. First, it reviewed the state-of-the-art demonstrators for multi-port feed-forward architectures, and feedforward/feedbackward arrangements. Although they represent implementations with moderate complexity up to 120 phase actuators, they are complex enough to demonstrate the benefits of programmable PICs.

Next, it provided a scalability analysis for finding the main potential performance limits in the future evolution of programmable PICs in general and waveguide mesh arrangements in particular. From it, it showed that insertion loss per TBU, footprint, optical crosstalk and power consumption per TBU are positioned as the most challenging obstacles to be overcome in future work.

We predict that the integration of 300 working phase actuators will become a reality in the short term and that the integration of 1,000 phase actuators will be exceeded by 2025. Note that some of these numbers have been already reached in the integration of optical switching matrices. However, although similar, the circuitry is not totally comparable since the cross-bar-only state relaxes the specifications of control system when compared to waveguide mesh arrangements that require both amplitude and phase tuning.

Finally, it described a disaggregated architecture based on a feedback loop to manage programmable PICs in general and with the description of the software framework and its main tasks.

The next chapters describe several applications where programmable PICs are essential. Chapter 7 covers their application in the synthesis of linear transformations for quantum circuits. Chapter 8 extends the analysis to the implementation of classical applications. Finally, Chapter 9 describes the next evolutionary step of programmable PICs, based on the combination of programmable high-performance building blocks and a reconfigurable optical core that, together with the same software and control environment, offers the highest versatility to date—the field-programmable photonic array.

REFERENCES

[1] D. Pérez, *Integrated Microwave Photonic Processors Using Waveguide Mesh Cores*. PhD Thesis, Universitat Politècnica de València, 2017.

[2] D. Pérez, I. Gasulla, J. Capmany and R.A. Soref, Reconfigurable Lattice Mesh Designs for Programmable Photonic Processors. (2016) *Opt Express* 24, 12093–106.

[3] D.A.B. Miller, Perfect Optics with Imperfect Components. (2015) *Optica* 2(8), 747–50.

[4] J. Mower, et al., High-fidelity quantum state evolution in Imperfect Photonic Integrated Circuits. (2015) *Phys Rev A*, 92: 032322.

[5] D. Pérez and J. Capmany, Scalable Analysis for Arbitrary Photonic Integrated Waveguide Meshes. (2019) *Optica* 6(1), 19–12.

[6] J. Carolan, et al., Universal Linear Optics. (2015) *Science* 349, 711.

[7] N.C. Harris, et al., Quantum Transport Simulations in a Programmable Nanophotonic Processor. (2017) *Nat Photonics* 11, 447–52.
[8] C. Taballione, et al., 8×8 Programmable Quantum Photonic Processor Based on Silicon Nitride Waveguides. (2018) *Frontiers in Optics/Laser Science*, paper JTu3A.58.
[9] A. Ribeiro, A. Ruocco, L. Vanacker and W. Bogaerts, Demonstration of a 4×4 -port Universal Linear Circuit. (2016) *Optica* 3(12), 1348–57.
[10] A. Annoni, et. al., Unscrambling Light—Automatically Undoing Strong Mixing between Modes. (2017) *Light Sci Appl* 6(12), e17110.
[11] Y. Shen, et al., Deep Learning with Coherent Nanophotonic Circuits. (2017) *Nat Photonics* 11, 441–6.
[12] H. Zhou, et. al., Self-learning Photonic Signal Processor with an Optical Neural Network Chip. (2019). Available at https://arxiv.org/abs/1902.07318
[13] P. Mennea, et. al., Modular Linear Optical Circuits. (2018) *Optica* 5, 1087–90.
[14] D. Zheng, et al., Low-loss Broadband 5×5 Non-blocking Si3N4 Optical Switch Matrix. (2019) *Opt Lett* 44(11), 2629–32.
[15] M. Reck, A. Zeilinger, H. J. Bernstein and P. Bertani, Experimental Realization of Any Discrete Unitary Operator. (1994) *Phys Rev Lett* 73(58), 58–61.
[16] L. Zhuang, et al., Programmable Photonic Signal Processor Chip for Radiofrequency Applications. (2015) *Optica* 2, 854–9.
[17] D. Pérez, et al., Multipurpose Silicon Photonics Signal Processor Core. (2017) *Nat Commun* 8(1), 636.
[18] D. Pérez, E. Sánchez and J. Capmany, Programmable True-Time Delay Lines Using Integrated Waveguide Meshes. (2018) *IEEE J Lightwave Tech* 36(19), 4591–601.
[19] D. Pérez, et al., Silicon Photonics Rectangular Universal Interferometer. (2017) *Laser Photonics Rev* 11, 1700219.
[20] D. Pérez and J. Capmany, Programmable Integrated Photonics: State of the Art and Future Trends. *Proceedings of Integrated Photonics Research, Silicon and Nanophotonics* (pp. ITh1I–2). 2–5 July 2018, Zurich, Switzerland.
[21] D. Pérez-López, A. M. Gutierrez, E. Sánchez, P. DasMahapatra, J. Capmany, Integrated photonic tunable basic units using dual-drive directional couplers, *Optics Express*, 27(26), 2019.
[22] W.R. Clements, et al., Optimal Design for Universal Multiport Interferometers. (2016) *Optica* 3, 1460–5.
[23] D. Celo et al., 32×32 Silicon Photonic Switch. *Proceedings of OptoElectronics and Communications Conference (OECC) & IEEE Photonics in Switching (PS)*. 3–7 July 2016, Niigata, Japan.
[24] A. Densmore, et al., Compact and Low-power Thermo-optic Switch Using Folded Silicon Waveguides. (2009) *Opt Express* 17(13), 10457–65.
[25] Fang, et al., Ultralow-power Silicon Photonics Thermo-optic Switch with Suspended Phase Arms. (2011) *IEEE Photonic Tech L* 23(8), 525–7.
[26] M. Rudé, et al., Optical Switching at 1.55 μm in Silicon Racetrack Resonators Using Phase Change Materials. (2013) *Appl Phys Lett* 103(14), https://doi.org/10.1063/1.4824714.
[27] J. Zheng, et al., GST-on-silicon Hybrid Nanophotonic Integrated Circuits: A Non-volatile Quasi-continuously Reprogrammable Platform. (2018) *Opt Mater Express* 8(6), 1551–61.
[28] http://www.europractice-ic.com/index.php
[29] T. Tekin, Review of Packaging of Optoelectronic, Photonic, and MEMS Components. (2011) *IEEE J Sel Top Quantum Electron* 17(3), 704–19.

[30] L. Carroll, et al., Photonic Packaging: Transforming Silicon Photonic Integrated Circuits into Photonic Devices. (2016) *Appl Sci* 6(12), 426.
[31] H.D. Thacker, et al., Flip-chip Integrated Silicon Photonic Bridge Chips for Sub-picojoule per Bit Optical Links. *IEEE Proceedings of the 60th Electronic Components and Technology Conference (ECTC)*, 240–6. 1–4 June 2010, Las Vegas, NV.
[32] M. Bahadori, et al., Thermal Rectification Of Integrated Microheaters for Microring Resonators in Silicon Photonics Platform. (2018) *J Light Technol* 36(3), 773–88.
[33] G. Cocorullo, F.G. Della Corte, I. Rendina and P.M. Sarro, Thermo-optic Effect Exploitation in Silicon Microstructures. (1998) *Sensor Actuat A Phys* 71(1–2), 19–26.
[34] N. Zecevic, M. Hofbauer and H. Zimmermann, Integrated Pulsewidth Modulation Control for a Scalable Optical Switch Matrix. (2015) *IEEE Photon J* 7(6), 1–7.
[35] Z. Sheng, et al., A Compact and Low-loss MMI Coupler Fabricated with CMOS Technology. (2012) *IEEE Photon J* 4(6), 22272–7.
[36] G.W. Cong, et al., Demonstration of a 3-dB Directional Coupler with Enhanced Robustness to Gap Variations for Silicon Wire Waveguide. (2014) *Opt Express* 22(2), 2051–9.
[37] N.C. Harris, et al., Efficient, Compact and Low-loss Thermo-optic Phase Shifter in Silicon. (2014) *Opt Express* 22(9), 10487–93.
[38] T.J. Seok, N. Quack, S. Han and M.C. Wu, 50 × 50 Digital Silicon Photonic Switches with MEMS-actuated Adiabatic Couplers. *Proceedings of the 2015 Optical Fiber Communications Conference and Exhibition (OFC)*, 1–3. 22–26 March 2015, Los Angeles, CA.
[39] T.Y. Yun and K. Chang, Analysis and Optimization of a Phase Shifter Controlled by a Piezoelectric Transducer. (2002) *IEEE Trans Microw Theory Tech* 50(1), 105–11.
[40] W.A. Zortman, D.C. Trotter and M.R. Watts, Silicon Photonics Manufacturing. (2010) *Opt Express* 18, 23598–607.
[41] C.J. Oton, et al, Silicon Photonic Waveguide Metrology Using Mach–Zehnder Interferometers. (2016) *Opt Express* 24, 6265–70.
[42] Z. Lu, et al., Performance prediction for silicon photonics integrated circuits with layout-dependent correlated manufacturing variability. (2017) *Opt Express* 25, 9712–33.
[43] D. Pérez, et al., Thermal Tuners on a Silicon Nitride Platform. (2016). Available at https://arxiv.org/abs/1604.02958.
[44] N.Y. Winnie, J. Michel and L.C. Kimerling, Athermal High-index-contrast Waveguide Design. (2008) *IEEE Photonic Tech L* 20(11), 885–7.
[45] D. Feng, et al., Compact Single-chip VMUX/DEMUX on the Silicon-on-insulator Platform. (2011) *Opt Express* 19(7), 6125–30.
[46] J. Song, et al., Fast- and Low-power Michelson Interferometer Thermo-optical Switch on SOI. (2008) *Opt Express* 16(20), 15304–11.
[47] D. Pérez and J. Capmany, Programmable Multifunctional Nanophotonics ICs: Architectures, Performance, and Challenges. In S. He and E. Lee (Eds), *Smart Photonic and Optoelectronic Integrated Circuits XXI*, vol. 10922, 1092213, Curran Associates, Inc., 2019.
[48] A. Prinzen, M. Waldow and H. Kurz, Fabrication Tolerances of SOI-based Directional Couplers and Ring Resonators. (2013) *Opt Express* 21(14), 17212–20.
[49] Mikkelsen, J. C. Sacher, W. D., and Poon J.K.S., Dimensional variation tolerant silicon-on-insulator directional couplers. (2014) *Opt Express* 22, 3145–50.
[50] D. Pérez, A. López, P. DasMahapatra and J. Capmany, Field-programmable Photonic Array for Multipurpose Microwave Photonic Applications, Presented at the *IEEE 2019 International Topical Meeting on Microwave Photonics (MWP2019)*, 7–10 October 2019, Ottowa, Canada.

[51] M.A. Nielsen and I.L. Chuang, *Quantum Computation and Quantum Information*, 2nd edn, Cambridge University Press, 2011.

[52] N. Thomas-Peter, et al., Integrated Photonic Sensing. (2011) *New J Phys* 13(5), 055024.

[53] C.K. Madsen and J.H. Zhao, *Optical Filter Design and Analysis: A Signal Processing Approach*, Wiley, 1999.

[54] J.S. Fandiño, D. Domenech, P. Muñoz and J. Capmany, A Monolithic Integrated Photonic Microwave Filter. (2017) *Nat Photon* 11, 124–9.

[55] R. Burgwal, et al., Using an Imperfect Photonic Network to Implement Random Unitaries. (2017) *Opt Express* 25, 28239.

[56] F. Flamini, et al., Benchmarking Integrated Linear-optic Architectures for Quantum Information Processing. (2017) *Sci Rep* 7, 15133.

[57] A. Mahendra, et al., Multiwavelength Stabilization Control of a Thermo-optic System with Adaptive Reconfiguration. (2017) *Appl Opt* 56(4), 1113–18.

[58] X. Leijtens, JePPIX: The Platform for Indium Phosphide-based Photonics. (2011) *IET Optoelectron* 5(5), 202–6.

[59] W. Wei, L. Yi, Y. Jaouën and W. Hu, Software-defined Microwave Photonic Filter with High Reconfigurable Resolution. (2016) *Sci Rep* 6, 3561.

[60] J. Mitola, The Software Radio Architecture. (1995) *IEEE Commun Mag* 33(5), 26–38.

[61] D. Pérez, I. Gasulla and J. Capmany, Software-defined Universal Microwave Photonics Processor. (2015) *Opt Express* 23(11), 14640–54.

[62] T. Korkmaz and M. Krunz, Multi-constrained Optimal Path Selection. In *Proceedings IEEE INFOCOM 2001. Conference on Computer Communications. Twentieth Annual Joint Conference of the IEEE Computer and Communications Society*, vol. 2, 834–43, IEEE, 2001.

[63] R.J. Trudeau, *Introduction to Graph Theory*, Dover, 1993.

[64] A. López, D. Pérez, P. DasMahapatra, and J. Capmany, Dynamic reconfiguration in field-programmable photonic arrays. Proceedings of the European Conference on Optical Communications ECOC 2019, 22–26 September 2019, Dublin, Ireland.

[65] P. Edinger, C. Errando-Herranz and K. Gylfason, Low-loss MEMS Phase Shifter for Large-scale Reconfigurable Silicon Photonics. *Proceeding of the 32nd IEEE International Conference on Micro Electro Mechanical Systems (MEMS)*, 27–31 January 2019, Seoul, Korea.

[66] International Telecommunication Union, *Series G: Transmission Systems and Media, Digital Systems and Networks: Transmission Media and Optical Systems Characteristics—Characteristics of Optical Systems. Recommendation ITU-T G.694.1, Spectral Grids for WDM applications: DWDM frequency grid*, In force 02/2012.

[67] D. Pérez, I. Gasulla and J. Capmany, Field-programmable Photonic Arrays. (2018) *Opt Express* 26, 27265–78.

[68] A. López, *Implementation of Self-reconfigurable Integrated Optical Filters based on Mixture Density Networks*, MSc Thesis, Universidad Politécnica de Madrid, 2019.

[69] A. López, D. Pérez, P. DasMahapatra and J. Capmany, Auto-routing algorithm for field-programmable photonic gate arrays, *Opt Express*, 28, 737–752 (2020).

[70] D. Zibar, et al., Machine Learning Under the Spotlight. (2017) *Nat Photonics* 11, 749–51.

[71] B. Rahmani, Multimode Optical Fiber Transmission with a Deep Learning Network. (2018) *Light Sci Appl* 7(69). doi:10.1038/s41377-018-0074-1

[72] C. Barth and C. Becker, Machine Learning Classification for Field Distributions of Photonic Modes. (2018) *Nat Commun* 1(58). doi:10.1038/s42005-018-0060-1

[73] J. Wass, et al., Gaussian Process Regression for WDM System Performance Prediction. In *Proceedings of the Optical Fiber Communications (OFC)*, 19–23 March 2017, Los Angeles, CA.

[74] D. Zibar, et al., Machine Learning-Based Raman Amplifier Design. In *Proceedings of the Optical Fiber Communications Conference*, 11–15 March 2018, San Diego, CA.
[75] A.Y. Piggott, et al., Inverse Design and Demonstration of a Compact and Broadband On-chip Wavelength Demultiplexer. (2015) *Nat Photonics* 9, 374–7.
[76] D. Melati, et al., Machine Learning Design of Subwavelength Integrated Photonic Devices. In *Proceedings of 2019 International Conference on Numerical Simulation of Optoelectronic Devices (NUSOD)*. 8–12 July 2019, Ottawa, ON, Canada.
[77] G. Choo, et al., Automatic Monitor-Based Tuning of Reconfigurable Silicon Photonic APF-Based Pole/Z. (2018) *J Light Technol* 36(10), 1899–1911.
[78] D.A.B. Miller, Self-configuring Universal Linear Optical Component. (2015) *Photonics Res* 1, 1–15.
[79] D. Pérez, Programmable Integrated Silicon Photonics Waveguide Meshes: Optimized Designs and Control Algorithms. *IEEE J. Sel. Top. in Quantum Electronics*, 26, 8301312. (2019).

7

Programmable Integrated Photonics for Quantum Systems

Programmable integrated photonics can find applications in myriad areas including the quantum information field, which encompasses communications, computing, sensing and tomography. Proof of many of these topics using lightwave technology require the use of reconfigurable linear optics transformations that, at earlier stages, were implemented via large-scale bulk optics setups, which prevented the development of more complex and scalable quantum optics configurations requiring tens or hundreds of modes. Furthermore, linear optic systems with the required fidelity require a strict control of interference through demanding phase stability mechanisms. Integrating a considerable number of photonic elements on a chip in order to implement multi-port interferometers has become the only viable technological path towards quantum information systems. Programmability brings the added value of enabling successive implementations of quantum optics experiments featuring random variation in the parameters of interest. This chapter introduces the applications of programmable photonics to quantum information systems and the general framework of a programmable quantum photonic system integrated on a chip. It then briefly describes the role of more external components such as sources and detectors, and covers the applications of reconfigurable integrated optic circuits to several topics, including linear optical quantum gates, quantum transport simulation, boson sampling and complex Hadamard and quantum Fourier transforms.

7.1 An Introduction to Programable Integrated Quantum Photonics

7.1.1 Introductory Remarks

Quantum photonics relies on the encoding of photons into input quantum states, the processing of these states through a reliable linear optics unitary transformation and subsequent measurement of the resulting output quantum state via optical detection [1]–[10].

Programmable Integrated Photonics. José Capmany and Daniel Pérez, Oxford University Press (2020). © José Capmany and Daniel Pérez.
DOI: 10.1093/oso/9780198844402.001.0001

Through this scheme, novel and powerful techniques for computation, measurement and communications can be developed that leverage the properties of quantum superposition and entanglement [3]. Many experiments over the last few years [5], [6] have demonstrated systems with increased complexity that have required a migration from bulk optics configurations to integrated optics implementations to provide a higher spatial mode count, phase stability and fidelity [7]–[9], [11], [12]. Integrated optics provides additional advantages that include low loss, potential for high-density circuit integration, strict optical path and interference control and strong photon interaction exploiting non-linearities. The merging of these two technologies has led to the new field of integrated quantum photonics (IQP) [8], [11].

The initial work on IQP focused on glass-based waveguide circuits, most notably by planar and laser-written silica (SiO_2) on silicon (Si) technologies [7], [8], which enabled mode field dimensions comparable to those of single-mode optical fibres. Quantum logic, bosonic sampling, teleportation, and multi-particle quantum walks were demonstrated among other functionalities. The low refractive index contrast nature of these technologies featured an excellent performance in terms of propagation and coupling losses limits at the same time as the high circuit footprint principally as a consequence of the high values of their minimum allowed waveguide-bending radius. Circuits with even modest complexity require footprints in the order of several cm, which precludes the use of this technology for scaling their complexity and functionality.

Overcoming this limitation requires use of high-index contrast material platforms like silicon-on-insulator (SOI) [11], [12], Si_3N_4[13], $LiNbO_3$[14] and InP, bearing in mind that reducing photon propagation losses is also a key requirement. For example, recent experimental results have shown [11] that SOI nanophotonic platforms can implement systems with hundreds of optical components in the same chip. We review some of these experiments later in the chapter. However, first we define a reference model for what is understood as an *integrated programmable photonic quantum system* and review the operation principle of its main constituents.

7.1.2 Reference Model for a Programmable Photonics Quantum System

Figure 7.1(a) shows a schematic diagram of a programmable photonics quantum processor, including its main relevant blocks.

The processor requires one or several *quantum sources*, mainly single photon, with different photon statistics to obtain the required input quantum states/modes. In many cases, quantum sources need a bright pumping lightwave to generate single photons and therefore, *highly selective tunable filters* will be required at their output to suppress the undesired remaining pump signal in order to avoid detector saturation. The input quantum states/modes are then processed by means of a $N \times N$ *programmable unitary transformer* based on a linear optics configuration. Either multi-port interferometers, described in Chapter 4, or waveguide meshes, described in Chapter 5, can be employed to implement this block, which includes tunable delay lines and switches. The outcome from the unitary transformer is an output quantum state from which we need to

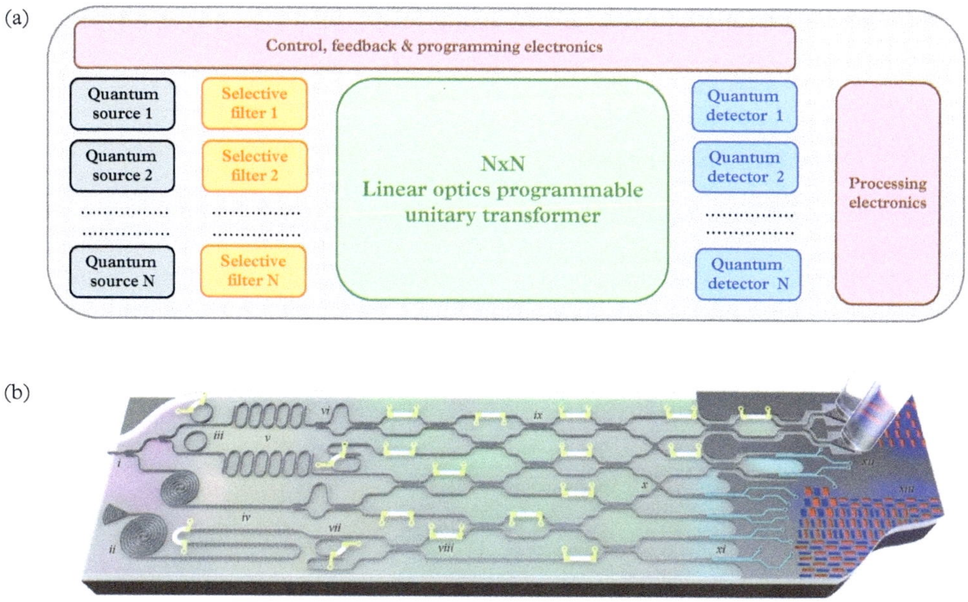

Figure 7.1 *(a) Schematic Bulk optics implementation of a programmable photonic quantum processor including all the relevant optical and electronic elements. (b) A proposed mock-up for implementation of the main parts of the processor in SOI technology (after [11]). From left to right: photon sources (magenta), pump-removal filters (yellow), passive and active optics (green), single-photon detectors (cyan), and control and feedback electronics (blue). Labels indicate: i. pump input and splitter, ii. Spiralled waveguide photon-pair source, iii. Ring resonator photon-pair source, iv. Bragg reflector pump removal filter, v. coupled-resonator optical waveguide (CROW) pump removal filter, vi. Asymmetric MZI wavelength division multiplexer (WDM), vii. Ring resonator WDM, viii. Thermal phase tuner, ix. Multi-mode interference waveguide coupler (MMI), x. Waveguide crossing, xi. Superconducting nanowire single-photon detector (SNSPD), xii. Grating based fibre-to-chip coupler, and xiii. Control and logic electronics.*

extract some classical information by means of measurement. This is achieved by using quantum or *single-photon* detectors. The classical information will need to be further processed by electronic circuits to extract the relevant content. Finally, if the output classical information is required outside the processor, then fibre-to-chip couplers can be employed for this task. To achieve full programmability and stable performance the processor needs an *electronic block for control, supervision and programming*. Control and supervision electronics are required first to track and keep the bias points of all programmed photonic components stable against environmental drifts and second, to supervise the result of any heralding operation. Programming electronics takes care of setting the bias points of all required photonic components to the values provided by the software.

All these blocks have been demonstrated separately and the main challenge is to integrate them in one or few interconnected photonic chips. To date, SOI technology appears to be the strongest technology contender. We now briefly review some relevant results reported in the literature for these blocks.

7.1.3 Quantum Sources

Quantum sources must be able to generate the required input quantum states to the programmable core. Single-photon sources are especially desired for this purpose [15], [16]. One simple alternative that can very well approximate a single-photon source and that is commonly employed in quantum key distribution systems is to strongly attenuate the light from a laser (i.e. a coherent state) [11]. In doing so, the probability that a pulse contains more than one photon decreases much faster than the probability of containing only one. Ideally, however, true single-photon emitters that produce single photons on demand are preferred. These photons must also be indistinguishable in all the relevant degrees of freedom, including central frequency, bandwidth, polarisation and spatial mode.

Reported techniques for the implementation of single-photon sources fall in two categories. The first is based on using single-emitter quantum systems, such as quantum dots and colour centres in diamond. In particular, quantum dots have been recently integrated with other on-chip photonic elements [17]. However, the difficulty in fabricating identical emitters at the nanoscale still challenges the production of highly indistinguishable photons from different sources. The second approach, which has received considerable attention, is to leverage on the non-linear properties of silicon to generate correlated photon pairs in silicon waveguide devices through spontaneous non-linear processes, namely spontaneous parametric down-conversion (SPDC) and spontaneous four wave mixing (SFWM) [16]. Since silicon is centrosymmetric, there is no second-order non-linearity; the dominant process is third-order non-linearity and SFWM is the preferred approach. In non-degenerate SFWM two photons from one bright pump laser source can be spontaneously shifted to different wavelengths (Figure 7.2(a)). These photons, known as signal (higher frequency) and idler (lower frequency) are quantum correlated. For SFWM to happen photons must fulfil both energy conservation and phase matching. Energy conservation is stated by $\omega_s + \omega_i = 2\omega_p$, where ω stands for the photon frequency and the subscripts s, i and p refer to the signal, idler and pump photons, respectively. Phase matching under waveguide propagation is defined in a similar way over the (collinear) momentum vectors, $k_s + k_i - 2k_p + 2\gamma P = 0$ where $\gamma = n_2 k_o / A$ is the non-linear parameter with n_2 the non-linear refractive index, k_o the vacuum wavenumber, A the effective mode area and P the optical pump power.

A second alternative is known as degenerate SFWM (Figure 7.2(b)). Here, two photons from two bright pump laser sources ω_{p1} and ω_{p2} are spontaneously shifted to the same wavelength $\omega_{s,i}$. Here, energy conservation dictates that $\omega_{p1} + \omega_{p2} = 2\omega_{s,i}$, while the phase matching condition is given by $2k_{s,i} + 2\gamma\,(P_1 + P_2) - k_{p1} - k_{p2} = 0$.

Both processes are ultrafast and therefore the emission of photons is quasi-instantaneous, which means that their temporal and spectral properties can be very well controlled, leading to highly indistinguishable single-photon emission.

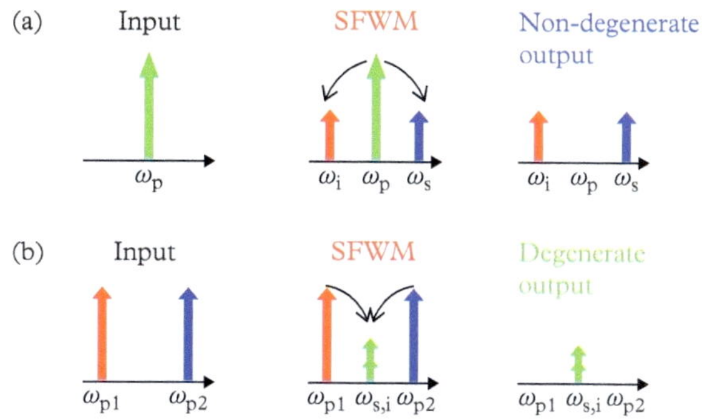

Figure 7.2 *(a) nondegenerate- and (b) degenerate-correlated photon-pair generation via SFWM for single-photon sources (after [16]).*

Non-degenerate SFWM can be exploited for single-photon heralded generation, where the detection of one of the two photons after separation by a spectral filter, heralds the existence of the other. Degenerate SFWM can be employed for indistinguishable photon pair generation.

One drawback of SFWM sources is that the photon emission is probabilistic. An ideal photon source is operated with thermal statistics so each time a pump pulse is used to seed the quantum source either no pair ($n = 0$) is produced, a single pair ($n = 1$) is produced or multiple pairs ($n > 1$) are produced. In fact, the probability to obtain n pair per pulse is given by [11]

$$p(n) = \frac{\mu^n}{(1+\mu)^{n+1}} \tag{7.1}$$

Where μ stands for the mean number or pairs created by pump pulse. In particular and under ideal conditions

$$\mu \approx \gamma^2 L_{eff}^2 P^2 \Theta^2 \tag{7.2}$$

where $L_{eff} = \left(1 - e^{-\alpha L}\right)/\alpha$ is the effective non-linear interaction length defined by the waveguide length L and the linear loss α, and $L_{eff} = \left(1 - e^{-\alpha L}\right)/\alpha$ and $\Theta = \mathrm{sinc}\,(\Delta kL)$ is the phase matching coefficient with Δk being the momentum mismatch between the four fields engaged in the process. In practice however other non-linear processes occur together with SFWM. Of concern, especially in silicon is two-photon absorption (TPA), which increases single-photon propagation loss. TPA happens when the material energy bandgap is such that it can absorb two photons to excite a valence electron. This absorption can involve two pump photons, or more importantly, one pump and one generated photon. Furthermore, excited electrons can also absorb incident photons introducing extra free-carrier absorption (FCA) losses. Under the effect of TPA and FCA, Eqn. (7.2) transforms to

$$\mu \approx \frac{\gamma^2 A^2 \Theta^2}{\alpha_2^2} \ln^2 \left(1 + \frac{\alpha_2 P L_{eff}}{A}\right) \tag{7.3}$$

where α_2 is the TPA coefficient. Importantly, Eqn. (7.3) indicates that TPA leads to saturation in the pair generation rate.

From Eqn. (7.1) we have:

$$\frac{p(1)}{p(n > 1)} = \frac{1}{\mu} \tag{7.4}$$

which means that we cannot increase the probability of getting one pair while simultaneously supressing the probability of producing multiple pairs. Eqn. (7.4) indicates that both $p(1)$ and $p(\mathrm{n} > 1)$ increase with μ, but $p(\mathrm{n} > 1)$ increases more rapidly.

The consequence is that SFWM sources must be operated in the regime given by $\mu \ll 1$, and hence $p(1) \ll 1$ and, therefore these sources do not seem to provide a clear advantage over attenuated laser-based sources. However, these sources can be employed to construct near-deterministic single-photon sources by combining the use of heralding and active optical switching in the so-called active spatial and temporal multiplexing. The basic principle is to use the knowledge of heralding photons to deterministically combine the outputs of a number of probabilistic sources, which are either spatially separated [18], [19] or different from each other in temporal modes [20]–[22]. These multiplexing techniques effectively decouple the probabilities of single- and multi-pair generation and, therefore, both probabilities increase with the number of multiplexed (spatial or temporal modes) while the ratio between them remains fixed.

Several approaches compatible with photonic integration have been reported in the literature for the implementation of non-degenerate SFWM quantum sources (Figure 7.3). These include straight and spiralled silicon nanowires [23], ring and micro-disk resonators [24], [25], slow light structures based on photonic crystal waveguides [26] and micro-ring and PhC coupled resonant optical waveguides (CROWs) [27], [28]. In short, nanowires implemented by means of 450 mn wide and 220 nm high silicon cores surrounded by SiO_2 cladding provide a small area and high index contrast that result in a very high non-linearity in the 1,550 nm region, enabling flexible SFWM over a broad bandwidth in only a few millimetres of length. In addition, they tolerate higher inaccuracies in the fabrication process. However, they require narrow bandpass filters to block the pump and usually are longer in footprint and require high peak power for the pulsed pumping signal. In micro-ring resonator-based sources pair generation scales as Q^2/R^2, where Q is the cavity quality factor and R is the cavity radius. These structures have a reduced footprint, but the sets of generated photon frequencies are restricted to the values of the cavity resonance frequencies. Finally, in slow-light SFWM sources the photon pair generation scales with S^4, where $S = n_g/n_o$ is the slow-down factor that depends on the pump field group index n_g and the bulk refractive index n_o. Xiong, Bell and Eggleton provide a detailed description of each approach [16].

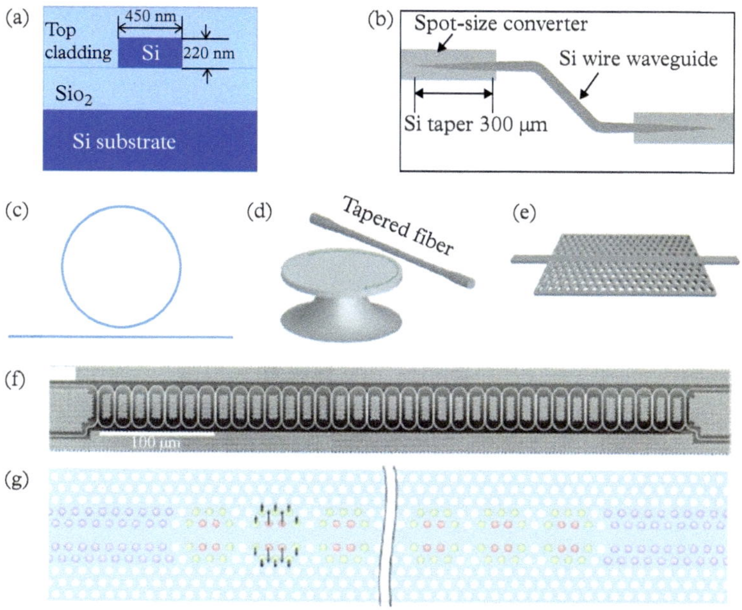

Figure 7.3 *Layouts for different implementation alternatives of SFWM single photon sources (after ([16]). (a) Cross section of a typical nanowire and (b) Nanowire with inversed tapers [16], (c) Integrated microring, (d) A microdisk [16], (e) A photonic crystal waveguide (PhCW) [26], (f) A microring CROW [27], and (g) A PhC CROW [28].*

Using data from Xiong, Bell and Eggleton [16], Table 7.1 and Table 7.2 provide a comparative analysis of these different approaches, which indicate that there is not a single structure that *outperforms* any other, but rather that, depending on the application, one is more *appropriate* than any other. A programmable photonic processor might therefore need to include more than one of these designs as high performance blocks.

7.1.4 Highly Selective Filters

Highly selective integrated optical filters are needed to supress on-chip the remaining pump light after the photon pair generation. This translates into a quite stringent selectivity requirement as the pump field is usually around 100 dB or more above the signal and idler intensities. For example, CROW filters can achieve these high rejection orders of magnitude [29]. Incoherently cascaded SOI lattice filters [30] are also capable of achieving over 100 dB suppression as recently reported by Piekarek and colleagues [30]. Harris and colleagues recently explored the combination of Bragg reflectors and ring resonators as another attractive alternative [31], [32]. Figure 7.4 illustrates this configuration where a pump field coupled into a CMOS silicon photonics chip is routed to a critically coupled tunable ring resonator, where photon pair generation via SFWM is accomplished. The pump field already attenuated by the SWFM process and the photon

Table 7.1 *Typical experimental values for performance parameter of silicon-based SFWM photon pair sources (after Xiong, Bell and Eggleton [16]).*

	Nanowires	Micro-rings	Micro-disks	PhC	Micro-ring CROWs	PhC CROWs
Non-linear coefficient ($\mathbf{W^{-1}m^{-1}}$)	300			4000	4100	9000
Device footprint ($\mathbf{mm^2}$)	5.3×10^{-3}	4.1×10^{-4}	7.5×10^{-5}	3.9×10^{-4}	-	1.8×10^{-3}
Coupled pump average power (mW)	0.18	0.0019	0.079	0.055	16	0.03
Collected photon bandwidth (GHz)	25	5.2	2.2	50	62.5	25
Brightness (pairs $\mathbf{s^{-1}mW^{-2}GHz^{-1}}$)	1.6×10^5	4.4×10^8	6.2×10^7	1.5×10^6	4.6	$8 - 8 \times 10^7$
Coincidence to accidental ratio	320	602	1386	330	10	2

Table 7.2 *Qualitative summary of performance metrics for different types of SFWM photon pair sources. More circles mean better performance (after Xiong, Bell and Eggleton [16]).*

	Nanowires	Micro-rings	Micro-disks	PhC	Micro-ring CROWs	PhC CROWs
Fabrication	ooo	o	o	o	o	o
Non-linearity	o	ooo	ooo	ooo	ooo	ooo
Footprint	o	ooo	ooo	ooo	o	o
Stability	ooo	o	o	oo	o	oo
Power	oo	ooo	ooo	ooo	o	ooo
Package	ooo	o	o	oo	oo	oo
Tuning bandwidth	ooo	oo	oo	o	oo	o

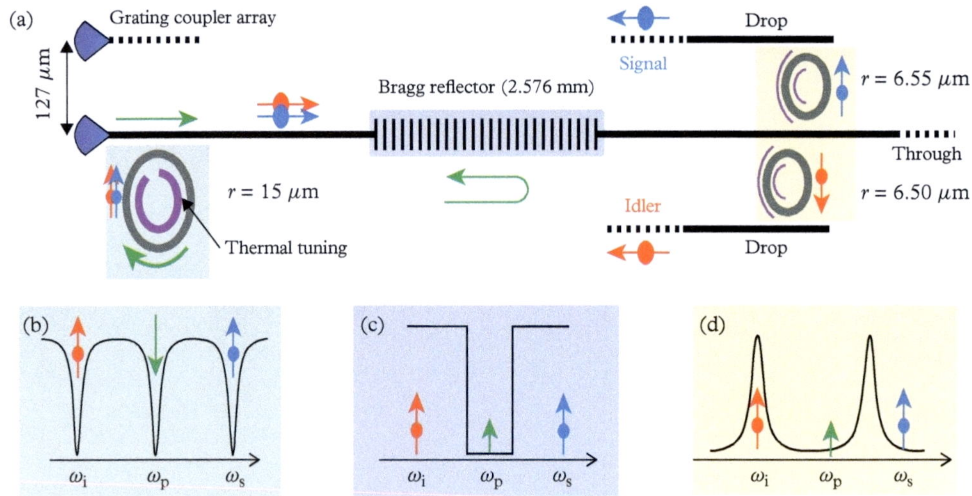

Figure 7.4 *(a) Layout of a photonic integrated circuit composed of a high-Q thermally tunable ring for efficient pair generation by spontaneous four-wave mixing, followed by a highly selective integrated photonic filter composed of DBR for pump rejection and the add-drop ring-resonator filters for the demultiplexing of signal and idler photons (after [32]). (b) Schematic transmission spectrum of the first ring around the pump wavelength ω_p.(c) Schematic transmission spectrum of the DBR with the stop band overlapping with the pump wavelength ω_p. (d) Add-drop filter spectrum tuned to route idler photons to the drop port.*

pairs then pass through a distributed Bragg reflector filter that provides a 65 dB rejection of the pump power. Subsequent demultiplexing of the signal and idler photons using tunable ring resonators provides an additional attenuation (over 35 dB) of the pump.

As pointed out by Silverstone and colleagues [11], high-extinction optical filters might require a separate placement as a high-performance block in a different chip from that of the linear optics core to avoid the limitation in their extinction ratio due to the possible leaking of the pump around the filter.

7.1.5 N × N Programmable Unitary Transformers

The core of the programmable integrated quantum photonics processor is the linear optics unitary transformer. Its implementation requires a multiple input/multiple output programmable integrated waveguide core. The main alternatives are multi-port integrated interferometers, considered in Chapter 4, and integrated waveguide mesh circuits, considered in Chapter 5. Both architectures can implement arbitrary unitary transformations. While multi-port interferometers are built to implement either the triangular or rectangular multi-port architectures, integrated waveguide meshes can be programmed to emulate any of the two. In the first case, promising results have been obtained, especially in silicon photonics. For instance, a large core with up to eighty-eight MZIs and the capability of providing twenty-six mode connections has been recently reported and the circuit micrograph is shown in Figure 7.5(a)[12]. Carolan

(a)

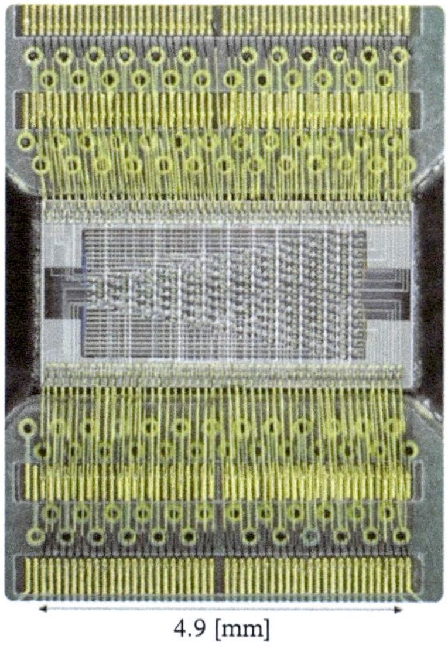

(b)

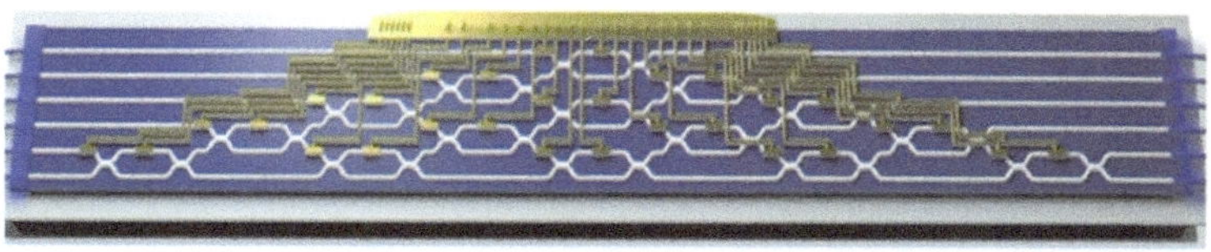

Figure 7.5 *(a) Optical micrograph of 26-mode, 88-MZI Silicon photonics programmable core (after[12]). (b) Germanium-doped glass six-mode, 15-MZI Silica on Silicon programmable core (after [10]).*

and colleagues [10] demonstrated a $U(6)$ unitary transformer based on a germanium-doped Silica PLC (Figure 7.5(b)), where phase shifters are implemented by thermal modulators. Both circuits have been employed in the demonstration of several quantum information processing tasks, as is highlighted later in this chapter.

In the second case, and as discussed in Chapter 5, an integrated waveguide mesh can be suitably programmed to emulate the operation of either triangular or square multi-port interferometers [33]. For instance, Figure 7.6 shows an emulation of a square interferometer implementing a $U(9)$ transformation by means of a hexagonal integrated waveguide mesh.

The reader can find a detailed treatment on the implementation of programmable integrated linear optics unitary and non-unitary transformers in Chapter 4 and Chapter 5.

7.1.6 Single Photon Detectors

The detection of single photons is the final, but essential, step in the programmable integrated quantum chip. Achieving compatibility with the wavelength region around

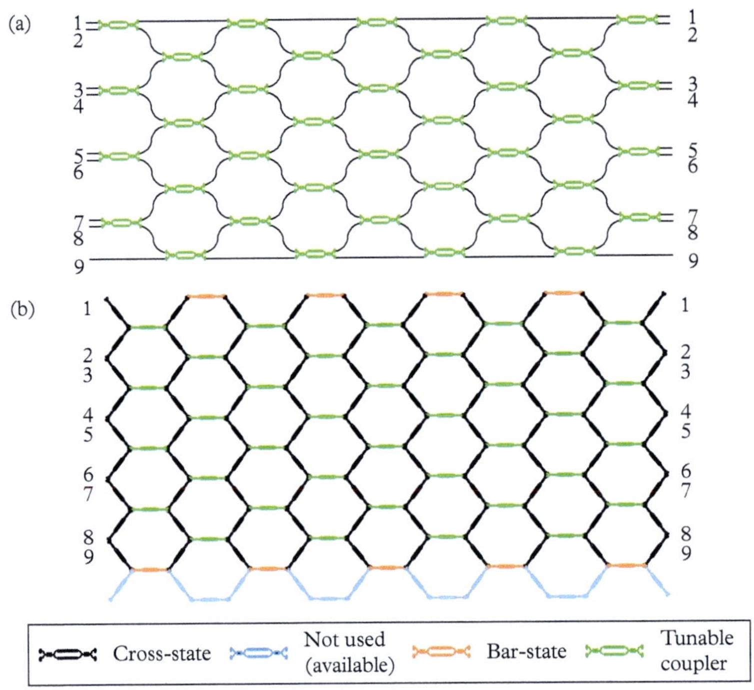

Figure 7.6 *(a) Rectangular arrangement of a 9x9 interferometer as proposed in [clements]. (b) Equivalent implementation using the hexagonal waveguide mesh (after [33]).*

1550 nm means that silicon avalanche photodiodes (APDs) must be discarded as their region of maximum absorption and APD sensitivity is around 800 nm. Single-photon detection can be performed off-chip using III–V InGaAs single-photon APDs and superconducting nanowire single-photon detectors (SNSPDs). The main problem with off-chip coupling is that it adds extra coupling losses and might require enhanced operating complexity in terms of wiring and time delay matching. Monolithic integration is therefore desirable to avoid these limitations and, furthermore, to allow path and gate reconfiguration in response to the detection of one or more heralded photons. As previously mentioned, this feature is critical for scalable quantum optic integrated circuits.

SNSPDs operate at cryogenic temperatures but are sensitive to a wide range of wavelengths, including 1550 nm [11], [12]. Several materials have been explored, mostly based on polycrystaline films of niobium nitride (NbN) [34] and amorphous semiconductors such as WSi [35], MoSi [36] and MoGe [37]. First steps towards monolithic integration have been reported based on the patterning of detectors on top of waveguides by interposing NbN film. For example, in a proof of concept experiment, Harris and colleagues [38] fabricated a set of 10 NbN SNSPDs, each one on its own Si_3N_4 membrane. After fabrication each SNSPD was evaluated in a cryostat and selected for high performance before being removed from the Si_3N_4 substrate and aligned and placed on top of a silicon waveguide in the quantum circuit. The detectors were integrated with beamsplitters to check their correct operation by means of second-order correlation function measurements.

So far, no waveguide-coupled single-photon detector has been demonstrated at or near room temperature. However, compact integrated optics can be fitted into a cryostat, including also the possibility of incorporating CMOS electronic control circuitry. The downside of this approach resides in the many challenges posed by this so-called *cold integrated photonic* approach [11], as all the components must operate at low temperatures, including thermal phase shifters and modulators, and also might require special packaging techniques.

7.2 Quantum Logic and Computing

Operations required in quantum computing can be implemented using two main approaches [39]. In the first, known as a *circuit model*, single or multi-qubit gates operate sequentially one after the other on qubits. In the second, known as a *cluster state* or *measurement-based model*, a large entangled state is first created with the aim of using it as a resource state and then single-qubit gates are performed to encode the processing.

Linear optics quantum gates acting over two qubits operate probabilistically and therefore the successful operation of the gate must be heralded by means of auxiliary or *ancillary* photons [4]. This means that even the simplest circuits must include extra input/output ports for the injection and measurement of these ancillary photons. Early works demonstrated the operation of simple one- and two-qubit non-programmable gates, e.g. the Hadamard, Pauli X, Y, Z and CNOT gates with no heralding modes [7], [40], [41]. For example, Figure 7.7 shows the layout and characterisation results of CNOT and CNOT plus Hadamard gates (i.e. 2×2 directional couplers) implemented using a silica-on-silicon chip.

Qubit encoding is usually performed by means of a *dual rail* configuration [3], [7], where two waveguides are employed. In this configuration, the state $|\mathbf{0}\rangle$ is represented by the presence of a photon in the upper waveguide and no photon in the lower waveguide, while the state $|\mathbf{1}\rangle$ is represented by the presence of a photon in the lower waveguide and no photon in the upper waveguide (Figure 7.7(b)). The same figure also provides the column vector description of each state. Single-qubit gates can then be described by means of 2×2 unitary matrices, while two-qubit gates will be represented by 4×4 unitary transformations. For instance, and referring to Figure 7.7(a) the ports C_0 and C_1 are used to encode the control bit, while ports T_0 and T_1 encode the signal qubit.

Figure 7.8(a)–(c) shows the diagram, circuit layout and measurement results of a heralded two-qubit logic designed to be scalable. The gate is implemented by programming part of the fifteen-MZI silica on silicon programmable core reported in [10]. The gate receives four input photons, where two are employed to encode the two qubits and the other two are employed as ancillas. Qubit inputs are labelled in Figure 7.8(b) as (c_0, c_1) and (t_0, t_1) respectively, while ancilla input ports are labelled by $|\mathbf{1}\rangle$. The correct

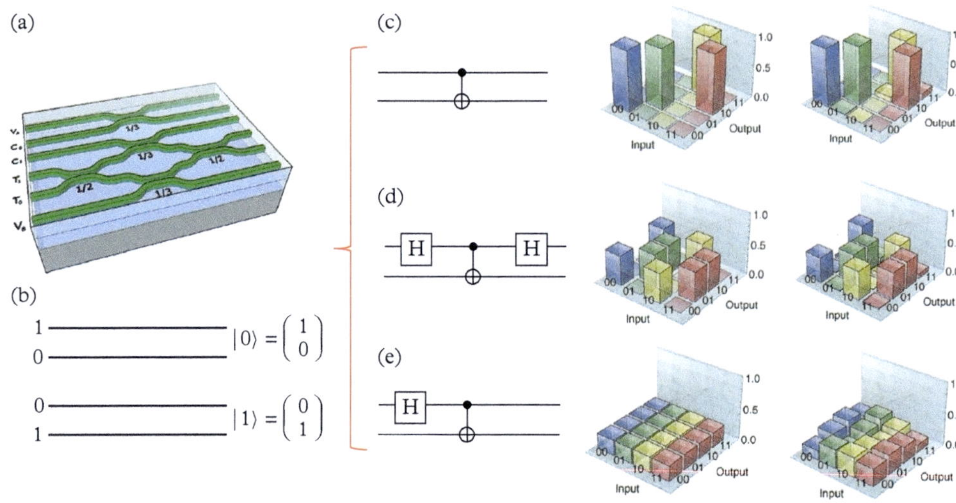

Figure 7.7 *(a) Layout of a Silica on silicon Chip reported in [7] for the implementation of a C-NOT quantum gate. (b) Scheme for the dual rail encoding of qubit input states. (c)-(e) Ideal and experimental results for different integrated quantum circuits built using the configuration in (a). In (d) and (e) the C-NOT gate is combined with several Hadamard gates implemented using 2x2 directional couplers.*

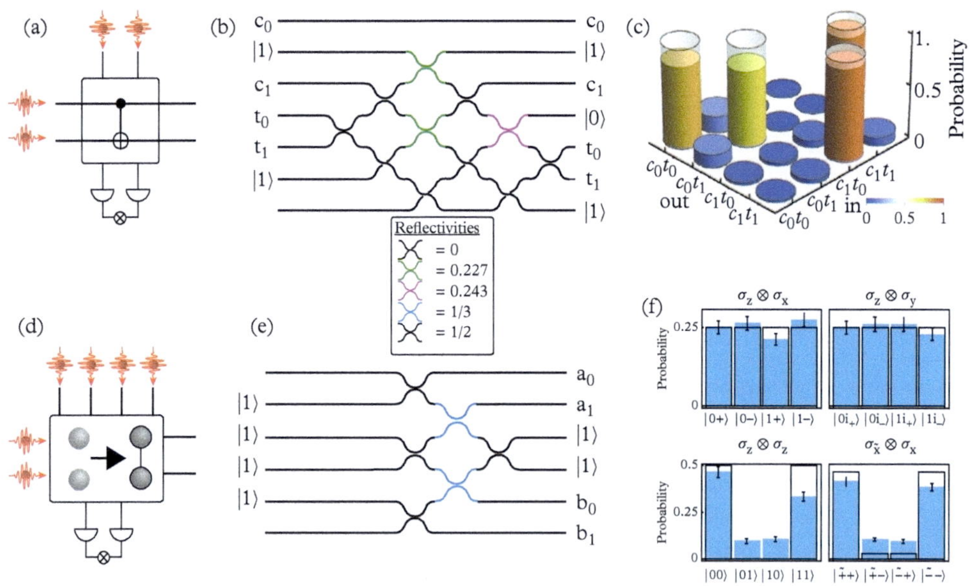

Figure 7.8 *7.8 (a)-(c) Diagram, circuit layout and measurement results of a heralded 2-qubit CNOT logic gate implemented using the programmable circuit reported in [10]. Values of the programmed MZIs are shown by a colour code in the inset. 7.8 (e)-(f) Diagram, circuit layout and measurement results of an entangled 2 qubit Bell state generator after reprogramming the same circuit.*

gate operation is guaranteed when a photon is detected in each of the two heralding modes signals, identified by the output ports marked as $|\mathbf{1}\rangle$. When this happens the gate implements a CNOT operation between the two photonic qubits and the output states are obtained from output ports (c_0,c_1) and (t_0,t_1) respectively. The measured results for the gate operation shown in Figure 7.8(c) do clearly reproduce the 4×4 linear transformation matrix expected for a CNOT gate with a reported fidelity of 0.93 ± 0.003 averaged over the different computational inputs.

Figure 7.8(d)–(f) shows an example of how this same circuit can be reprogrammed to implement a cluster state model operation. In this case it is a four-photon scheme that generates the maximally entangled states of two photonic qubits upon the successful detection of another two ancilla photons. Figure 7.8(d)–(e) show the circuit diagram and layout respectively, while Figure 7.8(f) provides the measured results when common and different bases are employed featuring a mean statistical fidelity of 0.966 ± 0.004. As expected, in the first case the measurements are correlated featuring an entanglement test measurement outcome of $0.673 \pm 0.031 > 0.5$, whereas in the second case, they are clearly uncorrelated.

As a final example we consider a large-scale silicon photonic circuit that realises a fully programmable two-qubit quantum processor [42]. The circuit is based on the implementation of a universal two-qubit 4×4 unitary $U \in SU(4)$ by means of a four-operator linear combination

$$U = \sum_{i=0}^{3} \alpha_i A_i \otimes B_i = \sum_{i=0}^{3} \alpha_i (P_1 \sigma_i Q_1) \otimes (P_2 \sigma_i Q_2) \tag{7.5}$$

where P and Q are single- (2×2) qubit gates, σ_i are the identity and the Pauli gates ($\sigma_0 = I; \sigma_1 = X; \sigma_2 = Y; \sigma_3 = z$) and α_i complex coefficients satisfying $\sum_i |\alpha_i|^2 = 1$. Figure 7.9(a) shows the layout of a probabilistic implementation of the transformation, $\sum_{i=0}^{k=2^n} \alpha_i \mathbf{V}_i$, which is implemented when all the n control bits are in the $|\mathbf{0}\rangle$ state. U_{LC} is a unitary matrix where the first row is given by $\{\alpha_0, \alpha_1, \rightleftharpoons \alpha_{n-1}\}$ and the rest are obtained in order to make U_{LC} unitary. For two-qubit circuits $n = 2$ and the success probability is $\frac{1}{4}$. Figure 7.9(b) shows the schematic of the silicon photonic chip, which is operated in combination with an external electrical control, laser input and fibre-coupled superconducting detectors. The chip footprint is 7.1 mm × 1.9 mm and consists of four spiral-waveguide SFWM photon-pair sources, four laser pump rejection filters, eighty-two multi-mode interferometer (MMI) beamsplitters and fifty-eight simultaneously running thermo-optic phase shifters. The device includes five functional parts. The first, third and fourth implement the required given $SU(4)$ operation, with the first part encoding the linear-combination coefficients α_i, the third implementing linear terms A_i and B_i, and the fourth realising the linear combination of terms $A_i \otimes B_i$ together with post-selection. The second part prepares arbitrary separable two-qubit states $|\varphi_{in}\rangle = |\varphi_1\rangle \otimes |\varphi_2\rangle$ as the input. The fifth part rotates the output state so that it can be measured at desired basis. The circuit was programmed to implement ninety-eight different two-qubit quantum logic gates and its operation verified by means of

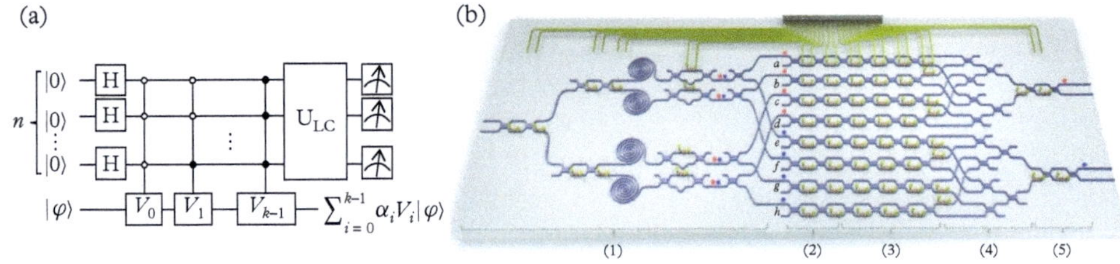

(c)

Gate	Fidelity (%)
CNOT	98.85 ± 0.06
CZ	96.90 ± 0.17
CH	97.57 ± 0.07
SWAP	95.33 ± 0.24
iSWAP	94.45 ± 0.27
$\sqrt{\text{SWAP}}$	92.41 ± 0.33

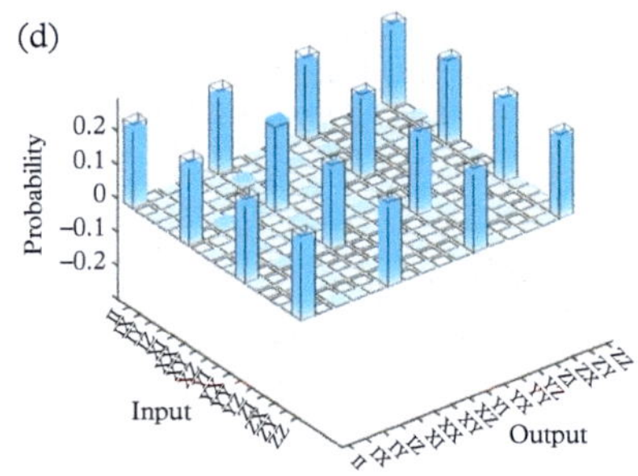

Figure 7.9 *(a) Layout of a probabilistic implementation of a linear combiner of operators. The linear combination is implemented when all n control qubits are measured to be 0. The success probability is 1/k. (b) Schematic of the silicon photonic chip, which is operated in combination with a external electrical control, laser input and fibre-coupled superconducting detectors. The chip details can be found in the text. (c) Fidelity results for the implementation of several SU(4) gates. (d) Tomography map for the implementation of the SWAP transformation (after [42]).*

tomography. Figure 7.9(c) shows the fidelity results for the implementation of several of these gates, while Figure 7.9(d) the measured real part of the tomography corresponding to the SWAP gate. Further programming of the chip allowed for the experimental demonstration of a two-qubit quantum approximate optimisation algorithm and efficient simulation of Szegedy directed quantum walks.

Linear transformations corresponding to quantum logic gates can also be implemented using waveguide mesh structures. For example, Pérez and colleagues [33] report a variety of 2×2, 3×3 and 4×4 configurations implemented over the same silicon photonic waveguide mesh configuration. Figure 7.10 shows the experimental results for different 4×4 linear transformations employed in quantum computation implemented over the same seven-cell waveguide mesh structure reported in Chapter 5. In this case, the waveguide mesh is programmed to emulate a 4×4 rectangular interferometer and the internal phase shifters are tailored in each case to obtain the desired transformation. In addition to the inherent advantage brought by programmability in terms of sharing the same hardware architecture, the use of reconfigurable waveguide meshes also allows leverage on unused resources (phase shifters) to implement, for example, extra connections required to enable input and output ports to enable flexible heralding operations. Furthermore, these connections enable *dynamic heralding* since they can be reprogrammed at the same time as when the main linear transformation is changed.

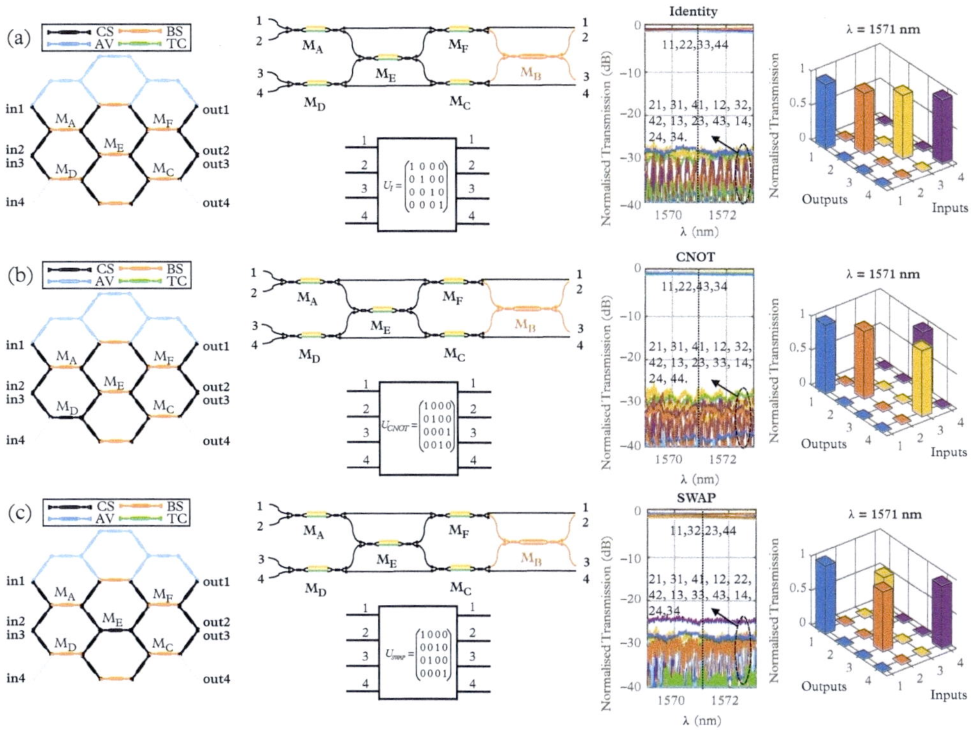

Figure 7.10 *Examples of 4 x 4 linear transformations for quantum gates implemented over the same 7-cell hexagonal waveguide mesh structure reported in Chapter 5, by programing its internal phase shifters (a) Identity transformation, (b) C-NOT transformation, (c) SWAP transformation. Column1: 7-cell configuration (CS = MZI in cross state, BS = MZI in bar state, TC = MZI in tunable coupler state, AV = MZI not employed), Column 2: circuit layout of the implemented interferometer, Column 3: Spectral measurement of all input/output port connections. Column 4: Normalized Bar diagram of the resulting unitary matrix for λ = 1571nm. (After [33]).*

7.3 Quantum Transport

Quantum transport (QT) phenomena are behind several problems of interest in different areas of physics such as solid state and biological systems, quantum computing and quantum search. In these scenarios environmental noise and disorder play critical roles that affect quantum particle and wave transport. For their simulation, understanding the role that static (time-invariant) and dynamic (time-varying) disorders play in the transport of coherent particles is of particular interest as it enables studies of fundamentally interesting and potentially useful QT phenomena.

Integrated optics is a particularly interesting platform for implementing QT simulation circuits. Both static and dynamic disorders have been introduced by fabricating circuits with random parameter variations or post-processing. Furthermore, linear photonic quantum walks along discrete lattice sites can be implemented by means of waveguides and beamsplitters in programmable nanophotonic circuits. In a recent experiment [43] Harris and colleagues reported a complex QT simulator based on a silicon photonics processor shown in Figure 7.11(a). The processor is composed of eighty-eight MZIs, twenty-six input modes, twenty-six output modes, 176 phase shifters and is capable of implement up to $U(9)$ linear transformations. The inset in Figure 7.11(a) shows the detail of one MZI including its two independent phase shifters. Figure 7.11(b) provides a circuit layout implemented using the silicon chip that includes the cross sites (black points) implemented via the MZIs and the connections between them implemented by means of integrated waveguides.

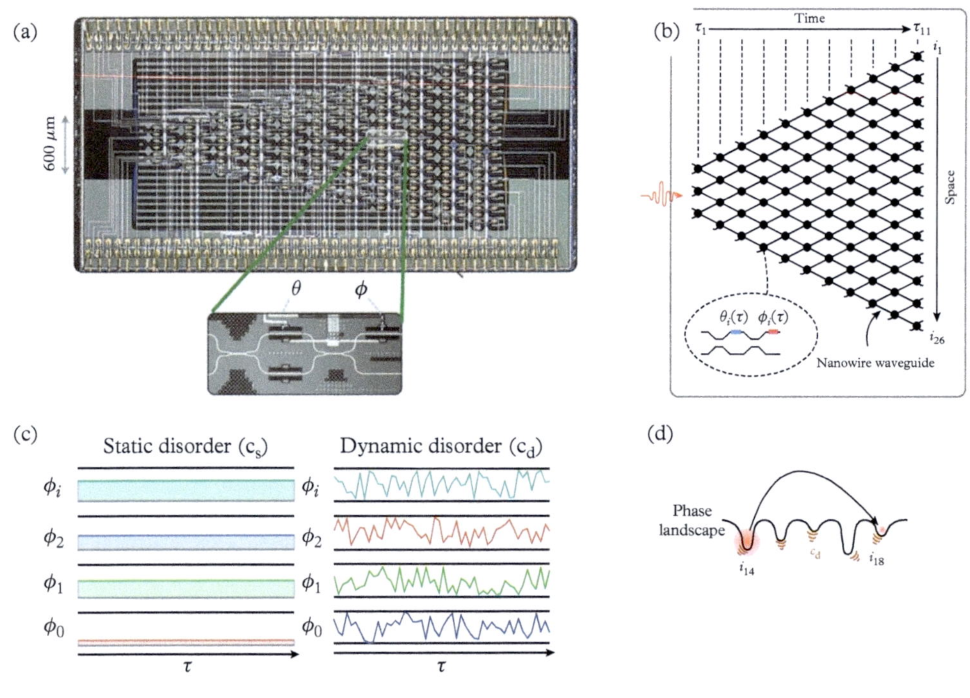

Figure 7.11 *(a) Silicon programmable processor reported in [43] to simulate discrete-site quantum transport experiments. The inset shows the layout and the internal and external phase shifters of a MZI unit. (b) Layout of the one photon quantum walker circuit implemented by the processor and coordinate system definition. Time (τ) is defined from left to right. Space (i) is defined from top to bottom. Dark lines are silicon nanowire waveguides; circles are programmable MZIs. (c) Phase vs time evolution graphs describing the static and dynamic disorders in the programmable circuit. (d) A stuck transport of quantum particles in strong, statically disordered system to external sites can be optimized by introducing dynamic disorder.*

The circuit operation features an input state at the left of the circuit that undergoes a quantum walk on a one-dimensional chain as it propagates in time to the right. Through the internal phase shifters of the MZIs (θ_i) it is possible to program the splitting ratios of the sites and therefore explore different QT scenarios implemented over a set of discrete sites. Harris and colleagues [43] explored a single quantum walker subject to both static and dynamic phase disorders. Initially, each MZI was set to implement a 50:50 splitting ratio but the external phase shifters (ϕ_i) were programmed to exhibit either static or dynamic phase variations (Figure 7.11(c)) or a combination of both. With this configuration the circuit implements a discrete-time balanced coin 1D quantum walk. If, for example, a static-only phase variation with sufficient dynamic range is implemented, then photons can be confined to local vicinity in similarity with the Anderson localisation (see i_{14} point corresponding to the output of waveguide 14 in the phase landscape of Figure 7.11(d)). On the other extreme, if strong dynamic phase variations are incorporated, then ballistic photon diffusion is enabled. The ratio between static and dynamic disorder can be optimised to favour a given ballistic diffusion to a given output waveguide (Figure 7.11(d)). The implemented programmable photonic circuit enabled up to 64,000 different and unique parameter settings of static and dynamic disorder, which correspond to different QT experiments.

7.4 Boson Sampling

The practical implementation of universal quantum computers is still far from feasible due to several reasons, one of the most important being the need to isolate the hardware from decoherence effects of the environment. Until these hurdles are overcome it is important to demonstrate intermediate schemes that can prove the enhancement provided by quantum computation under more relaxed experimental conditions. One of these intermediate scenarios is based on the computational complexity of simulating the output distribution of bosons propagating in linear optical interferometers. This task, known as *boson sampling* [10], has recently been proposed as a specific quantum computation that is more efficient than its classical counterpart. The distribution of bosons that have been subject to a unitary transformation is thought to be exponentially hard to sample from classically, but photonics provides affordable experimental conditions for boson sampling as sources of indistinguishable photons are available and linear transformations can be implemented by means of multi-port interferometers. An integrated platform adds the value of mode scalability and controlled environmental conditions to avoid decoherence. On top of this, programmability allows for the rapid reconfiguration of the circuit to sample from a user-defined operation.

Figure 7.12(a) shows the model of a boson sampling circuit [44]. It samples the output distribution of a multi-particle bosonic state $|\Psi_{out}\rangle$ prepared from an initial state $|\mathbf{I}\rangle$ and a linear transformation U. The initial Fock state is $|\mathbf{I}\rangle = |I_1, I_2 \rightleftharpoons I_M\rangle \propto \prod_{k=1}^{M} \left(\mathbf{a}_k^\dagger\right)^{T_k} |$

0⟩ that describes $N = \sum_{k=1}^{M} I_k$ bosons distributed in M modes. The output state is generated according to the linear map between input and output creation operators $\mathbf{a}_k^\dagger = \sum_{j=1}^{M} u_{kj}\mathbf{b}_j^\dagger$, where u_{ij} are the coefficients of the $M \times M$ matrix representing U. After sampling, the boson in each of the M output modes are counted and the probability of a particular outcome $|\,\mathbf{O}\rangle = |\,O_1, O_2 \rightleftharpoons O_M\rangle$ is given by

$$P(\mathbf{O}/\mathbf{I}) = \left|\left\langle \mathbf{O}\middle|\mathbf{I}\right\rangle\right|^2 = \frac{\left|\mathrm{Per}\left(U^{(\mathbf{O},\mathbf{I})}\right)\right|^2}{\prod_{k=1}^{M} O_k! \prod_{r=1}^{M} I_r!} \tag{7.6}$$

where $U^{(\mathbf{O},\mathbf{I})}$ is a $N \times N$ submatrix obtained by keeping O_k copies of the kth column and I_r copies of the rth row of$U$. Per(A) stands for the *permanent* of the matrix A and expresses the wave function of identical bosons, which are symmetric under exchange. The definition of permanent is similar to that of the determinant but without negative signs for odd permutations of matrix elements. Interestingly, despite the similarity in the definitions it turns out that the determinant and the permanent are quite different in terms of their computational complexity, as the first can be calculated in polynomial time while the second has been proved to be computationally #P-hard belonging to a class of intractable problems that include the more popular NP-hard problems.

Several boson sampling experiments using integrated photonics have reported in the literature for input Fock distributions ranging from one to four photons [10], [44], [45]. These low number count input populations allow comparing experimental results with those predicted by Eqn. (7.6) at a reasonable classical computational cost for permanents. These results provide good evidence that boson sampling provides an efficient and fast procedure for the computation of the output distributions that is expected to scale up with the number of input photons and ports. Figures 7.12(b)–(c) show the basic features of an integrated boson sampler reported by Crespi and colleagues [45]. The linear transformation is provided by a 5 × 5 3D integrated triangular interferometer where waveguides were fabricated using the femtosecond laser micromachining technique described in Section 4.4. In this case, the beamsplitters are implemented by TDCs, the operation of which has been described in Section 2.5.1. Tunability of the external phase shifts is achieved by deforming the S-bent waveguides at the input of each directional coupler in order to stretch the optical path (Figure 7.12(b)). Coupling constant tunability is performed by rotating one arm of the directional coupler out of the main circuit plane (Figure 7.12(b)). In this way, waveguide separation is changed without affecting the interaction length and therefore avoiding additional (undesired) phase shifting.

The chip was used in single-, two- and three-photon experiments in combination with several external elements already described in Section 7.1, e.g. input photon states to the interferometer where prepared using SPDC sources and output photons from the interferometer ports detected using avalanche photodiodes. Figures 7.12(d)–(e) illustrate the results obtained for the two-photon input case. Figure 7.12(d) represents the expected theoretical results according to Eqn. (7.6), while the measured results are displayed in Figure 7.12(e).

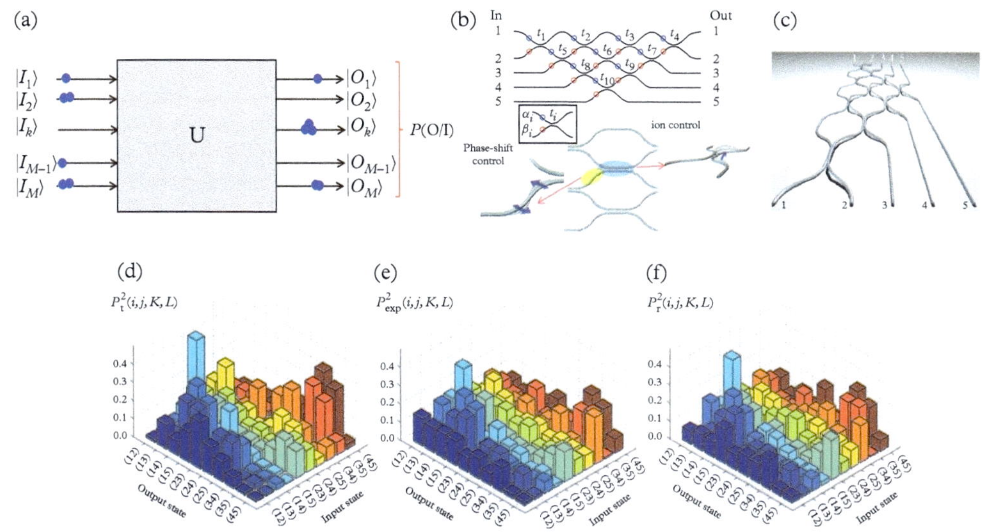

Figure 7.12 *(a) Schematic layout of the operation of a Boson Sampler. (b)-(f) Layouts (b),(c) and Results (d)-(f) for an integrated boson sampler implemented using a 3D integrated triangular interferometer (after [45]). (b) Triangular 5x5 interferometer layout and description of the phase and coupling constant tuning mechanisms. (c) Picture of the implemented interferometer. (d) Theoretical results predicted by (7.6) for an experiment with two input photon states, (e) Experimental results for the same case (f) Recovered unitary transformation U obtained from the experimental results.*

To quantify the agreement between theory and experiments in boson sampling experiments it is customary to calculate the so-called variation distance between two generic probability distributions p_k andq_k, which is defined as $d = (1/2)\sum_k | p_k - q_k |$. In this case, $d_{\exp,t} = 0.221 \pm 0.013$, which represents a quite good degree of agreement. Figure 7.12(f) depicts the reconstructed unitary transformation U obtained from the experimental data. Compared to the theoretical transformation, the degree of similarity was $d_{\exp,t} = 0.103 \pm 0.013$. The experiments up to three-photon input state confirmed the validity of Eqn. (7.6), but further work needs to be done to scale up the experiments to provide evidence of the advantage of quantum processing to simulate hard-to-compute problems.

Boson samplers have also been reported using planar integrated circuits [10], [44]. Spring and colleagues [44] reported a silica-on-silicon integrated photonic circuit fabricated using ultraviolet laser writing consisting of $M = 6$ input and output modes, as shown in Figure 7.13(a). This chip was employed in experiments featuring input states with three and four photons. For the three-photon experiments, photons were input in ports 2, 3 and 5 and the implemented moduli and phase components of the linear transformation are shown in Figure 7.13(b) and Figure 7.13 (c), respectively. Experimental versus theoretical results for the three photon experiments shown in Figure

7.13 (d) feature an excellent degree of similarity with $d_{\text{exp},t} = 0.094 \pm 0.013$, while experiments for the case of four-photons provided a value of $d_{\text{exp},t} = 0.097 \pm 0.004$.

While Spring and colleagues [44] and Crespi and colleagues [45] focused on fixed unitary transformations, experiments by Carolan and colleagues [10] reported a reconfigurable chip featuring a triangular multi-port interferometer shown in Figure 7.13(e) that was programmed to implement 100 different boson sampling routines. The programmed configurations were randomly chosen for the Haar measure implemented by direct parametrisation of the phase shifters in the triangular interferometer following the random distributions for the beamsplitter reflectivities and phase shifter values (Figure 7.13(f)). The experiments conducted for input number states from one to three photons. For the three-photon input case, detection events were computed for each of the twenty collision-free ways in which the photons can exit the six output ports of the device. From

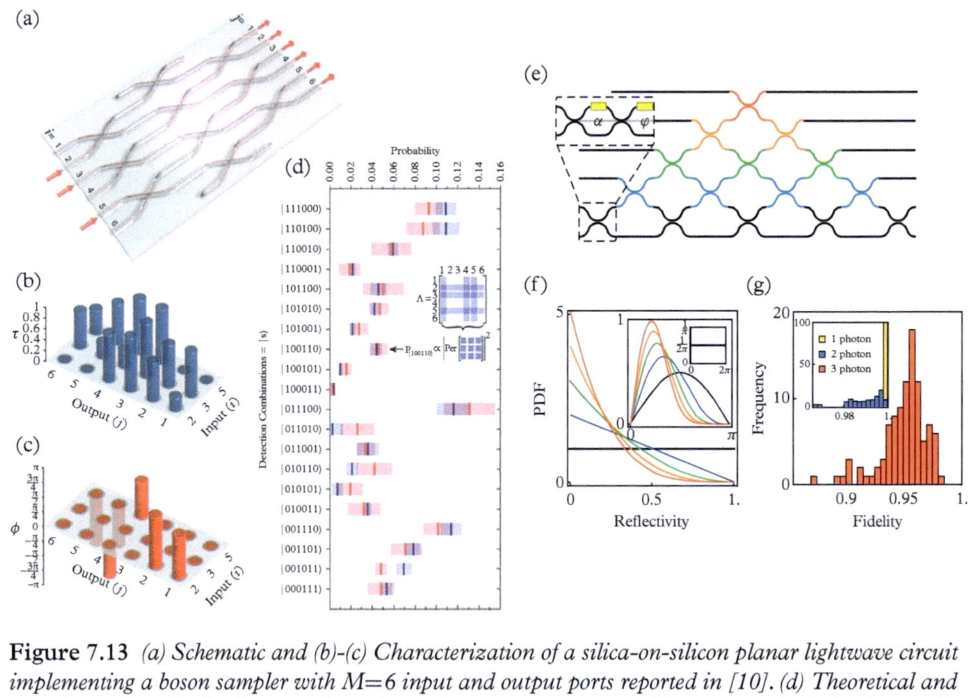

Figure 7.13 *(a) Schematic and (b)-(c) Characterization of a silica-on-silicon planar lightwave circuit implementing a boson sampler with M=6 input and output ports reported in [10]. (d) Theoretical and experimental results for the case of a 3 photon input state* | 010110⟩. *The measured relative frequencies* P^{exp} *of outcomes in which the photons were detected in distinct modes are shown in red, with statistical variations in counts shown by the red shaded bars. The theoretical distributions* P^{th} *(blue) are obtained from the permanents of sub-matrices constructed from the full transformation U, as depicted in the inset. The blue error bars arise from uncertainties in the characterization of U. (e) Schematic of a programmable bosn sampler based on a planar lightwave triangular interferometer reconfigured for implementing over 100 different Haar random unitaries directly implemented by choosing beam splitter reflectivities, and phase-shifter values from the probability density functions shown in (f). (g) Average fidelity results for 3 photon-experiments (results for 1 and 2 photons shown in the inset). After [44].*

these, statistics based on the computation of matrix permanents were performed and a final fidelity histogram provided, which is shown in Figure 7.13(g) featuring an average fidelity [46]$F = \sum_k \sqrt{p_k q_k}$ value of $F = 0.950 \pm 0.02$, demonstrating the remarkable chip performance over many different configurations.

7.5 Complex Hadamard Transformations

Further to the implementation of quantum logic gates, which in most of the cases need to be based on a combination of either 2×2 Pauli matrices and/or controlled-U gates, there is also the possibility of employing programmable photonic circuits to implement other linear operations. In particular, complex Hadamard transforms are of interest in the field of quantum computation. The Fourier transform and the ${F_6}^{(2)}\,(\theta_1,\theta_2)$ (a two-parameter set of six-dimensional matrices) are examples of this general class. Several studies [10], [33] have recently reported integrated configurations capable of performing these operations, although at a limited scale since issues related to losses and scalability pointed out in Chapter 6 need still a considerable amount of work to be, at least, partially overcome.

7.6 Chapter Summary

Quantum information processing requires compact and low-footprint devices to overcome the limitations that environmental conditions impose over decoherence. Integrated photonics and, more particularly, integrated programmable photonics enable optimum solutions for the implementation of large-scale processors provided that several demanding technical requirements affecting losses and scalability are overcome.

This chapter provided an introduction to the applications of programmable photonics to quantum information systems. The possibility of performing arbitrary and reconfigurable unitary transforms using multi-port interferometers and/or reconfigurable waveguide meshes that emulate these interferometers opens the path to the potential implementation of compact cores for generic quantum processors based on linear optics operations. While fundamental, the linear integrated photonic core is only one of the blocks required to construct an integrated generic quantum processor. Other high-performance components, or blocks, e.g. single-photon sources, single-photon detectors and very high rejection filters, are needed.

It introduced the general framework of a programmable quantum photonic system integrated on a chip and briefly described the role and state of the art of these high-performance blocks before it described the application of reconfigurable integrated optic circuits to several topics, including linear optical quantum gates, QT simulation, boson sampling, and complex Hadamard and quantum Fourier transforms. The experimental results reported so far are still quite modest, and the research area is still in its infancy, but clear progress has been made over the last few years and large-scale systems are expected to be demonstrated in the near future as propagation losses are reduced within the photonic core and high-performance external blocks are more efficiently coupled. Most probably before this is achieved programmable photonics will have been successfully introduced and employed in classical applications, some of which are reviewed in the next two chapters.

REFERENCES

[1] R. Loudon, *The Quantum Theory of Light*, 3rd edn, Oxford University Press, 2000.

[2] C.C. Gerry and P.L. Knight, *Introductory Quantum Optics*, Cambridge University Press, 2005.

[3] M.A. Nielsen and I.L. Chuang, *Quantum Computation and Quantum Information*, Cambridge University Press, 2000.

[4] E. Knill, R. Laflamme and G.J. Milburn, A Scheme for Efficient Quantum Computation with Linear Optics. (2001) *Nature*, 409, 46–52.

[5] J.L. O'Brien, A. Furusawa and J. Vučković, Photonic Quantum Technologies. (2009) *Nat Photonics* 3, 687–95.

[6] P. Kok, et al., Linear optical quantum computing with photonic qubits. (2007) *Rev Mod Phys* 79, 135–74.

[7] A. Politi, et al., Silica-on-silicon waveguide quantum circuits. (2008) *Science* 320, 646–49.

[8] A. Politi, J. Matthews, M.G. Thompson and J.L. O'Brien, Integrated Quantum Photonics. (2009) *IEEE J Select Top Quantum Electron* 15, 1673–84.

[9] M.G. Thompson, A. Politi, J.C. Matthews and J.L. O'Brien, Integrated Waveguide Circuits for Optical Quantum Computing. (2011) *IET Circuits Devices Syst* 5, 94–102.

[10] J. Carolan, et al., Universal Linear Optics. (2015) *Science* 349, 711–16.

[11] J.W. Silverstone, D. Bonneau, J. L. O'Brien and M.G. Thompson, Silicon Quantum Photonics. (2016) *IEEE J Select Top Quantum Electron* 22, 390–402.

[12] N. Harris, et al., Large-scale Quantum Photonic Circuits in Silicon. (2016) *Nanophotonics* 5, 456–68.

[13] M. Poot, et al., Design and Characterization of Integrated Components for SiN Photonic Quantum Circuits. (2016) *Opt Express* 24, 6843–60.

[14] M. F. Saleh, G. Di Giuseppe, B.E.A. Saleh and M.C. Teich, Modal and Polarization Qubits in Ti:$LiNbO_3$ Photonic Circuits for a Universal Quantum Logic Gate. (2010) *Opt Express* 18, 20475–90.

[15] M.D. Eisaman, J. Fan, A. Migdall and S.V. Polyakov, Invited Review Article: Single-photon Sources and Detectors. (2011) *Rev Sci Instrum* 82, 071101.

[16] C. Xiong, B. Bell and B.J. Eggleton, CMOS-compatible Photonic Devices for Single-photon Generation. (2016) *Nanophotonics*, 5 427–39.

[17] G. Reithmaier, et al., On-chip Generation, Routing, and Detection of Resonance Fluorescence. (2015) *Nano Lett* 15, 5208–13.

[18] A.L. Migdall, D. Branning and S. Castelletto, Tailoring Single-photon and Multiphoton Probabilities of a Single Photon-on-demand Source. (2002) *Phys Rev A* 66, 053805.

[19] J.H. Shapiro and F.N.C. Wong, On-demand Single-photon Generation Using a Modular Array of Parametric Downconverters with Electro-optic Polarization Controls. (2007) *Opt Lett* 32, 2698–2700.

[20] T. Pittman, B. Jacobs and J. Franson, Single Photons on Demand from Stored Parametric Down-conversion. (2002) *Phys Rev A* 66, 042303.

[21] E. Jeffrey, N.A. Peters and P.G. Kwiat, Towards a Periodic Deterministic Source of Arbitrary Single Photons. (2004) *New J Phys* 6, 100.
[22] J. Mower and D. Englund, Efficient Generation of Single and Entangled Photons on a Silicon Photonic Integrated Chip. (2011) *Phys Rev A* 84, 052326.
[23] S. Clemmen, et al., Continuous Wave Photon Pair Generation in Silicon-on-insulator Waveguides and Ring Resonators. (2009) *Opt Express* 17, 16558–70.
[24] J.E. Sharping, et al., Generation of Correlated Photons in Nanoscale Silicon Waveguides. (2006) *Opt Express* 14, 12388–93.
[25] K. Harada, et al., Frequency and Polarization Characteristics of Correlated Photon-pair Generation Using a Silicon Wire Waveguide. (2010) *IEEE J Sel Top Quantum Electron* 16, 325–31.
[26] C. Xiong, et al., Characteristics of Correlated Photon Pairs Generated in Ultra-compact Silicon Slow-light Photonic Crystal Waveguides. (2012) *IEEE J Sel Top Quantum Electron* 18, 1676–83.
[27] M. Davanco, et al., Telecommunications-band Heralded Single Photons from a Silicon Nanophotonic Chip. (2012) *Appl Phys Lett* 100, 261104.
[28] N. Matsuda, et al., Slow Light Enhanced Correlated Photon Pair Generation in Photonic-crystal Coupled-resonator Optical Waveguides. (2013) *Opt Express* 21, 8596–604.
[29] J.R. Ong, R. Kumar and S. Mookherjea, Ultra-high-contrast and Tunable-bandwidth Filter Using Cascaded High-order Silicon Microring Filters. (2013) *IEEE Photonic Tech L* 25, 1543–6.
[30] M. Piekarek, et al., Passive High-extinction Integrated Photonic Filters for Silicon Quantum Photonics. Proceedings of 2016 Conference on Lasers and Electro-Optics (CLEO), paper FM1N.6, 5–20 June 2016, San Jose, CA.
[31] N.C. Harris, et al., Integrated Source of Spectrally Filtered Correlated Photons for Large-scale Quantum Photonic Systems. (2014) *Phys Rev X* 4, 041047.
[32] N.C. Harris, et al., Large-scale Quantum Photonic Circuits in Silicon. (2016) *Nanophotonics* 5, 456–68.
[33] D. Pérez, et al., Silicon Photonics Rectangular Universal Interferometer. (2017) *Laser Photonics Rev* 11, 6.
[34] E.A. Dauler, et al., Review of Superconducting Nanowire Single-photon Detector System Design Options and Demonstrated Performance. (2014) *Opt Eng* 53, 081907.
[35] P. Rath, et al., 'Superconducting Single Photon Detectors Integrated with Diamond Nanophotonic Circuits. (2015) *Light Sci Appl* 4, e338. Available at https://www.nature.com/articles/lsa2015111.
[36] A. Beyer, et al., Waveguide-coupled Superconducting Nanowire Single-photon Detectors. CLEO: Science and Innovations 2015, paper STh1I.2, 10–15 May 2015, San Jose, CA.
[37] Y.P. Korneeva, et al., Superconducting Single-photon Detector Made of MoSi Film. (2014) *Supercond Sci Technol* 27, 095012.
[38] F. Najafi, et al., On-chip Detection of Non-classical Light by Scalable Integration of Single-photon Detectors. (2015) *Nat Commun* 6, 5873.
[39] N.C. Harris et al., Linear Programmable Nanophotonic Processor. (2018) *Optica* 5, 1623–31.
[40] H.F. Hofmann and S. Takeuchi, Quantum Phase Gate for Photonic Qubits Using Only Beam Splitters and Postselection. (2001) *Phys Rev A* 66, 024308.
[41] J.L. O'Brien, et al., Quantum Process Tomography of a Controlled-NOT Gate. (2004) *Phys Rev Lett* 93(8), 080502.
[42] X. Qian, et al., Large-scale Silicon Quantum Photonics Implementing Arbitrary Two-qubit Processing. (2018) *Nat Photon* 12, 534–9.

[43] N.C. Harris, et al., Quantum Transport Simulations in a Programmable Nanophotonic Processor. (2017) *Nat Photon* 11, 447–53.
[44] J.B. Spring, et al., Boson Sampling on a Photonic Chip. (2013) *Science* 339, 798–801.
[45] A. Crespi, et al., Integrated Multimode Interferometers with Arbitrary Designs for Photonic Boson Sampling. (2013) *Nat Photon* 7, 545–9.
[46] J. Mower, et al., High-fidelity Quantum State Evolution in Imperfect Photonic Integrated Circuits. (2015) *Phys Rev A* 92, 032322.

8

Programmable Integrated Photonics for Classical Applications

A second and probably wider area of application of programmable integrated photonics is in classical systems. While so far most of these systems have relied mainly on pure fixed application specific photonic circuits, the advantages brought by programmability can become decisive when these systems grow in complexity. This chapter reviews several application areas where programmable photonics can become ubiquitous. It begins with optical switching interconnection and routing, which is, arguably, the classical area of application where integrated photonics has been adopted at an earlier stage. Here programmable photonics provides myriad powerful solutions that encompass broadcasting and wavelength selective operation. A second emerging area of application is artificial intelligence (AI) and neurocomputing where it enables the implementation of reconfigurable analogue cores required for the operation of neural networks and neurophotonic systems. Microwave and analogue photonics are particularly appealing areas for programmable photonics as they can provide substantial cost reductions and broadband operation flexibility by enabling the possibility of implementing most of the required functionalities in a single chip. The use of these functionalities spans a huge area of applications, including 5G communications, satellite payloads, electromagnetic and photonic radar, the Internet of Things and autonomous driving. Finally, it discusses optical signal processing, including examples for the application of programmable photonics in optical mode conversion, photonic signal filtering and sensing, controllable Talbot effect and more speculative and futuristic applications such as synthetic domains. It aims to show, through different examples, the broad range of applications in which programmable photonics will play a relevant role in the near future.

8.1 Optical Switching, Interconnection and Routing

Optical switching and routing implemented by means of integrated photonic chips and controlled by electronic means has been proposed to replace partially or completely electronic switches in two application scenarios [1]–[3]: metro and long-haul communication networks [1] and short-reach communication networks within data centres and

Programmable Integrated Photonics. José Capmany and Daniel Pérez, Oxford University Press (2020). © José Capmany and Daniel Pérez.
DOI: 10.1093/oso/9780198844402.001.0001

high-performance computers [2], [3]. In the first case the features of dense photonic integration can be crucial for reducing the cost over current switching technologies. This is especially significant if silicon photonics technology is considered for leveraging large-scale integration. In the context of short-reach networks the use of photonic switching avoids the pin and power limitations imposed by electronic switches, eliminate costly and inefficient optoelectronic and electro-optic conversions and therefore enable higher bandwidth operation. In this case it is envisaged that optical switches will co-exist with conventional electronic switches.

8.1.1 Elementary Programmable Switch Cells

The core element of an optical switch is the 2×2 switch cell [1] (Figure 8.1(a)). As shown, it has two optical inputs $(a(t), b(t))$ and two optical outputs $(c(t), d(t))$. Multiple switch cells can be assembled to form a switch fabric that lies at the heart of a photonic switch module, as we shall shortly describe. The switch cell is operated in two states: bar (or$\|$) when input port 1 (port 2) is connected to output port 1 (port 2) and cross (or x) when input port 1 (port 2) is connected to output port 2 (port 1). If we assume $b(t) = 0$ then we can define the switching efficiencies for both states as

$$\eta_{\|} = \frac{|c|^2}{|a|^2}; \quad \eta_{\perp} = \frac{|d|^2}{|a|^2}; \tag{8.1}$$

While the extinction ratios (ERs) are given by

$$\varepsilon_{\|} = \frac{|d|^2}{|a|^2}; \quad \varepsilon_{\perp} = \frac{|c|^2}{|a|^2}; \tag{8.2}$$

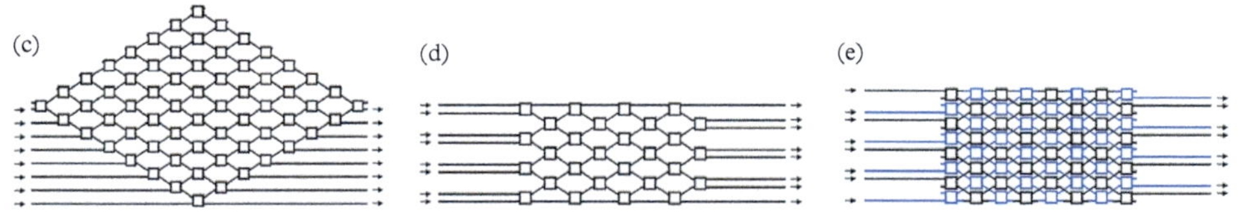

Figure 8.1 *(a) Schematic layout of a 2x2 switch cell including the driver for state programming [1]. (b) Configuration and operation of an elementary photonic MEMS switch cell consists of two crossed waveguides evanescently coupled to a third mechanically movable bent waveguide fabricated on an MEMS cantilever [13]. (c)-(e) Typical configurations of crosspoint switch fabrics [1]. (c)Triangular crosspoint. (d) Spanke-Benes (c) PILOSS.*

It is customary however to employ the worst values of Eqn. (8.1) and Eqn. (8.2) to quote the cell switching efficiency and ERs, i.e. $\eta = \min\{\eta_{\parallel}, \eta_{\perp}\}$ and $\varepsilon = \min\{\varepsilon_{\parallel}, \varepsilon_{\perp}\}$. Under general operation conditions $b(t) \neq 0$ and then both coherent and incoherent crosstalk terms have to be added to the transmission equations. For instance, under *bar* operation we have

$$\begin{aligned} |c|^2 &= \eta_{\parallel} |a|^2 + \xi_i(b) + \xi_c(a,b) \\ |d|^2 &= \eta_{\perp} |b|^2 + \xi_i(a) + \xi_c(a,b) \end{aligned} \tag{8.3}$$

Where ξ_i and ξ_c denote, respectively the incoherent and coherent crosstalk terms. Similar equations can be defined for the *cross* state of operation. Although both sources must be minimised for efficient operation, coherent crosstalk can be a major source of degradation as it creates beating distortion, which can be important if the frequencies of the interfering signals are within the bandwidth of the receiver. Additional metrics such as the operation bandwidth, switching time, power consumption and losses are required to fully characterise the 2×2 switch cell.

Several design approaches and technologies can be employed to implement integrated switch cells and switch fabrics and the reader can find a very complete description for instance in [1], [2]. For programmable photonics the most interesting are on one hand those that employ integrated interference-based switch cells based on integrated directional couplers [4], Mach–Zehnder interferometers (MZIs) [5]–[7] and ring resonator (RR) devices [8]–[10] and on the other those based on micro-electro mechanical (MEM) devices [11]–[14].

In the first case, switching cells have been reported both in InP and silicon photonics [2]. InP interferometric switches suffer from considerable losses and power ER and thus may require the use of auxiliary semiconductor optical amplifiers (SOAs) for improving operation, which increases power consumption and noise. Thus, the silicon photonics alternative is usually preferred [1]. Silicon MZI switching cells work in a similar way as the TBUs of a general programmable optical core and therefore their operation follows the same procedures as described in Chapter 2 and for the tuning of the required internal phase shifters one can resort to thermo-optic, electro-optic and other mechanisms described in Chapter 5. Their limitations and impairments are, again similar to those of TBUs and have been thoroughly described in Chapter 6. RR structures use their resonant complementary add/drop output ports to carry out a frequency-selective routing [8]. They have been employed to implement 2×2 switch cells but they are wavelength selective and very dependent on temperature. Finally, several silicon photonic horizontal and vertical MEMs switch cells have been reported in the last years.

An elementary photonic MEMS switch cell consists of two crossed waveguides [13] (Figure 8.1(b)), which can be evanescently coupled to a third mechanically movable bent waveguide fabricated on a MEMS cantilever. These cells use electrostatic actuators as moving parts are small and have response times in the order of microseconds. When the cell is on the off (or *through* propagation) state the cantilever is in an elevated position and the light entering the crossing passes through. When the switch is in the on (or *drop*) state the cantilever is lowered and coupled to the first two waveguides. Then the light entering the crossing is deflected by an angle of 90° by coupling in and out of the bent cantilever waveguide. Coupling to the cantilever can be lateral or vertical. In the first case [13] the operation bandwidth is limited as two directional couplers are required. Vertical coupling solves this problem by using a two-layer waveguide system and adiabatic coupling to an upper waveguide [14]. When comparing MZI switch cells with those based on RRs and MEMs, it is found that they feature intermediate value for figures of merit with insertion losses < 0.3 dB and ERs <-35 dB in the case of thermo-optically activated switches. These values are slightly worse when plasma dispersion activated switches are considered with 1 dB and -20 to -27 dB as standard values for the insertion losses and ERs, respectively. RR designs perform worse in all the metrics compared to MZIs with the exception of footprint. Finally, MEMs switch cells have ERs as low as -60 dB and insertion losses below 0.5 dB, however, these are asymmetric and correspond to the drop operation, while the thru operation results in insertion losses of around 0.04 dB. Interestingly, this asymmetry is also an undesired feature of RR switch cells and the reason why these and MEMs are usually only employed as building blocks for crossbar switch fabrics.

8.1.2 Programmable Crosspoint Switch Fabrics

Switch cells can be grouped and assembled to form switch fabrics in a variety of topologies. Some of these, like the Butterfly and the Benes architectures [15-17] are not particularly well suited for programmable integrated photonics implementations since they either do not provide path diversity result in a considerable level of crosstalk requiring, in addition, several waveguide crossings per established path. Cross-point switch matrix structures are much more amenable for integrated photonic implementations [10], [14], [18]-[20]. Figure 8.1(c) shows, for example a cross-point switch matrix featuring an architecture that is identical to the triangular multi-port interferometer presented in Chapter 4[21] if the 2×2 switch cells are implemented by means of MZI TBUs. This topology provides a strictly non-blocking operation and requires no waveguide crossings, however for a $N \times N$ fabric, it scales poorly with hop count $O(N)$ and also in overall switch count $O\left(N^2\right)$. Another drawback of this structure results from its high loss non-uniformity since the switch cell hop count can range from 1 to $2N-1$, although this can be compensated by including extra length waveguide sections at the input and output ports.

The former limitations regarding to switch count and footprint can be reduced to about half using a Spanke–Benes [22]–[24] permutation topology as shown in Figure 8.1(d). Note that this configuration is identical to the rectangular multi-port interferometer presented in Chapter 4[25] if the 2×2 switch cells are implemented by means of MZI TBUs. With the Spanke–Benes configuration loss uniformity is reduced although this structure is no longer strictly non-blocking but rearrangeably non-blocking. Switch hop count for a $N \times N$ fabric is $O(N)$, while the overall switch count is $O\left(N^2/2\right)$.

A third interesting option, which preserves the features of the Spanke–Benes configuration and regains the strictly non-blocking feature is the path independent loss (PILOSS) topology [26], [27] (Figure 8.1(e)). Here the upper and lower edges of the switch fabric are wrapped around in a cylinder and connected to each other while alternate first stage inputs are left unpopulated. This architecture requires waveguide crossings but loss uniformity is ideal since for a $N \times N$ fabric each path includes exactly $N-1$ cross switches, 1 bar switch and $N-1$ waveguide crossings. Its implementation however requires a novel type of 2D programmable photonic core allowing for waveguide crossings.

Several programmable cross-point switch fabrics have been reported in the literature both in InP as well as in silicon photonics. For instance, an 8×8 switch fabric arranged in an N-stage planar topology of 28 MZIs and shown in Figure 8.2(a)–(b) was recently reported [28] that constitutes the largest integrated MZI circuit in InP platform so far. This configuration suffered from various limitations that affect scalability. In first place, non-uniform polarisation behaviour degraded the crosstalk level from -20 dB to -11 dB for the TE polarisation. This non-uniformity can be corrected as it was attributed to fabrication variability. More important however is the limitation due to insertion losses, which can be compensated by combining the MZI-based switch cells with SOAs [29], where these elements help in addition to improve the crosstalk to values close to -40 dB.

The availability of these low-crosstalk cells shown in Figure 8.2(c) with almost negligible losses and crosstalk enables their intermediate assembly into dilated Benes 4×4 blocks shown in Figure 8.2(c) extending the scale of InP switch fabrics to up to 128×128 port counts [30]. A 32×32 monolithic silicon MZI PILOSS switch fabric shown in Figure 8.2(d) implementing 1024 MZIs switching cells and thermo-optic phase shifters (TO-PS) has been recently reported [31]. The fabric is electrically packaged by flip-chip attachment to a ceramic interposer with > 2000 bond pads. The TO-PS were operated by means of tunable pulse-width modulation signal generated by a field-programmable gate array. Chip metrics included a total on-chip loss of 6.4 dB, overall coupling losses from 2.8 to 6.8 dB, single path crosstalk around 20 dB below the signal level and a power consumption of 1.9 W with a reconfiguration speed of 30 microseconds.

A more recent design reported a 32×32 switch fabric based on 448 MZI cells activated by TO-PS [32]. The MZI devices were arranged twelve stages following a dilated topology. The chip also included 900 monitor photodiodes and its metrics included an on-chip loss range of 13–35 dB, single path crosstalk below -30 dB and power consumption below 20 mW. Designs incorporating electro-optic activated MZIs have also been recently reported. These have the potential for nanosecond scale reconfigurability. In this context carrier injection PIN junctions are the preferred option for phase shifters as they feature higher efficiencies in terms of nm/V leading to a much smaller footprint and lower operating voltage when compared to carrier depletion PN junctions. However, as compared to TO-PS, undesired electro-absorption is an undesired side effect that needs to be compensated. Reports exist of 16×16 configurations in wire-bonded packages and arranged in a Benes topology. These structures suffered from considerable loss and crosstalk as the effect of free carrier absorption compromised the achievement of ideal cross and bar states in the MZI cells. The high crosstalk effects can be partially mitigated by the use of push-pull configurations for the MZI phase shifters,

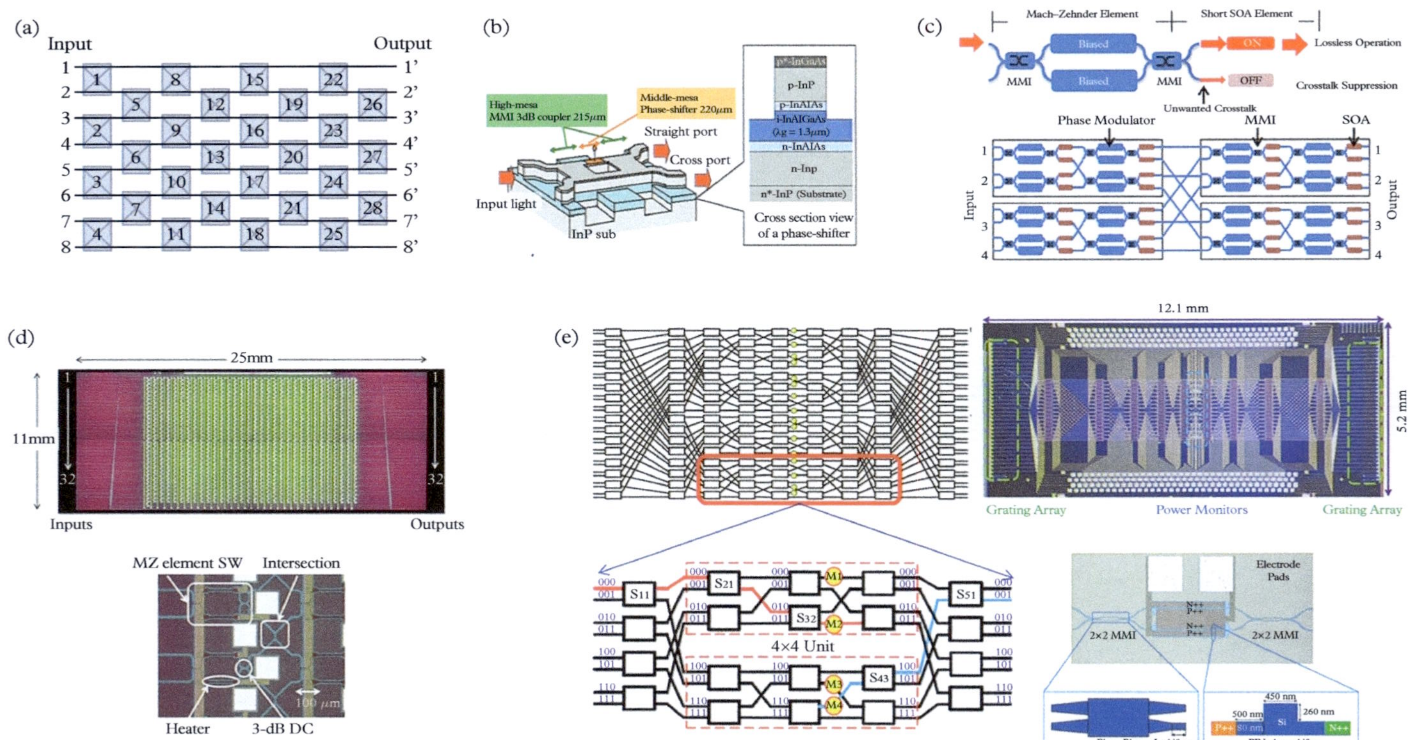

Figure 8.2 *(a) Layout of a passive 8×8 InP switch fabric (after [28]) (b) Detail of the elementary 2×2 switch cell corresponding to (a). (c) Active 2×2 interferometric InP switch cell combining an MZI and SOAs and construction of a 4x4 switch fabric by dilated combination of such cells (after [29]). (d) Photograph and detail of a 2×2 switch cell of a Silicon Photonics 32×32 switch fabric incorporating 1024 MZIs switching cells and thermo-optic phase shifters (after [31]). (e) Layout and chip photograph of a 32×32 switch fabric incorporating 144 MZIs switching cells and carrier injection PIN activated phase shifters. Also shown the basic switch cell unit with push-pull biasing and the 8x8 Benes building block (after [32]).*

which brings the additional advantage of reducing their footprint and by migration towards dilated topologies.

A 32 × 32 configuration has been recently reported employing the Benes topology [33]. The basic layout, chip photograph, configuration of the basic building 8 × 8 Benes fabric and detail of the push pull MZI units are shown in Figure 8.2(e). The chip has 144 phase-bias 2 × 2 MZI-switch units and 288 electrode pads; 16 bi-directional power taps were also implemented in the switch to detect the operation statuses of all 144 units. On-chip insertion losses were 12.9 to 16.5 dB, and the crosstalk ranged from −17.9 to −24.8 dB when all units were in the 'cross' status. The on-chip insertion losses were 14.4 to 18.5 dB, and the crosstalk ranged from −15.1 to −19.0 dB when all units were in the 'bar' status.

Switch fabrics have also been proposed using MEMs and RRs as basic building blocks (BBBs). The interested reader can find a detailed account of the most significant results reported so far in Lee and Dupuis [1] and Cheng, Rumley, Bahadori and Bergman [2].

8.1.3 Programmable Switch Modules

Despite playing a central role, a switch fabric is only a part of a complete switch module, as this system integrates other electrical and optical building blocks. This section describes the operational principles and structure of this module as a good example of a complete programmable photonics system. It is useful in understanding the design and operation principles of field programmable photonic gate arrays (FPPGAs) (Chapter 9).

The description of programmable photonic systems for switching applications takes a bottom-up approach, starting with the basic configuration of the programmable switching cell and from this it considers programmable switch fabric composed of interconnected cells. In turn, the switch fabric must be connected to two external elements; the electrical setting, monitoring and control interface and the optical interface to form a photonic switch module (Figure 8.3) [1].

The setting, monitoring and control interface enables electronic control signals to reach the analogue photonic components in the switch fabric with two purposes. The first is for programming purposes, i.e. to set their bias states according to the programmed switch configuration, the second for monitoring and control purposes, i.e. to collect feedback regarding their operation points and provide, if required, corrective signals in response to undesired deviations.

In this sense it intermediates between the specific current/voltages signals to set the state and monitor the conditions of the switch fabric and the standard communication protocols employed outside the element. Central to this part are the device drivers, which take input signals and produce/supply the corresponding current/voltage to the elements of the fabric. In the context of switching modules, input signals are usually digital, so drivers can be basically digital-to-analogue converters (DAC) or pulse width modulation generators. Drivers also are in charge of receiving the signals from the photonic elements in the switch fabric and sending them to the logic units for monitoring and control purposes and, in this case, they can be implemented by simple trans-impedance

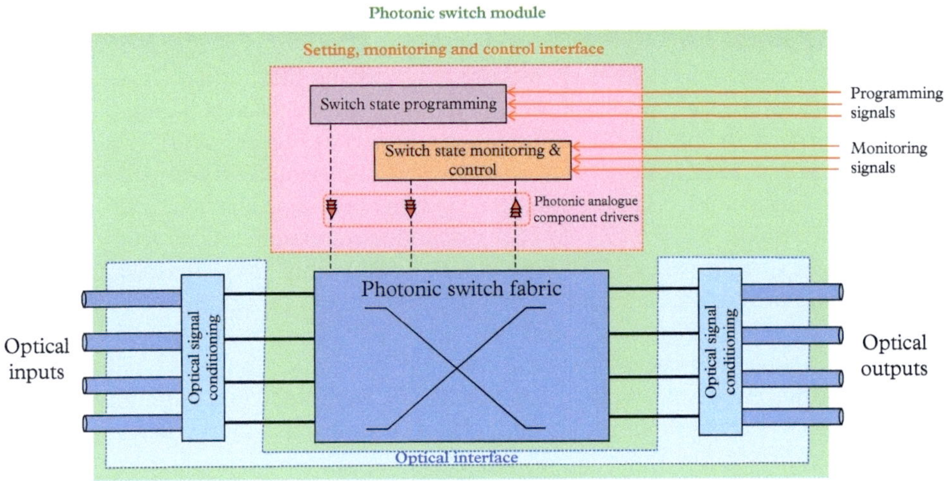

Figure 8.3 *Schematic diagram of a photonic switch module showing its main relevant units: the programmable switch fabric, the optical interface and the setting, monitoring and control interface (after [1]).*

amplifiers (TIAs) followed by and analogue-to-digital converter (ADC). The switch state programming unit receives the information from a high-level logic unit that includes the setting of the operating points of the photonic elements in the switch fabric required for a particular configuration and sends signals to the appropriate device drivers in order to configure the fabric. The switch state control and monitoring unit receives information regarding the physical status of the fabric elements and implements the necessary corrective algorithms for optimising the fabric operation according to the targeted programming.

The optical interface mission is principally to enable off-chip signals to be coupled in and out of the switch fabric and, in second place, if required, to provide suitable signal adaptation and conditioning. It interconnects external optical fibres with the internal switch fabric and this requires the means for efficiently coupling the signals in and out of the chip from and to the input/output fibres. This includes depending of the chip technology spot-size converters and edge couplers or grating couplers. In addition, this part may need to include optical signal conditioning elements, being the most common, optical amplifiers polarisation handling elements or wavelength division demultiplexers and multiplexers.

8.2 Artificial Intelligence and Neuromorphic Photonics

8.2.1 Neuromorphic Computing and Artificial Intelligence: Basic Concepts

The recent demise of different scaling laws that have ruled for several decades in micro-electronics is challenging the current dominating Von Neumann approach in computing, characterised by the principal role of a central processing unit (CPU) [34]–[42]. In first place, the breakdown of Moore's law by which the transistor count duplicates each two years is limiting the number of basic processing units per unit of surface. Second, the limitation in microprocessor clock rates caused by current leakage in nm transistors has frozen the growth rate in clock speeds. In third place, severe power consumption restrictions arise in nanometre-scale transistors as the well-known Dennard's scaling law (smaller transistors consume less power) no longer applies and hence the power density of microelectronics no longer stays constant as chips get denser. This is giving rise to the *dark silicon* phenomenon [35], by which a given percentage of an integrated circuit cannot be powered at the nominal operating voltage to meet power consumption restrictions. In 8-nm nodes this percentage is estimated to reach a fifty per cent.

All in all, the conclusion is that current CPUs have reached a limit in their operating speeds and, as a consequence, interest is shifting towards alternative computing paradigms and hardware architectures that can provide increased processing speed with a reasonable consumption [34].

A paradigmatic example [37] is the human brain, which is capable of a huge processing capacity (around 10^{20} MAC/s) that is around nine orders of magnitude beyond that of current supercomputers, which consume only 20 W. This is possible because it features a highly distributed processing architecture, where around 10^{11} of basic computing elements known as neurons are highly interconnected (around 10^4 connections per neuron). Inspired by this example, neuromorphic computing aims to produce large-scale hardware towards achieving efficiencies commensurable to those of the human brain.

The central hardware supporting a neuromorphic processor is a highly connected platform of neurons known as artificial neural network (ANN) [43]. An ANN is composed of two basic elements: a set of non-linear nodes or neurons and a network of reconfigurable (i.e. weighted) interconnections. These are bundled together into the neuromorphic processor itself that incorporates other external circuitry and systems in charge of implementing learning algorithms, controlling the interconnection weights, etc. (Figure 8.4).

The software part is not a set of programmed instructions but rather a resident knowledge that is stored in the interconnection weights and which is acquired by means of a *learning* or *training* process [43], in which the ANN is subject to different inputs and the weights are adjusted in order to obtain a given output for each one. AI based on machine and deep learning techniques is at the heart of the development of different efficient training and classification algorithms for ANN architectures.

Neuromorphic computing has been mainly developed in two principal directions. *Spike computing* [36] approaches, where the ANN processes spikes or pulses that encode information temporally. Reservoir computing [38] approaches, where the complex recurrent ANN performs a diversity of functions, from which linear classifiers extract the most useful information to perform a given algorithm.

While some efforts have been made to implement electronic neuromorphic systems, these systems are potentially limited by latency, speed and interconnection bandwidth. Thus, the attention has been turned to leverage on the properties of integrated photonics

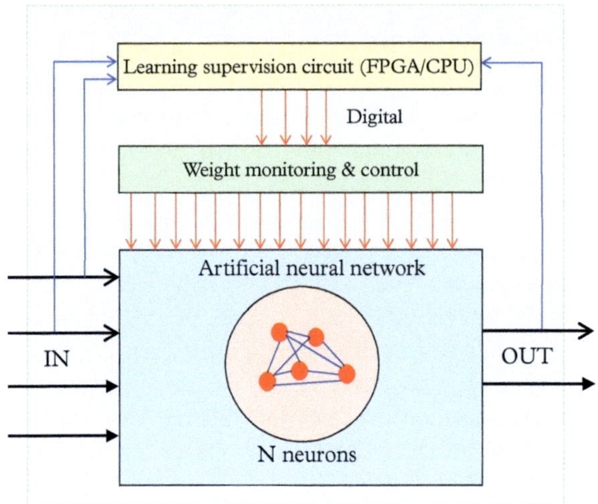

Figure 8.4 *Schematic description of a neuromorphic processor (after [37]). While two layers of electronics provide control, supervision and reconfigurability the programmable ANN operates as the core processor.*

in terms of high bandwidth, low propagation losses and low latency. It has been predicted that a neuromorphic approach based on integrated photonic chips could outperform their electronic counterparts by six to eight orders of magnitude in speed. Section 8.2.2 describes the BBBs of a photonics ANN.

8.2.2 Photonic Artificial Neural Network

An integrated photonic ANN is composed by a set of processing network nodes (PNNs), which can be either all optical or optical/electrical/optical (O/E/O) [34]–[37], and a network of reconfigurable interconnections implemented by means of integrated photonic waveguides.

PNNs aim to replicate the action of a neuron. Figure 8.5(a) shows a basic diagram of a neuron, and the layout illustrates a weighted and directed graph of connections or *synapses*. The input signal to the neuron is a weighted addition or linear combination of the outputs of other neurons connected to it

$$z_i = b_i + \sum_{k=1}^{N_{fi}} w_{ki} x_k \tag{8.4}$$

where b_i represents the bias point of the neuron, w_{ki} the connection weight between neuron k and neuron i, x_k the output of neuron k.

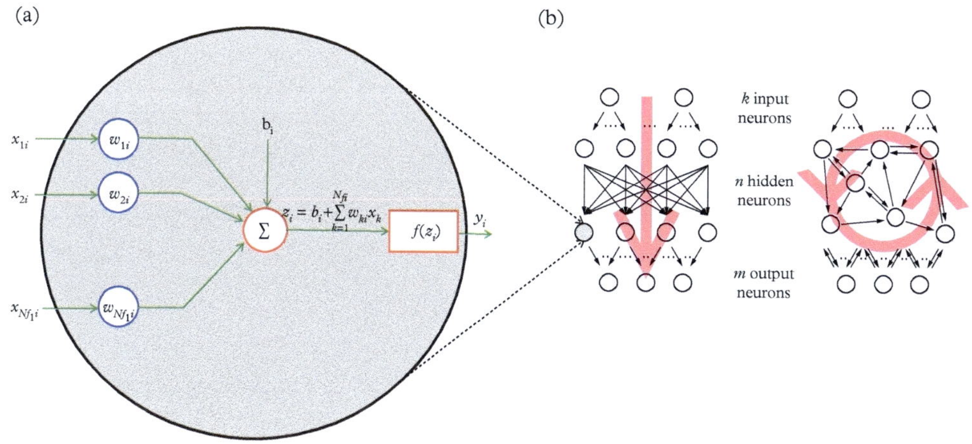

Figure 8.5 *(a) The schematic nonlinear model of a neuron. (b) ANN Architectures: feedforward (left handside) and recurrent (right handside), After [37].*

A neuron i takes into account this input signal and produces a nonlinear response according to a given monotonic and bounded *activation function* f

$$y_i = f(z_i) \tag{8.5}$$

Activation functions can be implemented directly in the optical domain, involving the use of either coherent or non-coherent injection techniques [35]. Stringent requirements in terms of phase control, operation wavelengths, pump powers, etc., need to be fulfilled in both schemes, which limits their possible applicability in terms of scalability. The overcoming of these restrictions is subject to current research. The second option is to implement the activation function in the electrical domain, which requires and optical to electrical conversion before the activation and an electronic to optical conversion after the activation. This approach opens a wide variety of physical mechanisms for the implementation of the nonlinear activation function that include; pure electronics, electro-optic physics in modulators and active laser behaviour. These mechanisms are reviewed in Ferreira de Lima and colleagues [34], Prucnal and Shastri [35] and Tait and colleagues [37].

Neurons can then be connected using different architectures. Figure 8.5(b) shows two of the most employed configurations. In the feedforward configuration neurons are grouped into several layers ($i = 1, 2 \ldots N$) and the output signals from layer $i - 1$ ($i > 1$) provide the inputs to the neurons in layer i after a linear matrix transformation. The signals thus propagate in a single direction from the input neuron layer ($i = 1$) towards the output neuron layer ($i = N$). This configuration is similar to that of multi-port interferometers [20], [21] considered in Chapter 4 and lends itself to an integrated photonics implementation using these configurations as will be shown later.

The second option, shown in the right part of Figure 8.5(b) is to employ a recurrent network topology [34], [43] in which any neuron output can reach any other neuron, including itself. Signals can thus propagate in forward and backward directions, enabling feedback. This configuration is similar to those of waveguide meshes [44] (Chapter 5) and lends itself to an integrated photonics implementation using this configuration as well as by means of programmable recirculating integrated interferometers.

Several proposals may benefit from the properties of programmable photonic circuits. In most of the cases, the work reported addresses node connection networks as the implementation of elements carrying activation functions in an integrated format amenable to scalability is still an unsolved issue.

8.2.3 Programmable Integrated Photonic Networks for ANN

In the context of photonic spike processing Tait and colleagues [37] proposed an integrated photonic neural networking configuration called *broadcast and weight* (Figure 8.6(a)) that employs wavelength-division multiplexing (WDM) to support a large number of reconfigurable analogue connections using silicon technology. This approach is named after the broadcast-and-select protocol employed in optical networks whereby all channels are broadcast in the network to all nodes and an active connection between a couple of nodes is established by tuning a filter at the receiver to drop the desired wavelength. In the broadcast and weight networks each neuron receives directed multiple inputs from other neurons (each one at a different wavelength) with a continuous range of analogue weights in between -1 and 1.

It is crucial to have therefore the ability to control each connection and each weight independently and the key element to achieving this objective is a programmable spectral filter bank placed at the input of each node. A very attractive option is the micro-ring resonator (MRR) weight bank [39]–[42] shown in Figure 8.6(b), which can be implemented either using a SCISSOR application-specific photonic integrated circuit (ASPIC) configuration or emulated using a waveguide mesh. The MRR weight bank is composed of N equal ring cavities connected by an input and an output common bus [39], [42]. Each ring is tuned to bypass/drop with a specified power ratio a given wavelength and bypass the rest. The bypassed and dropped signals are collected by the input and output common buses, respectively, and sent to a pair of photodiodes in a balanced configuration, which provides an electrical current proportional to their difference. Since the bypass/drop relationship can be independently programmed in each ring by acting over its specific couplers, in theory, any input weight ranging from -1 (all the signal is dropped) and 1 (all the signal is bypassed) can be continuously achieved. Of course, in practice there are accuracy and precision limits to the attainable weights and these limitations have been thoroughly addressed in the literature [34], [45]. It is no surprise that the main sources of limitation are precisely the same as those that affect general programmable circuits and waveguide meshes (Chapter 6). Filter weight banks based on MRRs have been implemented in silicon photonics and Figure 8.6(c) shows a specific example for a unit with four MRRs, although a scaling limit of around

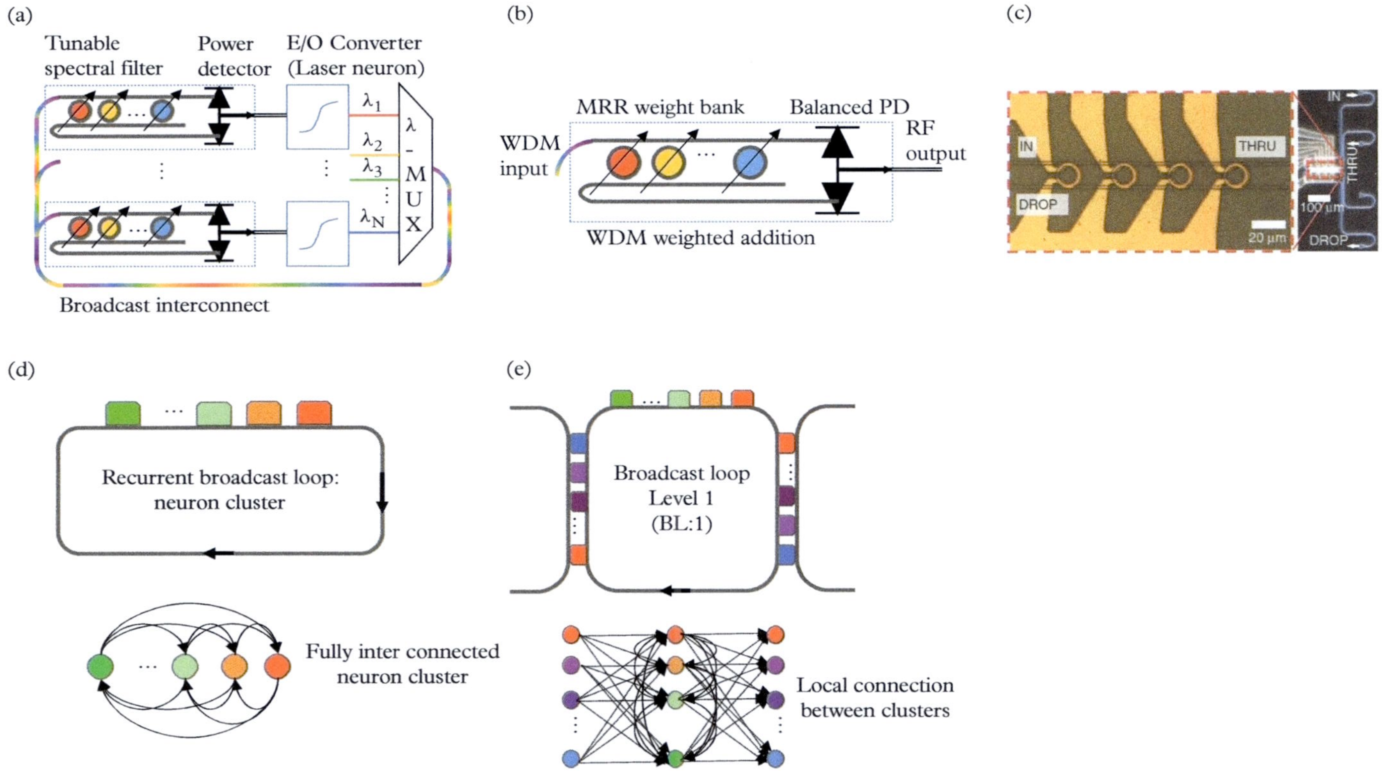

Figure 8.6 *(a) Broadcast and weight implementation of an optical ANN. (b) Tunable MRR weight bank. (c) Photograph of a tunable MRR filter bank fabricated in Silicon (c) Neuron cluster embedding a broadcast and weight ANN into a recurrent loop. (d) Multilayer architecture of fully interconnected neuron clusters embedding a broadcast and weight ANN into a recurrent loop and using spectral reutilization (after [35]).*

148 channels can be achieved [34]. These figures are impressive in terms of photonic integration but are far from state-of-the-art in the context of AI.

Since scalability is an important issue when using integrated photonic MRR filter banks, the broadcast and weight programmable architectures should allow for the implementation ANNs with extended number of neurons. This can be achieved by exploiting wavelength spectrum reuse strategies and spatial division multiplexing techniques. Figures 8.6(d)–(e) show two examples of how this can be achieved. In Figure 8.6(d) the PNNs are attached to a broadcast loop, which can emulate a fully interconnected neuron cluster. Figure 8.6(e) shows the extension of the former concept to achieve several layers of fully interconnected neuron clusters.

Recently Shen and colleagues [46] proposed a multi-port interferometer configuration for the implementation of N-layer feedforward optical ANNs. It is based on an integrated programmable nanophotonic processor featuring a cascaded array of fifty-six programmable MZIs in a silicon photonic integrated circuit (Figure 8.7(a)). It is composed of a set of input artificial neurons represented by circles, which are connected to at least one hidden neuron layer and an output neuron layer. Each neuron is then followed by the application of a non-linear activation function.

The internal operation of each ANN layer is shown in Figure 8.7(b). Each layer or instance basically consists of two stages. The first is an optical interference unit (OIU) where an arbitrary unitary linear transformation is performed over the input vector

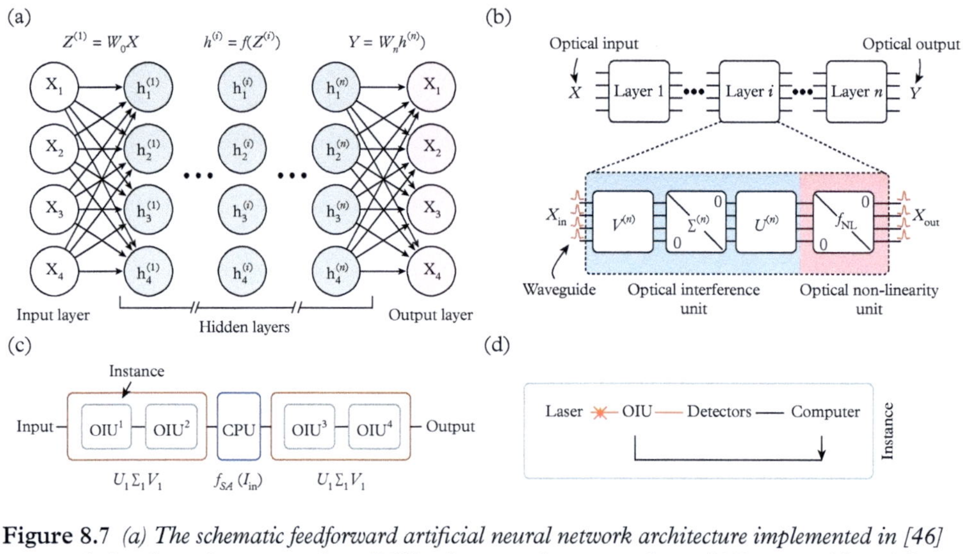

Figure 8.7 *(a) The schematic feedforward artificial neural network architecture implemented in [46] composed of an input layer, a number of hidden layers and an output layer. (b) Decomposition of the general neural network into individual layers with a detail of the optical interference and nonlinearity units that compose each layer of the artificial neural network. (c) Representation of the two-layer ONN experiment reported in [46]. (d) Experimental feedback and control loop used in the experiment. Laser light is coupled to the OIU, transformed, measured on a photodiode array, and then read on a computer.*

$\mathbf{X_i}$ by means of the singular value decomposition (SVD) technique (Chapter 2). This linear matrix transformation implements the weighted connections between neurons of adjacent layers. Once the linear transformation has been applied an intermediate vector $\mathbf{Z_i}$ is obtained, the elements of which are individually transformed by a second stage composed of optical non-linear units (ONUs) that provide the required activation functions and the output vector $\mathbf{Y_i}$.

In the experiment reported, the SVD matrix decomposition was implemented using two OIU stages connected by and intermediate layer of MZIs providing the diagonal matrix and the ONUs where implemented offline using a computer CPU. Figure 8.7(c) shows the operation of two instances in series and Figure 8.7(d) the internal operation of one instance. The ONUs where implemented using CPU that ran a program emulating the response of a saturable absorber.

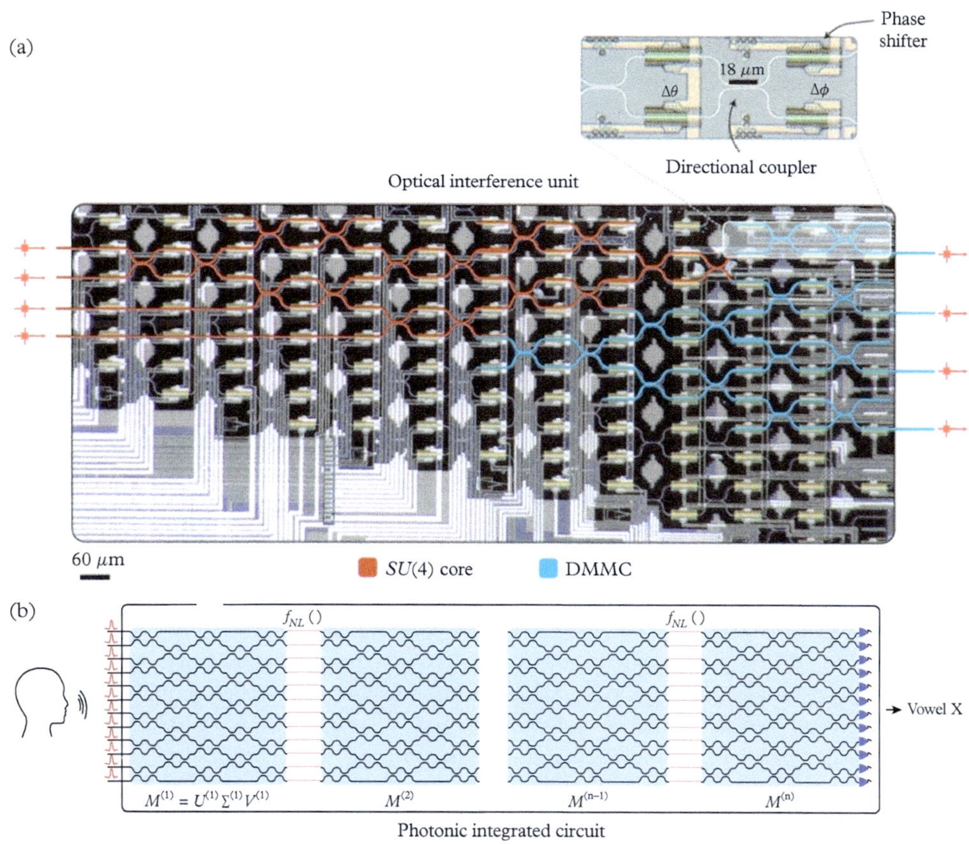

Figure 8.8 *(a) Optical micrograph illustration of the experimentally demonstrated OIU, which realizes both unitary matrix multiplication (highlighted in red) and diagonal matrix attenuation (highlighted in blue) required for SVD decomposition (after [46]). The inset shows a tunable Mach-Zehnder basic unit (b) Optically implemented ANN assembled for a vowel recognition experiment in [46].*

Figure 8.8(a) shows the layout of the silicon photonics programmable processor fabricated to implement an OIU. The processor implements a multi-port 4 × 4 triangular architecture and is composed of fifty-six programmable MZIs, each of which comprises a thermo-optic phase shifter between two adjacent fifty per cent evanescent directional couplers followed by a second phase shifter as shown in the inset. The red highlighted part corresponds to a 4 × 4 triangular interferometer implementing either the U or the V unitary matrix in the SVD decomposition, while the blue part corresponds to the linear array of MZI implementing the diagonal matrix interconnecting the U and V unitary matrices. Note that two chips are required to implement the full SVD decomposition. A two-layer neural network was then assembled using four OIU chips and trained for vowel recognition (Figure 8.8(b)).

The ONN was trained using a back-propagation algorithm and each vowel was encoded using 360 data points consisting in four log area ratio coefficients of one phoneme, which were then encoded in a vector with four coefficients. The reported experiment provided a reasonably good performance as compared to a 64-bit computer, which is a promising indicator of the possibilities of ANN implemented by means of integrated programmable photonics. Recurrent limiting factors arise again in connection to the precision attainable with the weight programming, the need to control and monitor the status of the matrix coefficients against drifts, etc. In addition, there is a concern related to the inefficient nature of backpropagation algorithms for training. A technique based on the feedforward calculation of gradient functions by means of a finite difference method has been proposed and, very recently Hughes, Minkov, Shi and Fan published an in-situ technique based on propagating the adjoint field and interfering its time reversed copy with the original field [47]. All in all, there is still much work to accomplish in hardware, supervision, learning and control software to make ONN practical. Future research is also expected in the implementation of the required activation functions in integrated version. Saturable absorbers and especially semiconductor optical amplifiers hold the promise for enabling these.

8.3 Microwave and Analogue Photonics

Although several programable circuits have been proposed for applications in microwave photonics [48], [49] (Chapter 3), the focus here is on a general-purpose photonic integrated processor using a reconfigurable optical core based on a hexagonal waveguide mesh. This configuration can be software defined to perform multiple applications required in microwave photonics MWP by suitable tuning of each element in the processor. Hence, the same hardware can be configured to perform the main functionalities in MWP: optical delay lines, radiofrequency (RF)-photonics filtering, optical generation of RF and millimetre waves, photonic-assisted RF-mixing, instantaneous frequency measurements, etc.

Figure 8.9 represents a processor architecture suitable for a wide range of RF-photonics applications. All the elements are connected to the reconfigurable optical core in such a way that not only they produce the desired filtering or delay schemes, but also

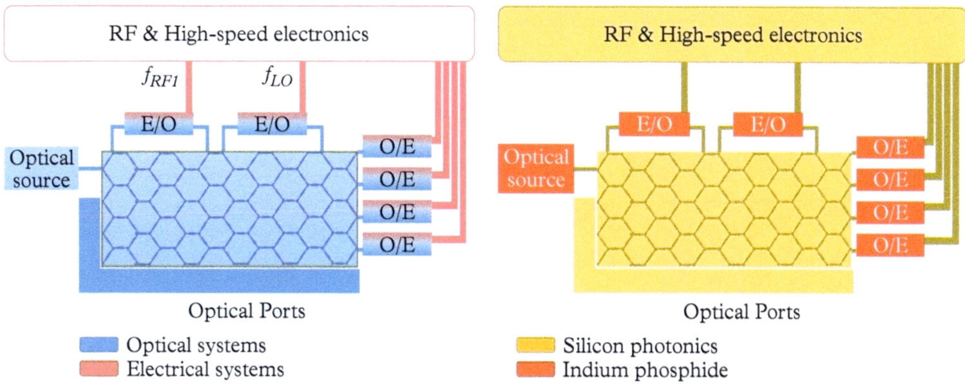

Figure 8.9 *General-purpose photonic integrated processor architecture and candidate fabrication platforms for each subsystem.*

they connect the internal and the external elements required for different functionalities. As highlighted in the right part of the figure, a hybrid design might be needed to achieve the most efficient performance. In this case, a natural choice can be silicon photonics platform (yellow colour) for the passive devices and indium phosphide (red colour) for the active devices.

Note that an array of optical amplifiers in this platform might be required to overcome the large conversion losses when moving from the radiofrequency to the optical domain. These losses are mainly related to the conversion efficiency of modulators and photodetectors as well as the propagation losses.

In order to test the benchmark of the multipurpose processor in terms of frequency response, we consider four different attainable values of basic unit lengths (BULs) for a silicon photonics platform with a group index of 4.18. $BUL_1 = 119.5\mu m$, $BUL_2 = 239.7\mu m$, $BUL_3 = 358.6\mu m$ and a larger one of $BUL_4 = 597.7\mu m$. Some of the MWP applications described here are limited to a set of free spectral range (FSR) values, a discrete frequency grid or delay given by the chosen BUL. For these examples featuring different BUL values, Table 8.1 shows the resulting FSRs of the filters (optical ring resonators (ORRs) and MZIs) that can be synthesised on the hexagonal mesh. The International Telecommunication Union (ITU) frequency grid has been highlighted for comparison. RF bands ranging from the K (20 GHz) up to the E (60 GHz) band are covered by all of the proposed examples. In addition, the Vernier effect [50] could be exploited to overcome this limitation and produce greater FSRs.

We now briefly review how this general processor structure can be employed to enable the most relevant functionalities required in MWP.

8.3.1 True-time Delay Lines

Together with tunable RF phase shifting, the tunable optical true time delay line is a key basic functionality required in complex signal processing applications. It constitutes an

Table 8.1 *Frequency grid associated to four different BULs. The interferometric structure (ring and/or MZI) available for the hexagonal core is also indicated.*

Interferometric length (BULs)	$BUL_1 = 119.5\,\mu m$ FSR(GHz)	$BUL_2 = 239.7\,\mu m$ FSR(GHz)	$BUL_3 = 358.6\,\mu m$ FSR(GHz)	$BUL_4 = 597.7\,\mu m$ FSR(GHz)
2	300.00	150.00	**100.00**	60.00
4	150.00	75.00	**50.00**	30.00
○ 6	**100.00**	**50.00**	33.33	20.00
8	75.00	37.50	**25.00**	15.00
○ 10	60.00	30.00	20.00	12.00
○ 12	**50.00**	**25.00**	16.67	10.00
○ 14	42.86	21.43	14.29	8.57
○ 16	37.50	18.75	12.50	7.50
○ 18	33.33	16.67	11.11	6.67
○ 20	30.00	15.00	10.00	6.00
○ 22	27.27	13.64	9.09	5.45
○ 24	**25.00**	12.50	8.33	5.00
○ 26	23.08	11.54	7.69	4.62
○ 28	21.43	10.71	7.14	4.29
○ 30	20.00	10.00	6.67	4.00

BBB in a considerable number of filtering and optical beamforming schemes that are described in following subsections. It is worth noticing that the first optical true-time delay lines were implemented using optical fibres of different lengths and, therefore, resulted in bulky devices. During the last twenty years, however, different schemes of integrated delay lines have been reported [51], [52] where some of them enable tunable delays. The objective is to achieve long tunable time delays while maintaining high-bandwidth operability, reduced footprint and reduced losses per delay unit.

Using optical switches, discretely tunable time delays can be implemented selecting integrated waveguides with different lengths. They enable large bandwidth operation that is typically independent of the selected delay value. Nevertheless, some applications would benefit from continuously tunable delay lines.

An alternative approach to achieve continuous time delay relies on overlapping the group delay of cascaded optical filters, e.g. ORRs loaded with a phase shifter. In this case, there is a trade-off between the bandwidth of the delay line and the time delay provided. In order to achieve longer continuous delays without reducing the bandwidth, a continuous delay ranging from 0 to τ seconds in series with switched discrete delay lines offering τ, 2τ, and 4 τ can provide an increased continuous delay line ranging from 0 to 8τ, [52]. In this case, since waveguide-based discrete delay lines enable large bandwidth transmission, the delay line bandwidth will be mainly limited by the coupled

ORRs maximum bandwidth provided by the first continuous delay stage. In general, demonstrated losses per delay value are between 3.3 and 60 dB/ns, depending on the selected optical delay line scheme.

Another approach enabling continuous delays is the separate carrier tuning (SCT) technique [53], [54]. It consists of two independent blocks: a tunable delay line for the modulation sideband and a separate carrier tuning phase shifter for the optical carrier that extrapolates the linear phase characteristic of the sideband delay line. Varying the group time delay while independently applying a full phase shift in the separate carrier allows higher bandwidths. This technique has demonstrated delays up to 400 ps for small signal *bandwidth* × *delay*. The hexagonal waveguide-mesh is a natural discrete optical delay line. If by switching different TBU we can set a light path corresponding to N BULs, then, the corresponding delay τ is as

$$\tau = \frac{n_g \cdot N \cdot BUL}{c}. \tag{8.6}$$

The basic unit delay for BULs values considered above and a group index of 4.18 are: $\tau_1 = 1.66$ ps, $\tau_2 = 3.34$ ps, $\tau_3 = 5.00$ ps and $\tau_4 = 8.33$ ps. The losses per delay value figure can be obtained as the ratio between the TBU insertion losses and its associated basic unit delay.

The techniques enabling continuously tunable delays reviewed here can be programmed as well in the hexagonal waveguide mesh. The upper part of Figure 8.10 illustrates the group-delay overlapping of the cascaded ORRs approach. In the first case, each tunable coupler that connects the rings to the bus waveguide should have a small/moderate coupling constant maintaining a trade-off related to resonant insertion losses and group delay. Next, a phase shift is applied to each RR to obtain the desired group delay overlapping profile for each wavelength. Note that the more overlapped they are, the higher the group delay and the smaller the bandwidth achieved.

The lower part of Figure 8.10 depicts the implementation of a tunable delay line based on the SCT technique, where the first two rings correspond to the carrier tuning and the two remaining provide the subcarrier phase shift. Finally, the waveguide mesh includes an optical filter based on a doubled loaded MZI to suppress one of the modulated sidebands. The signal bandwidth for this application will be limited by the passband response of this optical filter or the ORR stages corresponding to a few GHz for the selected BULs (Table 8.1).

8.3.2 Radiofrequency Tunable Filters

All-optical microwave signal filtering is probably the most widespread application in integrated MWP processors. Specifically, this application brings the possibility of designing fully integrated tunable photonic devices to meet flexible wideband spectral processing requirements for actual and future RF communication bands.

The general scheme of a MWP system consists of an optical source modulated by an external modulator with the RF signal to be processed, followed by an optical

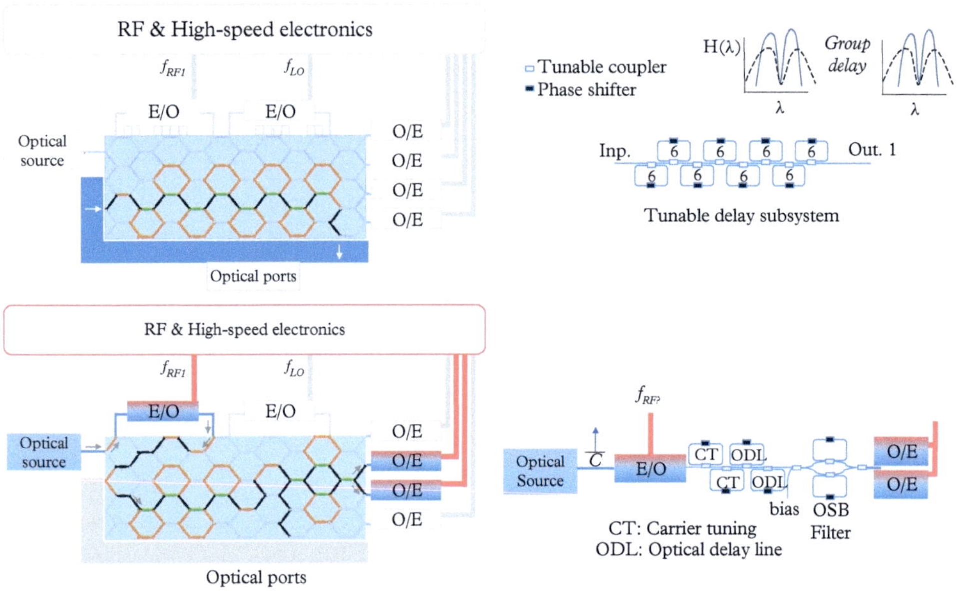

Figure 8.10 *True Time delay line implementation for two different techniques: (top) Overlapping group delays of 7 cascaded ORRs, (bottom) SCT technique. The left part depicts the processor configuration and the right part shows the targeted scheme.*

core where the RF signal, up-converted to the optical region of the electromagnetic spectrum (at hundreds of THz), is processed by optical techniques and components, such as filters and amplifiers. Finally, the signal can be distributed to large distances or can be downconverted back to the RF domain with a high-speed photodetector resulting in the processed signal in the electrical domain. This scheme leverages on the unique characteristics inherent to photonics, e.g. high bandwidth, low losses, tunability, reconfigurability.

Depending on the operation regime, the filter can be based on a summation of optical intensities (ruling out the optical phases) in the case of incoherent operation, or a summation of electric fields, where optical phases are key, e.g. in the coherent regime case. In the first, the shortest delay experienced by the signal inside the filter is much longer than the coherence time of the optical source. It is commonly implemented with discrete and large optical components, where the coefficients are set by optical amplification or attenuators and the delay is given by tunable delay lines to obtain tunable frequency responses. Filters built using optical discrete delay lines and fibre Bragg gratings reach up to 40 dB of ER and quality factors up to 325 with a high number of taps and, hence, require a considerable number of components. Some of them have demonstrated reconfigurability and tunability up to around tens of GHz, but it is quite difficult to maintain a stable temperature in the resulting bulky system. Implementation of integrated incoherent MWP is not easy on a chip scale as the inherent small footprint delays call for the use of broadband optical sources to meet the incoherent operation condition.

On the other hand, coherent filters, where the longest delay experienced by the signal is much shorter than the coherence time of the optical source, are amenable for integration. Integrated optics circuits have been demonstrated recently for the implementation of tunable coherent optical filters [55]–[59].

Optical single-side band modulation (OSSBM) is an interesting particular approach for the implementation of coherent MWP filters as it allows the direct transmission of the transfer filter implemented by the optical core to the RF region of the spectrum. In this case, the detected photocurrent is then proportional to the frequency response of the optical filter at the carrier frequency multiplied by the optical filter response at the RF sideband, mapping the shape of the optical filter transfer function into the electrical domain [60]

$$i(t) \propto s\,|H(f_o)|\cdot|H(f_o+f_{RF})|\cdot\cos(2\pi f_{RF}-\angle H(f_o)+\angle H(f_o+f_{RF})), \tag{8.7}$$

where $H(f)$ is the optical filter response at a given RF frequency f.

Furthermore, different modulation and detection schemes can be employed to increase the tunability range, the dynamic range, the gain and the noise figure of the whole system, such as self-heterodyning systems [61], where full-FSR frequency response tunability can be achieved by adding a phase shift into the optical filter and employing a copy of the carrier that was sent to a different path prior modulation.

Coherent filters are usually based on optical filters built from the concatenation of single unit cells such as MZIs (zeros), RRs (poles) and ring-loaded MZIs (Chapter 3). As shown for true-time delay lines, tuning can be achieved by adding a simultaneous phase to all the filtering structures. Optimum passband filters like Chebyshev, Butterworth, and elliptic are possible, where the filter order is related to the number of RRs in the structure [62]. The concatenation of unit cells increases the insertion losses of the whole device (if each cell has no zero insertion losses), but, on the other hand, can produce more selective filters as the result of multiplication of their individual frequency responses. Integrated solutions outstand due to their versatility and low SWAP, offering at the same time similar figures of ER and selectivity comparing to bulk optics designs.

Figure 8.11 illustrates an application example of the multi-purpose processor hexagonal core for the implementation of MWP filters. It corresponds to a RF-photonic filter employing a self-homodyne modulation/detection scheme. The optical filter stage synthesises, in this case, a sixth-order optical filter based on a SCISSOR structure. For the case of RF-photonics filters, the limitation will depend on the modulation scheme: double-side band modulations would have the half-FSR of the optical filter whereas single-side band modulations map the full FSR of the optical filter to the electrical domain.

Following the values provided in Table 8.1, the optical filter corresponds to an FSR associated to ORRs of 6 BULs of 100, 50, 33.33 and 20 GHz for each selected BUL. A wide variety of optical filters can be synthesised on the hexagonal waveguide core and which are amenable to be translated into the RF bands by suitable downconversion.

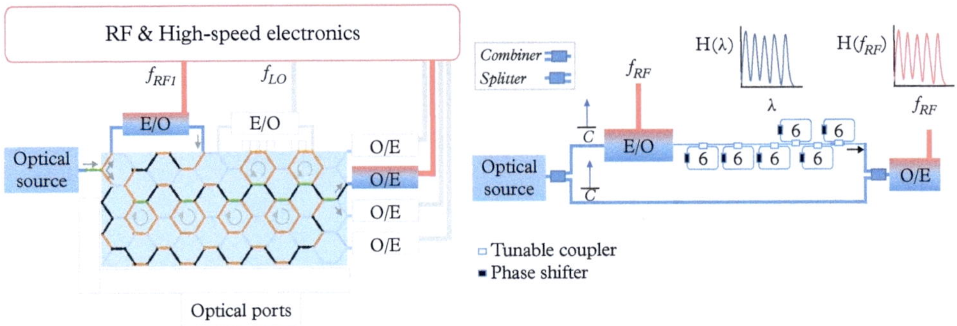

Figure 8.11 *General-purpose signal processor configuration for RF filtering implementation (left) based on a self-homodyne modulation/detection scheme (right). The optical filter is composed of six cascaded ORRs defined by a cavity length of 6 BULs.*

When compared to its ASPIC counterpart, the programmable processor suffers from extra excess losses due to the waveguide lattice mesh, reducing the total RF gain. To overcome this limitation, the integration of optical amplifiers must be considered. In contrast, the reconfigurability of the filter is greatly improved, resulting in a more powerful and versatile device.

8.3.3 Optical Generation of Continuous Microwave and Millimetre Wave Signals

The optical generation of high-quality CW microwave and millimetre wave (mm-wave) signals can be applied in many fibre-supported microwave and mm-wave systems (including antenna remoting and phased array antennas) with a special interest in the upcoming 5G radio access networks [63], [64]. This functionality allows the generation and distribution of high-frequency RF signals by taking advantage of inherent properties of optics, e.g. low propagation losses of standard fibres and the availability of fibre amplifiers. Actually, the purely electrical approach faces problems in the generation of signals above 25 GHz, while their distribution is highly challenging due to the propagation losses of coaxial cables. In addition, the increasing bandwidth demand in communications has driven the attention towards the mm-wave band because of the wide bandwidth available for high-data-rate wireless transmission where propagation losses are even higher for purely electrical/RF systems. The main performance figures of photonic-generated RF signals systems are the frequency range where the signal can be generated and the linewidth of the tone. As in every photonic integrated circuit (PIC), it is also important to evaluate the power consumption and the potential integration percentage of the system.

There are different reported approaches to generate microwave and mm-wave signals, but most of them are based on optical heterodyning in which two optical waves of different wavelengths beat at a photodetector generating a tone up to the THz band with a frequency coinciding on the difference between the two tones. The phase of the resulting tone will be the difference between the two phases. Thus, in order to generate low-phase noise, the use of correlated optical waves is advisable.

The maximum achievable frequency is mainly limited by the photodetector bandwidth (among other specific limits related to each approach). Figure 8.12 shows the performance results for the demonstrated devices for each technique reported during the last years. They can be classified into six groups:

1. optical injection locking
2. optical phase locking
3. optical injection phase locking
4. external modulator based
5. optoelectronic oscillator
6. dual parallel MZMs.

The first three have not been able to break the 42-GHz barrier and need two lasers to be working at the same time.

The approach employing an external modulator takes advantage of the frequency upconversion inherent when a continuous wave signal provided by a laser is modulated by an RF tone. By employing different modulation biasing points, the input frequency can be doubled or even quadrupled by suppressing the odd- or even-order modes, (together

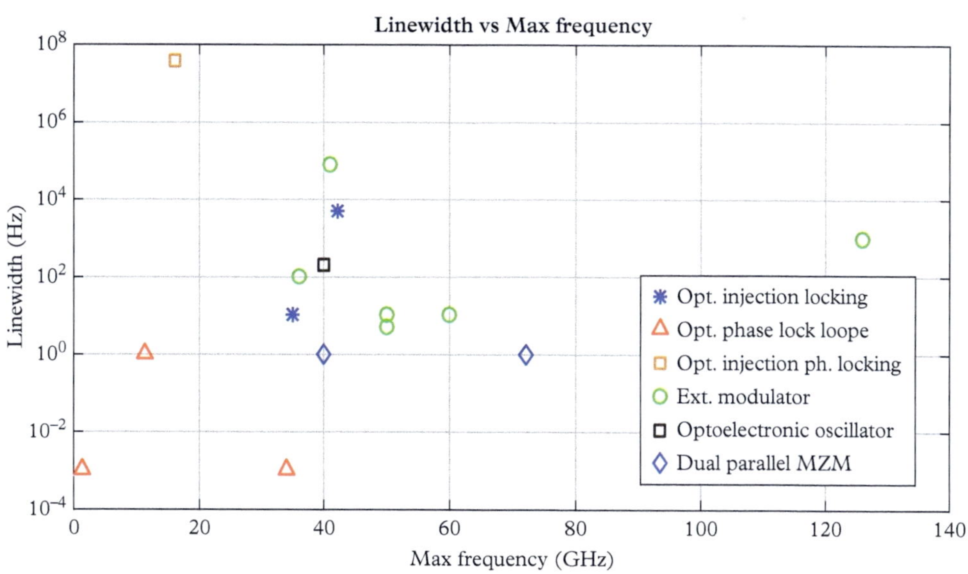

Figure 8.12 *Linewidth vs maximum frequency of the different demonstrated techniques for optical generation of microwave and millimetre-wave signals using application-specific systems, [64].*

with the use of an optical filter to suppress the optical carrier). The linewidth is limited by the input signal and can be as low as 5 Hz. The lowest frequency is limited by the bandwidth of the filter employed to suppress the carrier, while the highest achievable frequency is limited by the photodetector bandwidth and the modulator bandwidth. This technique has reached tone generation up to 60 GHz. The use of a phase modulator instead of an intensity modulator avoids the bias drift problem that would need a control circuit in order to increase the robustness [65].

An interesting alternative that has achieved an outstanding integration degree is opto-electronic oscillation. The oscillator operates by modulating a continuous wave signal from a laser with an external intensity modulator. The signal goes through a high-Q optical storage stage (Fabry–Pérot filter or optical delay line) in order to suppress the unwanted modes and provide a cavity round-trip delay. The photo-detected signal is injected again into the modulator after the electrical stage as a feedback loop. This architecture has provided a linewidth under 200 Hz and tones ranging from 10 up to 40 GHz [66].

Finally, the use of a dual parallel Mach–Zehnder modulator (DPMZM) maintains a good trade-off between linewidth obtained, (in the range of tens of Hz), and achievable frequency (that ranges from very low frequencies up to 60 GHz and is actually limited by the photo-detector and the modulator bandwidth). In this case, an MZM is connected at each arm of an outer MZM forming a third interferometric structure. This structure can provide optical carrier suppressed modulations without using an optical filter. The generated tone is then the difference between the optically modulated tones. Regarding power consumption, this structure only needs an optical source and an optical amplifier to increase the conversion gain.

None of the aforementioned analysed structures has been completely integrated on a complex PIC. However, Tang and colleagues [67] recently reported an integrated opto-electronic oscillator (OEO), where the optical components, including a directly modulated laser source, a spiral-shaped optical delay line and a high-speed photodetector, were fabricated on an indium phosphide substrate.

The multi-purpose programmable MWP processor architecture enables synthesis of different schemes for generating continuous wave RF signals. For example, Figure 8.13 shows the processor configurations implementing different approaches: (a) the external modulator approach, (b) the optoelectronic oscillator approach. Note that if a DPMZM were included in the architecture, this technique could be integrated as well. The achievable frequencies would mainly depend on the modulator and photodetector performance. Particularly, the bandwidth limitation in (a) due to the optical filter might be determined by the UMZI FSR of $\Delta L = 2$ BULs, ranging from 300 up to 60 GHz for the selected BUL examples.

8.3.4 Optical Mixing: Radiofrequency Upconversion/Downconversion

Frequency up- and/or downconversion of RF signals is usually required in many radio-over-fibre and intermediate-frequency-over-fibre communication systems. Different

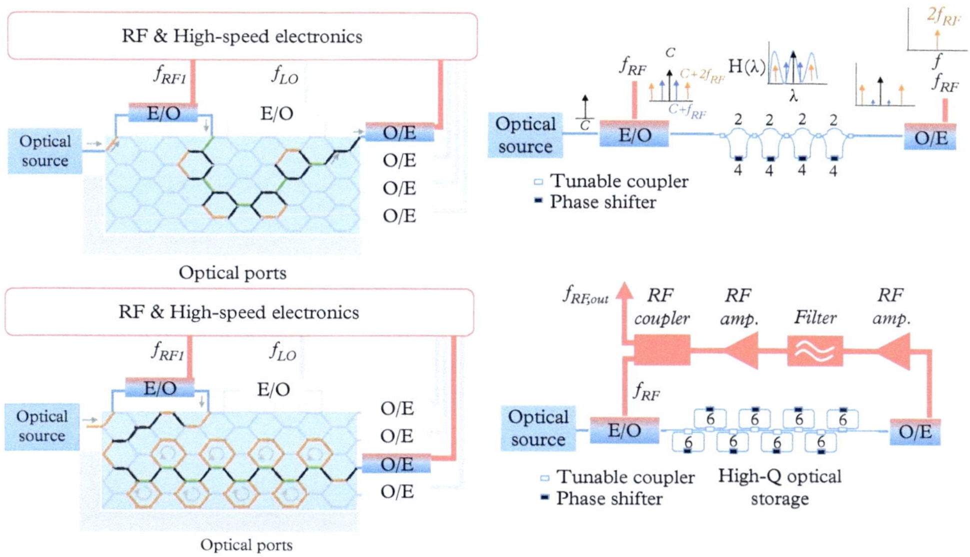

Figure 8.13 *General-purpose signal processor configuration for microwave and mm-wave tone generation based on (upper) External modulator approach, (bottom) Optoelectronic oscillation approach. The right figures illustrate the targeted configuration schemes.*

architectures have been proposed focusing on the modulation scheme of both the laser and the local oscillator signal, i.e. direct or external modulation [68]. The use of an optical modulator is equivalent to the use of an electrical mixer, but without the limitation to an intermediate frequency bandwidth of less than a few GHz. The use of an external modulator for both the RF input frequency and the local oscillator (LO) increases their maximum frequency at both system inputs.

DPMZMs [69] can upconvert signals to high frequencies as well. They can double the frequency by biasing the DPMZM to obtain optical double side band-carrier suppressed (ODSB-CS) modulation without employing optical filters providing destructive interference for the carrier.

The main figures of merit to be evaluated in these systems are the RF gain (conversion losses), the noise figure and the spurious free dynamic range (SFDR). The approach based on two cascaded intensity modulators with frequency doubling [70] has worse figures of merit than the one built upon the DPMZM. The latter improves the noise figure (NF) 10 dB, near 20 dB the gain and 10 $dB \cdot Hz^{2/3}$ the third-order SFDR ($SFDR_3$) [69].

For this application, it might be more efficient to have a DPMZM instead of two E/O intensity modulators in the photonic processor. Nevertheless, we consider the two modulators approach here. Figure 8.14 illustrates the processor configuration for optical mixing and the targeted scheme. An optional optical filter is employed in this case to suppress the possible intermodulation contributions prior to the second modulator input. An optical amplifier array prior to the photodetection stage would improve the overall

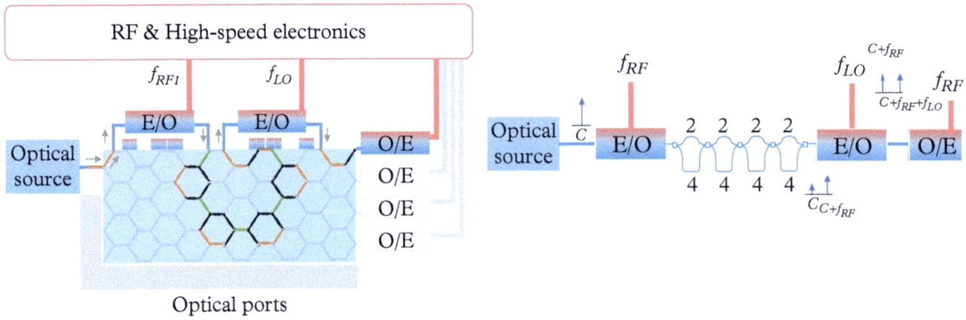

Figure 8.14 *General-purpose signal processor configuration for RF-mixing scheme example (left) for a targeted configuration scheme composed of 2 electrooptic modulators and an optical filter implemented by four cascaded MZIs (right).*

gain of the RF signal output, which is particularly necessary to overcome the electrooptic and optoelectronic conversion losses.

Another approach employing a pure-optical local oscillator would require the integration of another optical source. Also, the LO signal can be introduced externally by the optical ports.

The maximum bandwidth will be given again by the modulation and photodetector bandwidths. When compared to its ASPIC counterpart, only the programmed optical filter will decrease the RF gain of the system. However, this filtering stage can benefit from a higher reconfigurability degree and increase the system linearity.

8.3.5 Arbitrary Radiofrequency Waveform Generation

This application pursues generating ultra-broad bandwidth RF waves with arbitrary and reconfigurable phase or amplitude characteristics. Applications like pulsed radar, medical imaging, modern instrumentation systems and ultra-wideband (UWB) communications need a system for the generation of arbitrary waveforms. Again, the electronic counterpart is limited in frequency and bandwidth to few GHz. A solution is to take advantage of the high speed and broadband characteristics offered by photonic devices and generate the microwave arbitrary waveform in the optical domain. There are four main techniques:

1. direct space-to-time pulse shaping
2. spectral shaping and wavelength-to-time mapping
3. temporal pulse shaping
4. microwave signal processing shaping.

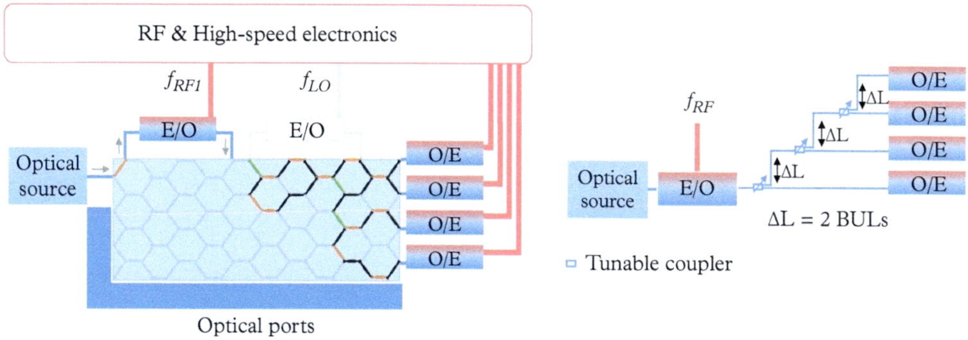

Figure 8.15 *General-purpose photonic processor settings (left) for the implementation of beamforming networks based on discrete optical delay lines with a differential length of* ΔL *(right).*

Lin, McKinney and Weiner [71] have demonstrated MWP techniques in bulky devices, lacking on full programmability and reconfigurability that is essential in arbitrary waveform generation.

The use of an optical filter to implement a spectral shaper followed by a dispersive element to implement the wavelength-to-time mapping is a promising solution and can be integrated on a chip [72]. However, the main difficulty remains that the dispersive element is not easy to integrate in a small footprint while simultaneously providing long differential delays.

Arbitrary waveform generators should have a wide range of tunability and reconfigurability and the hexagonal core enables this capability. Ultra-reconfigurable filter topologies and design parameters can be tuned to produce the spectral shaper followed by a dispersive component that can be programmed with cascaded lattice filters of unbalanced MZIs creating the desired dispersion profile, as shown in Figure 8.15. In contrast, more insertion losses will be added when compared to an ASPIC solution. If a certain application needs a higher differential delay, the optical ports can be employed to connect dispersive fibre reels.

For the BULs selected as an example, the FSR limitation of the optical filter will be again between 100 and 20 GHz for each BUL, limited by the shortest FSR that, in this case, is determined by the ORRs.

8.3.6 Beamforming Networks

In phased array antennas, the circuit that feeds the radiating elements is known as the beamforming network. It is possible to modify the radiation characteristics of an array of equal radiating elements simply by changing their electrical excitations. The beam is then steered by adjusting the phase relationship between the different feeding paths. The angle of transmission or reception θ is given by

$$\theta = \arcsin\left(\frac{\lambda \cdot \Delta\phi}{2 \cdot \pi \cdot d}\right), \tag{8.8}$$

where d is the distance between array radiating elements, λ the optical wavelength and $\Delta\phi$ is the differential phase shift introduced between the different array elements. There are two alternatives to perform the phase shift. The first directly applies the differential phase by an integrated phase shifter, where the pointing angle varies with the input RF frequency. The second approach uses tunable delay lines where the pointing angle θ remains independent of f_{RF} offering a squint-free approach suitable for wideband operation [73].

This field has raised a considerable attention for the last thirty years, moving from the military field to civil radars, remote sensing platforms, satellite or airborne communications, air traffic control, radio astronomy and flexible cellular wireless communications. From the first bulky fibre-based devices, integrated MWP approaches have been proposed based on ORRs as tunable optical delay lines elements [74]. They can provide up to 2.5-GHz bandwidth, limited by the ring performance. Furthermore, a more complex integrated beamforming based on SCT to provide delay has been proposed for a 2D array system of sixteen antennas [73].

Figure 8.15 shows a possible implementation of a beamformer based on the natural discrete delay lines of the hexagonal core. In this case, the bandwidth is wide and will mainly be limited by the modulator or photodetector bandwidths. The pointing angle can be varied modifying the differential length between the paths feeding the photodetectors (ΔL). The tilt angle tuning will be limited to discrete values.

Figure 8.16 shows the tilt angle for the BULs taken as an example at the beginning of this section and a group index of 4.18. Larger BULs produce less angle resolution. Negative waveguides lengths refer to the case where the lower antennas have a shorter feeding path. Note that continuously tunable delays can be added to overcome this problem, as described in the previous subsection *optical delay lines*.

If a certain application requires more antennas to enable 2D tunability, more multipurpose chips can be employed and connected between their optical ports. In addition, pure optical beamforming, useful in light detection and ranging (LIDAR) and autonomous driving applications can be done as well employing exclusively the optical core.

8.3.7 Instantaneous Frequency Measurement

The ability of measuring in real-time applications the frequency of microwave signals is enabled by photonic-assisted devices. Integrated photonics devices take advantage not only of low cost, weight and power consumption, but also of the reduced measurement latency due to the short optical paths when comparing to bulk-optics-based-systems. Large bandwidths up to 40 GHz (limited among others by the bandwidth of the modulator), and high accuracy (represented by an error below one per cent in the measured range), have been demonstrated increasing the range and capacity of electronic-based IFM systems [75], [76].

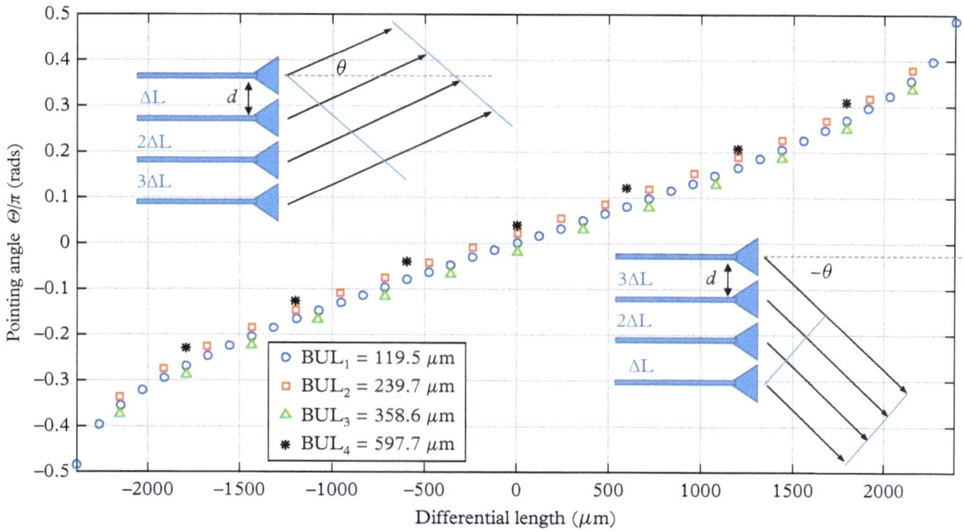

Figure 8.16 *Beamforming networks based on discrete delay line implementation for broadband operation. A differential of 0.04 rads/π on the y-axis has been applied to improve readability. For this example, it is considered a distance between radiating elements d = 1 cm. Spatial aliasing might be compromised.*

Demonstrated devices follow two main approaches clearly differentiated. On the one hand, the basic principle is to map the unknown microwave frequency to an optical power ratio by establishing a unique relationship, which is commonly referred as the amplitude comparison function (ACF) that is independent of both the laser and input RF powers. The reported devices have been partially fabricated on silicon nitride [77], InP [75] and silicon [78]. On the other hand, four wave mixing (FWM) approaches map the input frequency to the power of an idle tone generated by a non-linear process like the Kerr effect on thick silicon or ChG waveguides [76]. Neither of them includes the optical source nor the modulator and the detector on-chip. Thus, a maximum twenty per cent of the components have been integrated. The ones that employ interferometric structures are limited in range to the quality factor and the free spectral range, while the FWM-based IFM depends on obtaining a high on-chip non-linearity coefficient in order to avoid the use of fibre. Figure 8.17 reports the error and achievable bandwidth demonstrated by different IFM devices. Obviously, a low measurement error for a wide bandwidth is desired. The best results exhibit sub-one per cent error rates for up to 34 GHz for both approaches.

The approach that employs interferometric structures like ORRs exhibits a trade-off between the measured frequency range and the error while maintaining a real-time process due to the FSR limitation in frequency and the quality factor. Some of them use a coarse and a fine measure after setting a narrower range as an effective method to increase the accuracy while giving up a faster response time. All the processing is taken in the

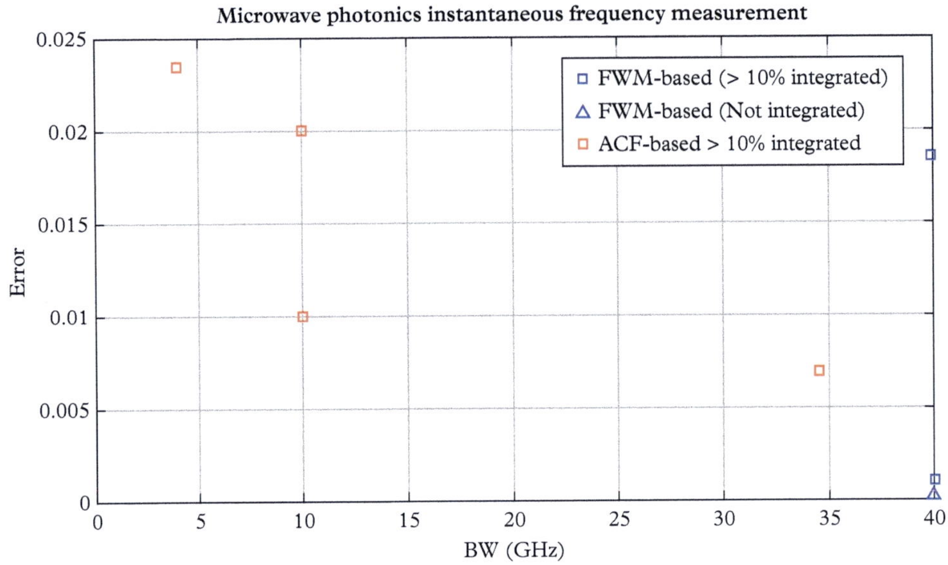

Figure 8.17 *Bandwidth vs Error (%) for partially integrated IFM systems demonstrated in the literature.*

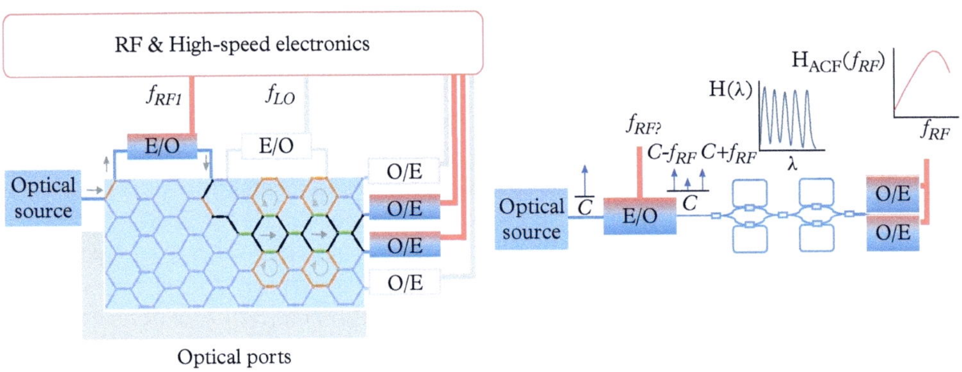

Figure 8.18 *General-purpose photonic processor settings for the implementation of instantaneous frequency measurement based on ACP function mapping (left) and targeted system scheme (right).*

photonic domain but an integrated electrical circuit is needed to map the optical power versus frequency function. These approaches open the door to highly accurate IFM systems with a reduced quantity of active components. A high-performance IFM has been demonstrated with DSB-CS based on the use of only one non-tunable optical source, an intensity modulator, a simple ORR-based add-drop and two photodetectors [78].

Figure 8.18 illustrates the configuration of the generic-purpose photonic integrated processor. In this case, the optical source, one electro-optic modulator (EOM) producing a carrier-suppressed double side-band modulation and two photodetectors are enabled.

The reconfigurable core leads the signal from the optical source to the modulator and performs the complementary function on the interferometric filter structure for the ACP function. In this case, it consists of two stages of double-loaded MZIs. Considering the previous example BULs, the frequency range of this device would be limited to a range up to of 50, 25, 16 and 10 GHz, respectively, which is half the FSR of the 6-BULs ORR.

A key advantage of the reconfigurable processor is that the ACF can be made reconfigurable. Sharper functions lead to an increase in the resolution at the expense of reducing the frequency range, so multiple measurements can be done while reconfiguring an ACP function for each frequency range of measurement. In contrast, the increment in the optical filter insertion losses might degrade the system sensitivity.

8.4 Optical Signal Processing

Direct signal processing in the optical domain avoids optical to electrical and electrical to optical conversions required to carry intermediate electronic signal processing and it is advantageous in many applications that call for broad bandwidth and high-speed reconfigurable analogue operation, which do not require sampling and digitisation. These functionalities include, for instance, optical mode conversion, optical filtering, programmable dispersion devices, advanced time/frequency multiplication techniques based on the Talbot effect and emerging areas such as synthetic domains. Integrated programmable photonics offers a unique platform for the implementation of these systems with unprecedented levels of reconfigurability and footprint. We review here some of the most relevant results reported so far.

8.4.1 Optical Mode Conversion

Silicon feedforward waveguide meshes have been reported for the implementation of mode conversion operations [79], [80]. Ribeiro and colleagues [70] reported the implementation and demonstration of a 4 × 4-port universal linear optical circuit based on Khope and colleagues [20] triangular multi-port interferometer arrangement where the silicon photonic circuit, combined with electronic control and software feedback, can perform any linear operation between its input and output ports. The circuit layout (Figure 8.19 (a)) and its microscope image (Figure 8.19 (b)) consisted of a network of thermally tunable symmetric MZIs with phase and amplitude control, in-circuit optical power monitors and local software-controlled feedback loops.

The chip was automatically configured to perform two different operations. The first was as a switch matrix providing the input/output port connections: IP#1->OP#4, IP#2->OP#3, IP#3->OP#2, IP#4->OP#1 and the second was a universal beam coupler as proposed originally by Miller. Circuit programming, control and testing were carried in a laboratory environment using the configuration displayed in Figure 8.20(a). The circuit is first configured by means of a training phase in which the individual MZIs are optimised stage by stage using their local feedback loops composed of optical monitoring elements and feedback electronics (the reader can find a detailed description of the

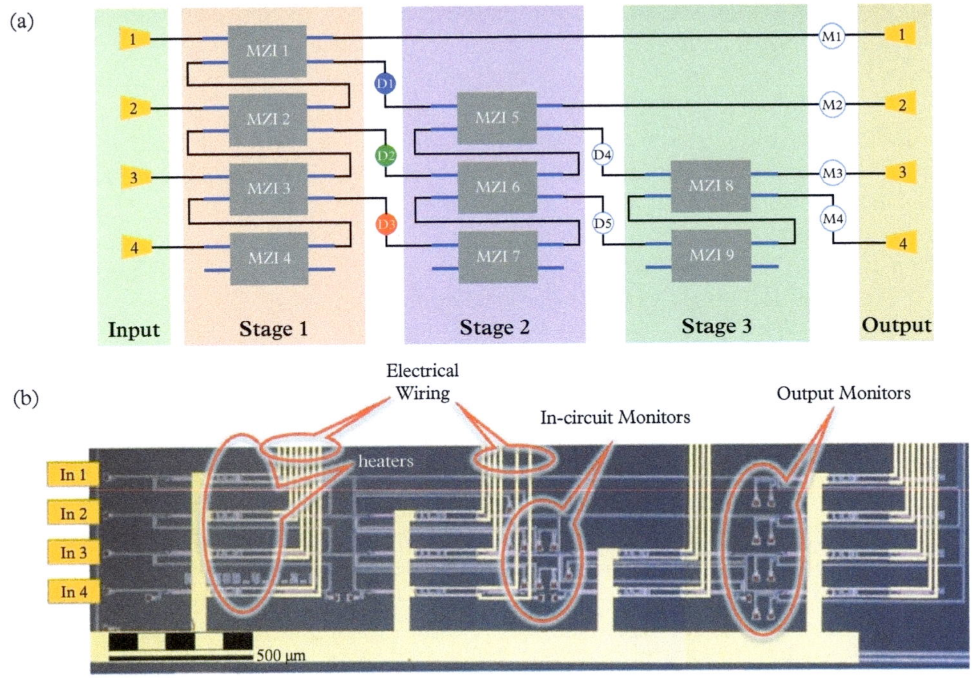

Figure 8.19 *Schematic of the integrated implementation of the 4x4 linear mode converter using MZIs and reported in [79]. The optical input and the output are implemented using grating couplers. Detectors D1 to D5 are implemented using optical taps connected to a grating coupler, for monitoring using an IR camera. The same approach was used with output monitors M1 to M4, which are used to monitor the optical power at the output of the circuit. (b) Microscope image of the fabricated device.*

training phase in Ribeiro, Ruocco, Vanacker and Bogaerts [79]). Once the training phase is completed and the circuit is operational, the feedback loops were used to keep the operation of the circuit stable against temperature variations.

The optical monitoring elements were implemented by means of grating couplers, which tapped a small proportion of light from the integrated waveguides. The reading of these monitoring signals was carried by means of an infrared camera and the obtained signals employed to control and supervise the status of the MZI phase shifters in the circuit.

Figures 8.20(b)–(e) show the results after the training phase for the operation of the circuit as a switch matrix. The results show a typical variation with wavelength that is due to the wavelength dependence of input and output grating couplers used to inject the input light from an external laser and the output light from the chip to an output fibre.

More recently, Annoni and co-workers [80] reported a 4×4 fibre mode unscrambler. In waveguides and optical fibres where the loss is essentially the same in all the different propagating modes, the scattering between the modes can be undone by means of a

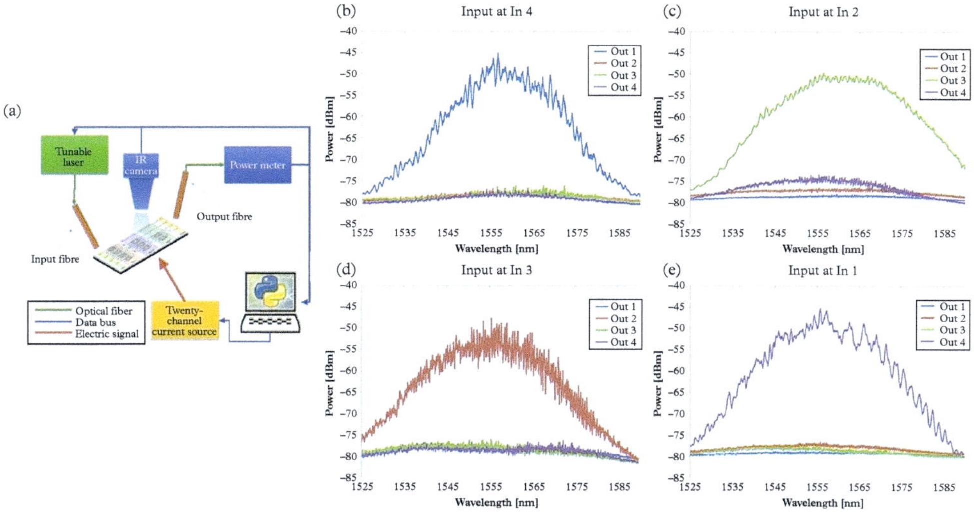

Figure 8.20 *(a) Schematic diagram of the experimental setup and control elements for the 4x4 mode converter reported in [79]. (b)-(e) Input/Output transmission spectra for input at (a) In 4, (b) In 2, (c) In 3, and (d) In 1.*

unitary linear processor. Annoni and colleagues [80] developed a linear unitary processor based on the triangular multi-port interferometer architecture [20] (see also Chapter 4). Figures 8.21(a)–(b) show the schematic concept of the N × N unscrambler and the guided-wave layout.

The interferometer was fabricated in silicon photonics technology and included in addition to the MZIs, transparent contactless integrated photonic probe (CLIPP) detectors that supplied monitoring information to the control system in charge of running progressive self-configuring algorithms required to simultaneously unscramble, sort and track four different mode beams. Figure 8.21(c) shows the silicon photonics four-mode unscrambler made of six thermally actuated MZIs individually monitored by transparent CLIPP detectors [81]. As a distinctive feature, this module incorporated a CMOS ASIC (Figure 8.21(d)) in charge of the configuration and control of the photonic chip. The authors demonstrated the separation and reconstruction of four optical beams after these were completely and arbitrarily mixed in a multi-mode waveguide coupler.

The chip training followed the sequential orthogonal mode sorting procedure into different spatial output ports (Chapter 4) for triangular interferometers. In fact, being a linear transformer, the chip can be configured so the lumped action of its transfer function $\mathbf{H_{mesh}}$ and that of the mode mixer $\mathbf{H}$ can be programmed to provide any arbitrary unitary matrix. Figure 8.22 shows the main features and results of the experimental setup used for the chip testing.

Figure 8.22(a)–(b) display the experimental concept and the actual laboratory setup respectively. Four different spatial modes (A, B, C and D) each one carrying a 10 Gb/s PRBS signal were generated and separately input to four different waveguides in the

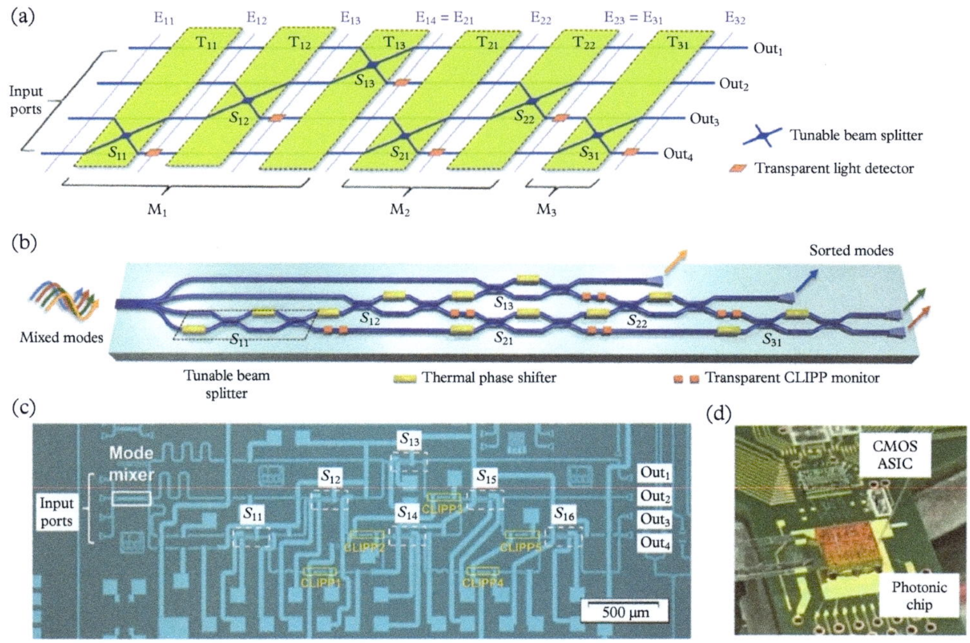

Figure 8.21 *(a) Schematic concept of an $N \times N$ ($N = 4$) triangular mesh of tuneable beam splitters implementing any arbitrary transformation on N-dimensional input vectors. Transparent detectors at the output port of each beam splitter monitor the evolution of the optical field along the entire mesh enabling local control operation on each beam splitter individually. (b) Guided-wave implementation of the mesh through a lattice of two-port cascaded MZIs realizing the tuneable beam splitters controlled through a pair of integrated phase shifters. (c) Silicon photonic four-mode unscrambler consisting of six thermally actuated MZIs individually monitored by transparent CLIPP detectors. Mode scrambling is induced on chip through a multimode waveguide section (mode mixer). Self-configuration and stabilization of the circuit is performed through a CMOS ASIC (d) bridged to the silicon chip (after [80]).*

silicon chip. Inputs were then combined and scrambled by a multimode interference coupler and the outputs fed to the unscrambler. After training the waveguide mesh was programmed to sort the different modes (A, B, C and D) to each of its four outputs with almost negligible impact due to the crosstalk arising from the rest of the modes. Figure 8.22(c) shows for example how mode D is completely recovered at output 1. The overall penalty for a BER of 10-9 was below 2 dB as it can be appreciated in Figure 8.22(d). The chip could also successfully sort out the modes when the scrambling conditions were changed dynamically.

8.4.2 Optical Filters and Tunable Dispersive Elements

Optical filtering is an essential operation in many applications supported by photonics, both ASPIC and programmable filters based on waveguide mesh arrangements have

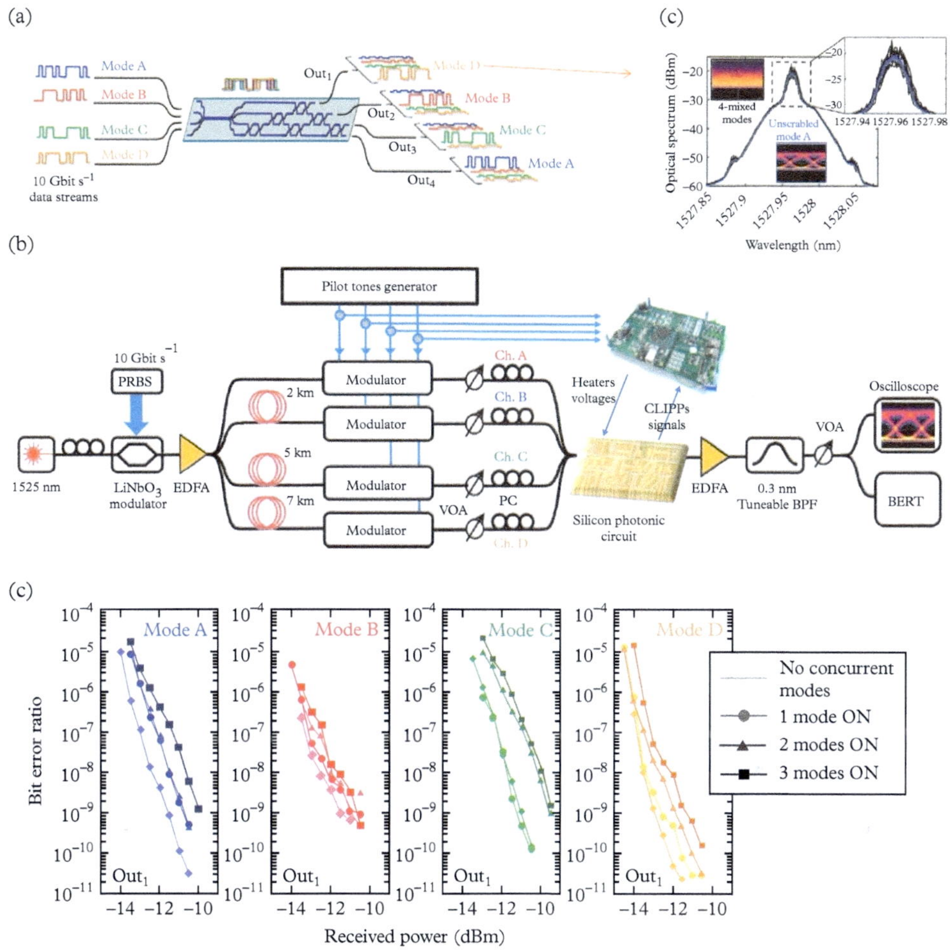

Figure 8.22 *On-chip unscrambling of 4 MDM optical channels. (a) Descriptive chip operation. (b) Experimental setup assembled in the laboratory. (c) Mode D reconstruction by the unscrambler at the output port 1. (d) BER measurements for each recovered mode demonstrating that information encoded in each channel can be retrieved with a very small power penalty independent of the number of mixed modes (after [80]).*

been reported for band selection and band rejection applications and some of the main results reported in the literature have already been considered in Chapter 3 and Chapter 5, respectively.

Programmable photonic filters can also be designed to exploit their phase, group delay or dispersion characteristics, e.g. a programmable tunable group delay module reported by Notaros and colleagues [82]. The module is based on the cascade of N single-cavity RRs (Figure 8.23(a)).

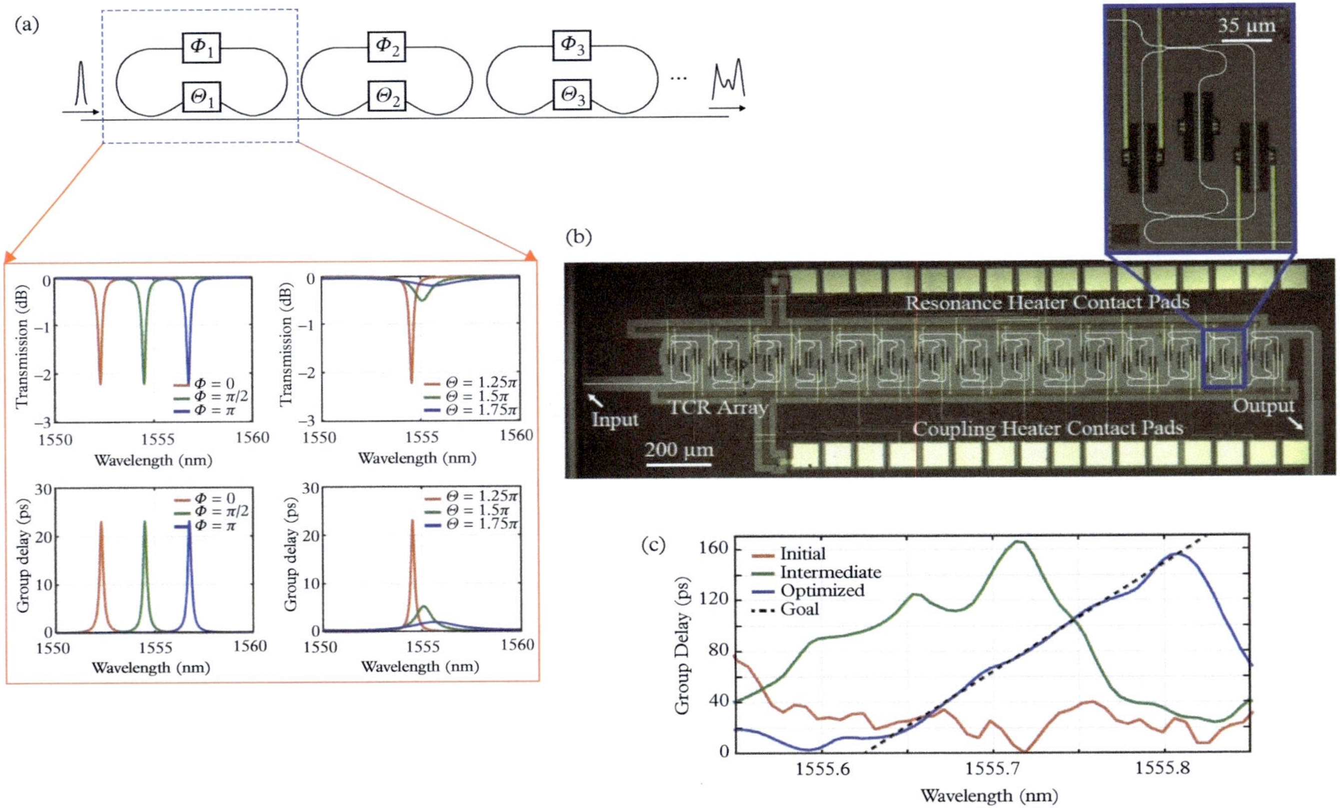

Figure 8.23 *Salient results of a programmable group delay module integrated in silicon and reported in [82]. (a) Schematic of the tunable-coupling ring array (TCRA) and operation of an individual cell. (b) Fabricated 15-ring tunable-coupling ring array device and inset showing a single tunable-coupling ring within the 15-ring array. (c) Optimization of the 15-ring tunable-coupling ring array device for an example goal group delay function (shown as a dashed line). The measured group delay spectra is shown in its initial state before the optimization has begun (solid red line), at an intermediate point within the optimization (solid green line), and in the final optimized state (solid blue line).*

Each ring cavity is closed by a 3dB-MZI tunable coupler and provides two degrees of freedom. In the inset of Figure 8.23(a), the phase Φ_i adjusts the position of the filter resonance within the available FSR, and the phase Θ_i adjusts the coupler value and hence, the cavity quality factor. Following the matrix descriptions provided in Chapter 2 for the BBBs of photonic integrated couplers and phase shifters one has for a given module 'k' the field transfer function and group delay are given by

$$T_k(\omega) = \frac{1 - e^{j\Theta_k} + 2e^{i[\beta(\omega)L_k + \Phi_k + \Theta_k]}\alpha_k}{2 - e^{i[\beta(\omega)L_k + \Phi_k]}\alpha_k + e^{i[\beta(\omega)L_k + \Phi_k + \Theta_k]}\alpha_k}$$
$$\tau_k(\omega) = -\frac{d}{d\omega}\left[\angle T_k(\omega)\right] \tag{8.9}$$

Where L_k is the length of the cavity, a_k represents the waveguide propagation losses and β the waveguide propagation constant for the fundamental mode. By cascading N of such elements, the result is

$$T_N(\omega) = \prod_{k=1}^{N} \frac{1 - e^{j\Theta_k} + 2e^{i[\beta(\omega)L_k + \Phi_k + \Theta_k]}\alpha_k}{2 - e^{i[\beta(\omega)L_k + \Phi_k]}\alpha_k + e^{i[\beta(\omega)L_k + \Phi_k + \Theta_k]}\alpha_k}$$
$$\tau_N(\omega) = -\frac{d}{d\omega}\left[\angle T_N(\omega)\right] = -\frac{d}{d\omega}\sum_{k=1}^{N} \angle T_k(\omega) \tag{8.10}$$

If losses and cavity lengths can be considered equal in all the modules, the group delay characteristic of Eqn. (8.10) is controlled by $2N$ degrees of freedom, which can be programmed to achieve a target group delay characteristic τ_{target}. Notaros and colleagues [82] reported that these coefficients are obtained by means of a non-linear optimisation algorithm for minimising the mean square error$\left|\tau_N - \tau_{\text{target}}\right|^2$. Figure 8.23(b) displays the optical micrograph of the device fabricated on a 200mm SOI wafer with 220 nm device layer thickness and 2 μm buried oxide thickness. A 248 nm photolithography was employed to define the resist patterns, based around 500 nm wide waveguides. Oxide was deposited on the top of the chip, aluminium routes are defined through the oxide to access active devices on the device layers, and aluminium contact pads were laid on the top oxide for contact to wire bonds or probes. Figure 8.23(c) shows as an example the optimisation procedure of the fifteen-ring tunable-coupling ring array device for an example goal group delay function, in this case a chromatic dispersion module featuring a dispersion parameter > 930 ps/mn over a bandwidth of around 0.15 nm, limited by the cavity length (430 μm).

Similar devices can also be programmed over a waveguide mesh. Figure 8.24 shows the results corresponding to a parabolic dispersion module based on a two-cavity SCISSOR emulated in a hexagonal waveguide mesh [44]. Figure 8.24(a) and Figure 8.24(b) show respectively the module layout and its emulation using the hexagonal waveguide mesh. Figure 8.24(c) shows the measured experimental results, which demonstrate a module providing 3.4 nsec/nm over a bandwidth of 5 GHz.

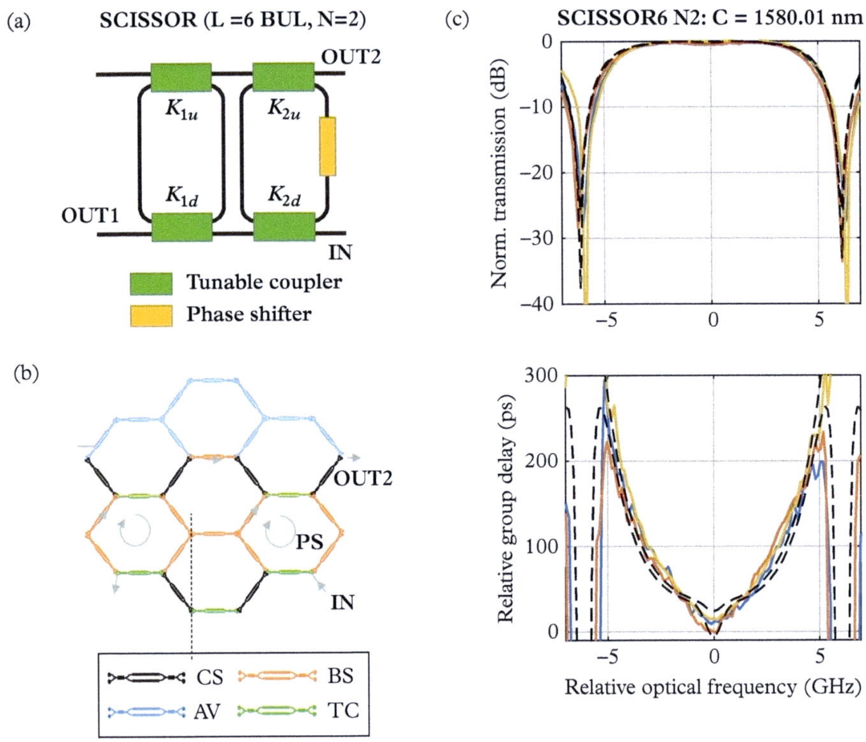

Figure 8.24 *Salient results of a programmable parabolic dispersion module integrated in silicon and reported in [44].(a) Circuit layout of the module (b) Circuit emulation on a 2d hexagonal integrated waveguide mesh (PS: Phase shifter, TC: 3dB MZI unit programmed as Tunable coupler, CS: 3dB MZI unit programmed as cross switch, BS: 3dB MZI unit programmed as Bar switch, AV: Available or unused 3dB MZI unit). (c) Transmission and group delay characteristics of the module for a given value set of the programmable parameters.*

8.4.3 Talbot Effect Processors

The temporal Talbot effect is a simple and efficient technique to change the repetition rate of a periodic train of optical pulses that has enabled a completely new area within the field of optical signal processing [83], [84]. It can be employed in an extensive list of applications that span laser physics, signal processing and telecommunications [85]–[88].

If we consider a periodic train of pulses (with repetition period given by T), its representation in the dual Fourier domain (i.e. the frequency spectrum) is a frequency comb with linear line spacing given by $f_r = 1/T$. The temporal Talbot effect is achieved when this comb spectrum is subject to the action of a frequency-domain parabolic phase filter [84]

$$H_T(\omega) = e^{j\frac{\theta\omega^2}{2}} \tag{8.11}$$

For the observation of the Talbot effect the parameter θ has to be given by [84]

$$2\pi \mid \theta \mid = \frac{p}{q} T^2 \tag{8.12}$$

where p and q are two mutually prime natural numbers. The transfer function given by Eqn. (8.11) has the form of that of a dispersive medium and this is the reason behind the fact that many of the practical implementations of the temporal Talbot effect employ optical fibres, where $q = \beta_2 z$, where β_2 is the first order dispersion coefficient of the fibre and z is the fibre length. A solution of Eqn. (8.12) with $q = 1$, also known as integer Talbot effect, results in output pulse trains corresponding to perfect reconstruction of the input train as if it never underwent dispersive propagation. If, however, $q > 1$ then we have a fractional Talbot effect in which output pulse trains are produced, where individual pulses are reconstructed but with a reduced pulse repetition period T/q. In summary, there is a repetition rate multiplication of the input pulse train and output pulses are referred to as temporal Talbot sub-images. Due to the requirement of energy conservation these output pulses have their energy content reduced by a factor of q. The basic equations for the temporal Talbot effect are the following. Given an input periodic pulse train (represented without loss of generality by a train of Dirac delta functions)

$$\psi(t) = \sum_{n=-\infty}^{\infty} \delta(t - nT) \Longleftrightarrow \tilde{\psi}(\omega) = \frac{2\pi}{T} \sum_{k=-\infty}^{\infty} \delta\left(\omega - \frac{2\pi k}{T}\right) \tag{8.13}$$

Then, the output signal from a medium providing a transfer function given by Eqns. (8.11)–(8.12) is given by [89]

$$\psi'(t) = \frac{e^{-j\pi c}}{\sqrt{q}} \sum_{n=-\infty}^{\infty} \delta\left(t - n\frac{T}{q} + p\frac{T}{2}\right) e^{-j\pi \frac{s}{m} n^2} \Longleftrightarrow \tilde{\psi}'(\omega) = \frac{2\pi}{T} \sum_{k=-\infty}^{\infty} \delta\left(\omega - k\frac{2\pi}{T}\right) e^{j\pi \frac{p}{q} k^2} \tag{8.14}$$

where s and c are given by Table 8.2:

Table 8.2 *Values of the s and c parameters in Eqn. (8.14). The operator [ ] stands for the integer part (after Romero Cortes [89]).*

Parity of p,q	s	c
p is Even and q Odd	$p\left(\left[\frac{1}{p}\right]_{\mathrm{mod}\,q}\right)^2$	$\frac{q-1}{4} + \frac{1-(p/q)}{2}$
p is Odd and q Odd	$8p\left[\frac{1}{2}\right]_{\mathrm{mod}\,q}\left(\left[\frac{1}{2p}\right]_{\mathrm{mod}\,q}\right)^2$	$\frac{q-1}{4} + \frac{1-(p/q)}{2}$
p is Odd and q Even	$p\left(\left[\frac{1}{p}\right]_{\mathrm{mod}\,q}\right)^2$	$-\frac{p}{4} - \frac{1-(p/q)}{2}$

Eqn. (8.14) and Table 8.2 include both the cases of integer and fractional Talbot effect. Note by comparing Eqn. (8.15) and Eqn. (8.14) that, while temporal Talbot sub-images are rate-multiplied replicas of the input pulse train, their power spectra is identical to that of the input train—a logical consequence of the fact that the involved transformation is simply a modification of the input phase distribution along the frequency domain. What is not so obvious, but however was demonstrated by Romero Cortes [89], is that the temporal Talbot sub-images acquire a temporal phase variation that follows a Talbot condition.

For the implementation of a system providing temporal Talbot we do not actually need a dispersive medium. It is enough to design a structure capable of providing a frequency-dependent parabolic phase shift as given by the spectral domain description of Eqn. (8.14). This can be readily be implemented by using a programmable integrated circuit, which can be reconfigured, if required, to provide different values of the multiplication factor q. Figure 8.25(a) shows a possible implementation based on a SCISSOR structure. The input periodic repetition pulse sequence is input on the upper waveguide bus that connects $N+1$ identical ring cavities, each one featuring a round-trip delay T and including an internal phase shifterϕ_i. Output signals from the rings are collected by the lower bus, which provides additional phase shifters φ_i and combines all the outputs towards the device exit. The operation of the device is as follows. Each ring cavity implements a band-pass spectral resonance between the upper and lower bus within the available free spectral range given by FSR $= 1/\text{T}$. The internal phase shifter in each cavity is tuned to select one and only one of the carriers in the optical comb of the input signal (Figure 8.25(b)). This carrier is passed onto the lower bus, while the rest are not affected by the cavity filter and proceed to the next ring cavity. In order to sequentially select a different carrier of the frequency comb, the phase shifter in cavity $i = k, k = -N, -n+1, \ldots 0, 1, N$ has to be programmed to provide a phase shift given by

$$\phi_k = \frac{2\pi}{N}k \tag{8.15}$$

The phase shifters in the lower bus have to be programmed to: (a) provide the required Talbot phase shift for each filtered carrier in the comb as given by Eqn. (8.14); (b) compensate for the inter-cavity phase shifts Δ in the upper bus and (c) to compensate for the action of the phase shifters for other frequencies in the lower bus. This can be achieved by setting the following values

$$\varphi_k = \pi\,(p/q)\,k^2 - \sum_{r=-N/2}^{k-1} \varphi_r - \left(\frac{N}{2} + k\right)\Delta \tag{8.16}$$

Note that the filtered and phase shifted carriers in the lower bus will be bypassed by all the cavities encountered upon propagation from right to left. At the left end of the lower bus a combined signal with spectrum given by Eqn. (8.14) will be collected.

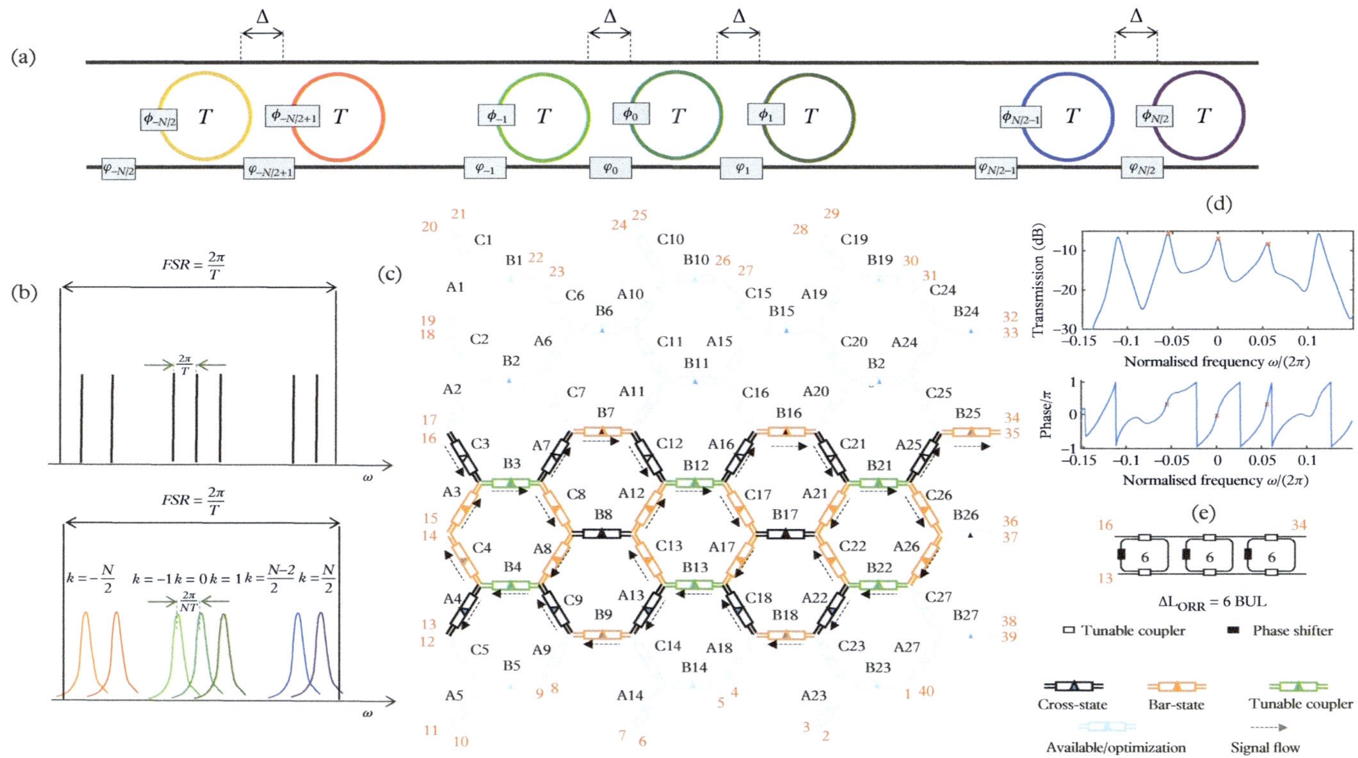

Figure 8.25 *Implementation of the temporal Talbot processor using programmable photonic circuits. (a)an implementation based on a N+1 SCISSOR. (b) Frequency selection of the input comb lines using programmable in-cavity phase shifters. (c) Temporal Talbot processor using anhexagonal waveguide mesh emulating a 3 cavity SCISSOR. (d) Simulation results with a three comb input signal showing the discrete parabolic phase response. The SCISSOR filter is emulated using an hexagonal waveguide mesh.*

An interesting feature of this configuration is that the pulse repetition rate can be changed by programming a new value for q in the phase shifters of the lower bus. It is also possible to emulate this temporal Talbot processor by means of an integrated waveguide mesh, as these are capable (see Chapter 5) of emulating SCISSOR structures. Moreover, the external TBUs connecting the cavities can be programmed to provide the required phase shifting values in the lower bus. Figure 8.25(c) shows an example of a possible configuration for a temporal Talbot processor using a hexagonal waveguide mesh. Experimental results for this application have not yet been proposed at the time of writing, but a simulation of a designed suture for processing a three-comb input signal using an hexagonal waveguide mesh shown in Figure 8.25(d) shows that it is possible to achieve the required parabolic phase dependence required by Eqn. (8.14).

8.4.4 Emerging and Future Applications

The availability of programmable integrated photonic processors opens interesting and exciting research avenues in many other classical fields, some of which are yet to be proposed and explored. One example is in the area of biophotonic sensors, where the capability of supporting both of simple and multiple input/multiple output interferometric structures for lab-on-a-chip and multi-parameter-sensing applications enable the future implementation of multi-parameter integrated photonic sensing [90], [91].

A second topic of interest is the emulation of complex physical systems with discrete energy levels. These are analogous to artificial atom- and molecule-like systems activated by light, which can enable the coherent and dynamic control of the frequency, amplitude and phase of photons, with applications in the understanding of both in classical and quantum systems. Recent experimental research [92] has demonstrated a photonic molecule with two different energy levels based on coupled lithium niobate MRRs controlled by a microwave signal. Waveguide mesh structures might provide the perfect platform for the emulation of even more complex molecular structures and assorted dynamic processes including Stark shifts, Rabi oscillations and Autler–Townes splitting.

In this same context of advanced physics concepts, programmable integrated photonics can provide a compact platform to implement programmable 2D topological systems such as multi-ring cavity structures to support research in synthetic dimensions and devices based on topological insulator [93]–[96] and inverse design [97], [98] principles.

8.5 Chapter Summary

Programmable integrated photonics has a considerable number of applications in classical physics and engineering systems. This chapter reviewed some of the most important fields of application. Due to space restrictions and the emerging nature of programmable photonics as a topic, the chapter sticks to fields in which there is at least some representative literature.

First it reviewed applications to optical switching, which is a key technology in telecommunication systems and also in emerging data centre communications and multiprocessor interconnection. Switching is perhaps the field where more contributions have been published that are based on programmable photonic chips. While the topics are very broad, this chapter investigated a selected set of approaches and results that may profit from generic chip programmability. It then described the application to an emerging field where integrated programmable photonics will play a significant role: AI and neurophotonics. The distributed nature of connections that can be enabled by waveguide mesh and/or ring based photonic chips and the analogue nature of signal processing provides a unique platform for the future implementation of neural networks with distinctive properties compared to electronic approaches. Here programmability brings the significant added value of enabling weight reconfigurability in the links connecting artificial neurons. A third area of application is in microwave or RF photonics, where programmability provided by photonic integrated circuits combined with efficient RF and electronic interfaces and again the analogue nature of the involved signals provides the possibility of programming the different required functionalities in a single chip. This can be a clear advantage for the implementation of compact and versatile RF-photonic transceivers, called to play a significant role in future 5G and beyond communications.

It also described several applications that fell into the realm of the so-called optical signal processing area. The functionalities visited here included optical channel filtering, optical mode transformation and unscrambling, tunable dispersion modules and temporal Talbot effect processors.

Many more applications can be envisaged and only time will tell in which sense programmable PICs will be efficiently employed for their implementation. Many of these applications could be supported using a common type of integrated photonic hardware that would be programmed to enable the user to employ different topologies and architectures. If this could be achieved in addition in a compact and application-agnostic configuration for a photonic integrated processor it could result in a paradigmatic change in integrated photonics. The field programmable photonic gate array (FPPGA) introduced in Chapter 9 might well be the suitable device to carry this revolution.

REFERENCES

[1] B.G. Lee and N. Dupuis, Silicon Photonic Switch Fabrics: Technology and Architecture. (2019) *IEEE J Lightwave Technol* 37, 6–20.

[2] Q. Cheng, S. Rumley, M. Bahadori and K. Bergman, Photonic Switching in High-performance Datacenters. (2018) *Opt Express* 26, 16022–43.

[3] A. Wonfor, et al., Large Port Count High-speed Optical Switch Fabric for Use within Datacenters. (2011) *J Opt Commun Netw* 3(8), A32–9.

[4] K. Hamamoto, et al., First 8×8 Semiconductor Optical Matrix Switches Using GaAs AlGaAs Electro-optic Guided-wave Directional Couplers. (1992) *Electron Lett* 28(5), 441.

[5] J. Van Campenhout, et al., Low-power, 2×2 Silicon Electro-optic Switch with 110-nm Bandwidth for Broadband Reconfigurable Optical Networks. (2009) *Opt Express* 17(26), 24020–9.

[6] L. Soldano and E. Pennings, Optical Multi-mode Interference Devices Based on Self-imaging: Principles and Applications. (1995) *IEEE J Lightwave Technol* 13, 615–27.

[7] N. Dupuis, et al., Design and Fabrication of Low-insertion-loss and Low-crosstalk Broadband 2×2 Mach–Zehnder Silicon Photonic Switches. (2015) *IEEE J Lightwave Technol* 33, 3597–606.

[8] A.W. Poon, et al., Cascaded Microresonator-based Matrix Switch for Silicon on-chip Optical Interconnection. (2009) *Proc IEEE* 97, 1216–38.

[9] X. Luo, et al., Silicon High-order Coupled-microring-based Electrooptical Switches for On-chip Optical Interconnects. (2012) *IEEE Photon Technol Lett* 24, 821–3.

[10] P. DasMahapatra, et al., Optical Crosspoint Matrix Using Broadband Resonant Switches. (2014) *IEEE J Sel Top Quantum Electron* 20, 1–10.

[11] M.-C.M. Lee and M.C. Wu, Tunable Coupling Regimes of Silicon Microdisk Resonators Using MEMS Actuators. (2006) *Opt Express* 14, 4703–12.

[12] J. Yao, et al., Silicon Microtoroidal Resonators with Integrated MEMS Tunable Coupler. (2007) *IEEE J Sel Top Quantum Electron* 13, 202–8.

[13] S. Han, et al., Large-scale Silicon Photonic Switches with Movable Directional Couplers. (2015) *Optica* 2, 370–5.

[14] T.J. Seok, et al., Large-scale Broadband Digital Silicon Photonic Switches with Vertical Adiabatic Couplers. (2016) *Optica* 3, 64–70.

[15] L. Lu, et al., 16×16 Non-blocking Silicon Optical Switch Based on Electro-optic Mach–Zehnder Interferometers. (2016) *Opt Express* 24, 9295–307.

[16] L. Qiao, et al., 32 × 32 Silicon Electro-optic Switch with Built-in Monitors and Balanced-status Units. (2017) *Sci Rep* 7, 1–7, 42306.

[17] Z. Guo, et al., 16×16 Silicon Optical Switch Based on Dual-ring-assisted Mach–Zehnder Interferometers. (2018) *IEEE J Light Technol* 36, 225–32.

[18] S. Han, et al., Large-scale Polarization-insensitive Silicon Photonic MEMS Switches. (2018) *IEEE J Light Technol* 36, 1824–30.

[19] H.Y. Hwang, et al., Flip Chip Packaging of Digital Silicon Photonics MEMS Switch for Cloud Computing and Data Centre. (2017) *IEEE Photon J* 9, 1–10.

[20] A.S.P. Khope, et al., Elastic WDM Optoelectronic Crossbar Switch with On-chip Wavelength Control. Proceedings of Advanced Photonics 2017, paper PTh1D.3, 24–27 July 2017, New Orleans, LA.

[21] M. Reck, A. Zeilinger, H.J. Bernstein and P. Bertani. Experimental Realization of Any Discrete Unitary Operator. (1994) *Phys Rev Lett* 73, 58–61.

[22] R.A. Spanke and V.E. Benes, N-stage planar optical permutation network. (1987) *Appl Opt* 26, 1226–9.

[23] T. Zhou, et al., Rearrangeable Nonblocking Five-port Silicon Optical Switch for 2D-mesh Network-on-chip. (2018) *IEEE Photon J* 10, 6601208.

[24] J. Xing, et al., Nonblocking 4×4 Silicon Electro-optic Switch Matrix with Push-pull Drive. (2013) *Opt Lett* 38, 3926–9.

[25] W.R. Clements, et al., Optimal Design for Universal Multiport Interferometers. (2016) *Optica* 3, 1460–5.

[26] K. Suzuki, et al., Ultra-compact 8×8 Strictly-non-blocking Si-wire PILOSS Switch. (2014) *Opt Express* 22, 3887–94.

[27] K. Tanizawa, et al., Non-duplicate Polarization-diversity 8×8 Si-wire PILOSS Switch Integrated with Polarization Splitter-rotators. (2017) *Opt Express* 25, 10885–92.
[28] S.K. Hiroki Kouketsu, et al., High-speed and Compact Non-blocking 8×8 InAlGaAs InAlAs Mach–Zehnder-type Optical Switch Fabric. *Proceedings of Optical Fiber Communication Conference 2014*, paper M2K.3, 9–13 March 2014, San Francisco, CA.
[29] M. Ding, et al, Hybrid MZI-SOA InGaAs/InP Photonic Integrated Switches. (2018) *IEEE J Sel Top Quantum Electron* 24, 3600108.
[30] S. Liu, et al., 'Low Latency Optical Switch for High Performance Computing with Minimized Processor Energy Load. (2015) *J Opt Commun Netw* 7, A498.
[31] K. Tanizawa, et al., Ultra-compact 32×32 Strictly Non-blocking Si-wire Optical Switch with Fan-out LGA Interposer. (2015) *Opt Express* 23, 17599–606.
[32] D. Celo, et al., 32×32 Silicon Photonic Switch. Proceedings of the 21st OptoElectronics and Communications Conference (OECC), 3–7 July 2016, Niigata, Japan.
[33] L. Qiao, W. Tang and T. Chu, 32×32 Silicon Electro-optic Switch with Built-in Monitors and Balanced-status Units. (2017) *Sci Rep* 7, 42306.
[34] T. Ferreira de Lima, et al., Progress in Neuromorphic Photonics. (2017) *Nanophotonics* 6, 577–99.
[35] P.R. Prucnal and B.J. Shastri, *Neuromorphic Photonics*, CRC Press, 2107.
[36] P.R. Prucnal, et al., Recent Progress in Semiconductor Excitable Lasers for Photonic Spike Processing. (2016) *Adv Opt Photon* 8, 228–99.
[37] A.N. Tait, et al., Photonic Neuromorphic Signal Processing and Computing. In M. Naruse (Ed), *Nanophotonic Information Physics*, 183–222, Springer, 2014.
[38] K. Vandoorne, et al., Experimental Demonstration of Reservoir Computing on a Silicon Photonics Chip. (2014) *Nat Commun* 5, 3451.
[39] A.N. Tait, et al., Continuous Calibration of Microring Weights for Analog Optical Networks. (2016) *IEEE Photon Technol Lett* 28, 887–90.
[40] A.N. Tait, et al., Multichannel Control for Microring Weight Banks. (2016) *Opt Express* 24, 8895–906.
[41] H. Peng, et al., 'Neuromorphic Photonic Integrated Circuits. (2018) *IEEE J Sel Top Quantum Electron* 24(6), Art no. 6101715.
[42] A.N. Tait, et al., Feedback Control for Microring Weight Banks. (2018) *Opt Express* 26, 26422–43.
[43] D.E. Rumelhart, G.E. Hinton and R.J. Williams, Learning Internal Representations by Error Propagation. In D.E. Rumelhart and J.L. McClelland (Eds), Parallel Distributed Processing, 318–62, MIT Press, 1986.
[44] D. Pérez, et al., Multipurpose Silicon Photonics Signal Processor Core. (2017) *Nat Commun* 8, 636.
[45] T.F. de Lima, et al., Machine Learning with Neuromorphic Photonics. (2019) *J Light Technol* 37, 1515–34.
[46] Y. Shen, et al., Deep Learning with Coherent Nanophotonic Circuits. (2017) *Nat Photonics* 11, 441–7.
[47] T.W. Hughes, M. Minkov, Y. Shi and S. Fan, Training of Photonic Neural Networks through In-situ Backpropagation and Gradient Measurement. (2018) *Optica* 5, 864–71.
[48] D. Marpaung, et al., Integrated Microwave Photonics. (2013) *Laser Photon Rev* 7, 506–38.
[49] D. Marpaung, J.P. Yao and J. Capmany, Integrated Microwave Photonics. (2019) *Nat Photonics* 13, 80–90.

[50] R. Boeck, N.A.F. Jaeger, N. Rouger and L. Chrostowski, Series-coupled Silicon Racetrack Resonators and the Vernier Effect: Theory and Measurement. (2010) *Opt Express* 18, 25151–7.

[51] D. Pérez, I. Gasulla and J. Capmany, Software-defined Microwave Photonics Processor: Concept, Core Design, Equations and Performance Analysis. (2015) *Opt Express* 23(11), 14640–54.

[52] M. Rasras, et al., Integrated Resonance-enhanced Variable Optical Delay Lines. (2005) *Photon Tech Lett* 17, 834–6.

[53] M. Burla, et al., Separate Carrier Tuning Scheme for Integrated Optical Delay Lines in Photonic Beamformers, *2011 International Topical Meeting on Microwave Photonics (MWP 2011) jointly held with the 2011 Asia-Pacific Microwave Photonics Conference (APMP 2011).* 18–21 October 2011, Singapore.

[54] D. Marpaung, et al., On-chip CMOS Compatible Reconfigurable Optical Delay-line with Separate Carrier Tuning for Microwave Photonic Signal Processing. (2011) *Opt Express* 19, 21475–84.

[55] E.J. Norberg, et al., Programmable Photonic Microwave Filters Monolithically Integrated in InP–InGaAsP. (2011) *J Light Technol* 29, 1611–19.

[56] H. Chen, et al., Integrated Microwave Photonic Filter on a Hybrid Silicon Platform. (2010) *IEEE Microw Wirel Compon Lett* 58, 3213–19.

[57] R.S. Guzzon, et al., Integrated InP-InGaAsP Tunable Coupled Ring Optical Bandpass Filters with Zero Insertion Loss. (2011) *Opt Express* 19(8), 7816–26.

[58] L. Zhuang, et al., Programmable Photonic Signal Processor Chip for Radiofrequency Applications. (2015) *Optica* 2, 854–9.

[59] J.S. Fandiño, P. Muñoz, D. Doménech and J. Capmany, A Monolithic Integrated Photonic Microwave Filter. (2017) *Nat Photon* 1, 124–9.

[60] M. Sagues, *Aplicaciones de la correspondencia entre el dominio óptico y eléctrico establecida por la modulación óptica en banda lateral única.* PhD Thesis (in Spanish) Universidad Pública de Navarra, (2009).

[61] D. Pérez, et al., Figures of merit for self-beating filtered microwave photonic systems. (2016) *Opt Express* 24, 10087–102.

[62] C.K. Madsen and J.H. Zhao, *Optical Filter Design and Analysis: A Signal Processing Approach*, John Wiley & Sons, Inc., 1999.

[63] D. Pérez, I. Gasulla, J. Capmany and R.A. Soref, Reconfigurable Lattice Mesh Designs for Programmable Photonic Processors. (2016) *Opt Express* 24, 12093–106.

[64] D. Perez, I. Gasulla and J. Capmany, Software-defined Universal Microwave Photonics Processor. (2015) *Opt Express* 23, 14640–54.

[65] G. Qi, et al., Generation and Distribution of a Wide-band Continuously Tunable Millimeter-wave Signal with an Optical External Modulation Technique. (2005) *IEEE Microw Wirel Compon Lett* 53, 3090–7.

[66] L. Maleki, Sources: The Optoelectronic Oscillator. (2011) *Nat Photon* 5, 728–30.

[67] J. Tang, et al., Integrated optoelectronic oscillator. (2018) *Opt Express* 26, 12257–65.

[68] K.-P. Ho, S. Liaw and C. Lin, Efficient Photonic Mixer with Frequency Doubling. (1997) *IEEE Photon Tech L* 9, 511–13.

[69] Y. Gao, et al., An Efficient Photonic Mixer with Frequency Doubling Based on a Dual-parallel MZM. (2014) *Opt Commun* 321, 11–15.

[70] K.-P. Ho, S. Liaw and C. Lin, Efficient Photonic Mixer with Frequency Doubling, (1997) *IEEE Photon Tech L* 9, 511–13.

[71] I.S. Lin, J.D. McKinney and A.M. Weiner, Photonic Synthesis of Broadband Microwave Arbitrary Waveforms Applicable to Ultra-wideband Communication. (2005) *IEEE Microw Wirel Compon Lett* 15, 226–8.
[72] M.H. Khan, et al., Ultrabroad-bandwidth Arbitrary Radiofrequency Waveform Generation with a Silicon Photonic Chip-based Spectral Shaper. (2010) Nat Photon 4, 117–22.
[73] M. Burla, *Advanced Integrated Optical Beam Forming Networks for Broadband Phased Array Antenna Systems*. PhD Thesis, University of Twente, 2013.
[74] L. Zhuang, et al., Single-Chip Ring Resonator-Based 1×8 Optical Beam Forming Network in CMOS-compatible Waveguide Technology. (2007) *IEEE Photon Tech Lett* 19, 1130–2.
[75] J.S. Fandiño and P. Muñoz, Photonics-based Microwave Frequency Measurement Using a Double-sideband Suppressed-carrier Modulation and an InP Integrated Ring-assisted Mach–Zehnder Interferometer Filter. (2013) *Opt Lett* 38, 4316–19.
[76] M. Pagani, et al., Instantaneous Microwave Frequency Measurement Using Four-wave Mixing in a Chalcogenide Chip. (2016) *Opt Commun* 373, 100–4.
[77] D. Marpaung, On-Chip Photonic-Assisted Instantaneous Microwave Frequency Measurement System. IEEE Photon. Tech. Lett., 25, 837–40 (2013).
[78] H. Shao, et al., Large Bandwidth and High Accuracy Photonic-assisted Instantaneous Microwave Frequency Estimation System Based on an Integrated Silicon Micro-resonator, in *Proceedings of the 2014 IEEE 11th International Conference on Group IV Photonics (GFP)*, 27–29 August 2014, Paris, France.
[79] A. Ribeiro, A. Ruocco, L. Vanacker and W. Bogaerts, Demonstration of a 4×4-port Universal Linear Circuit. (2016) *Optica* 3, 1348–57.
[80] A. Annoni, et al., Unscrambling Light—Automatically Undoing Strong Mixing Between Modes. (2017) *Light Sci Appl* 6, e17110.
[81] F. Morichetti, et al., Non-invasive On-chip Light Observation by Contactless Waveguide Conductance Monitoring. (2014) *IEEE J Sel Top Quantum Electron* 20, 292–301.
[82] J. Notaros, et al., Programmable Dispersion on a Photonic Integrated Circuit for Classical and Quantum Applications. (2017) *Opt Express* 25, 21275–85.
[83] T. Jannson and J. Jannson, Temporal Self-Imaging Effect in Single-mode Fibers. (1981) *J Opt Soc Am* 71, 1373–6.
[84] J. Azaña and M.A. Muriel, Temporal Self-Imaging Effects: Theory and Application for Multiplying Pulse Repetition Rates. (2001) *IEEE J Sel Top Quantum Electron* 7, 728–44.
[85] H. Guillet de Chatellus, et al., Theory of Talbot Lasers. (2013) *Phys Rev A* 88, 033 828.
[86] J. Azaña and M.A. Muriel, Temporal Talbot Effect in Fiber Gratings and its Applications. (1999) *Appl Opt* 38, 6700–4.
[87] J. Azaña and L.R. Chen, General Temporal Self-Imaging Phenomena. (2003) *J Opt Soc Am B* 20, 1447–58.
[88] D. Pudo, M. Depa and L.R. Chen, Single and Multiwavelength All-optical Clock Recovery in Singlemode Fiber Using the Temporal Talbot Effect. (2007) *J Lightwave Technol* 25, 2898–903.
[89] L. Romero Cortes, *Generalized Talbot Effect Theory and Application to Advanced Optical Wave Processing*. PhD Thesis, INRS, 2018.
[90] M.C. Estevez, M. Alvarez and L. Lechuga, Integrated Optical Devices for Lab-on-a-chip Biosensing Applications. (2012) *Laser Photon Rev* 6, 463–87.
[91] R. Heideman, M. Hoekman and E. Schreuder, TriPleX-based Integrated Optical Ring Resonators for Lab-on-a-chip and Environmental Detection. (2012) *IEEE J Sel Topics Quant Electron* 18, 1583–96.

[92] M. Zhang, et al., Electronically Programmable Photonic Molecule. (2019) *Nat Photon* 13, 36–40.
[93] T. Ozawa, et al., Synthetic Dimensions in Integrated Photonics: From Optical Isolation to Four-dimensional Quantum Hall Physics,' Phys Rev A, 93, 043827 (2016).
[94] L. Yuan, Q. Lin, M. Xiao and S. Fan, Synthetic Dimensions in Photonics. (2018) *Optica* 11, 1396–405.
[95] G. Harari, et al., Topological Insulator Laser: Theory. (2018) *Science* 359, DOI: 10.1126/science.aar4003.
[96] M.A. Bandres, et al., Topological Insulator Laser: Experiments. (2018) *Science* 359, DOI: 10.1126/science.aar400.
[97] S. Molesky, et al., Inverse Design in Nanophotonics. (2018) *Nat Photon* 12, 659–70.
[98] N. Estakhri, B. Edwards and N. Engheta, Inverse-designed Metastructures that Solve Equations. (2019) *Science* 363, 1333–8.

9

Field Programmable Photonic Gate Arrays

9.1 An Introduction to Field Programmable Gate Arrays

Electronic field programmable gate arrays (FPGAs) were introduced in 1984 by XilinX [1]–[3] a company founded by Ross Freeman, James Barnet and Bernie Voderschmitt. In the original patent application filed by Ross Freeman [4] the device was described as a configurable electrical circuit having configurable logic elements (CLEs) and configurable interconnects. Figure 9.1 depicts an original drawing of the cited patent application for a particular FPGA configuration using nine logic elements shown as black boxes and the corresponding interconnection circuitry. Interestingly, the term FPGA was introduced in 1988 by the competing firm Actel, and not by XilinX, which instead used configurable logic array (CLA) [3].

The process involving the proposal of the FPGA and its subsequent patenting followed after a period of forty years where electronics experienced two notable ground breaking milestones (Figure 9.2): Shockley, Bardeen and Brattain's invention of the transistor in 1947, and Kilby and Noyce's proposal in 1959 of the first integrated monolithic circuits that enabled the emergence of a twenty-year period where application-specific integrated circuits (ASICs) dominated the research and commercial areas of electronics.

Reid [5] suggests that the origin of programmable electronics can be traced back to the late sixties; by that time, Noyce was worried about the rapid proliferation of different integrated circuits, each designed for its own special purpose. Looking ahead, Noyce saw that the solution to proliferation of special-purpose integrated circuits would be the development of general-purpose chips that could be manufactured in huge quantities and adapted ('programmed') for specific applications. Nevertheless—and despite the fact that this period ushered in the first microprocessors—the proposal for a more general programmable device with applications beyond mere computing would need to wait for almost two decades.

9.1.1 Advantages of FPGAs over ASICs

The reasons for the success of programmable over electronic ASICs can be better understood if the advantages of FPGAs over ASICs are analysed in retrospective [2], [3].

Programmable Integrated Photonics. José Capmany and Daniel Pérez, Oxford University Press (2020). © José Capmany and Daniel Pérez.
DOI: 10.1093/oso/9780198844402.001.0001

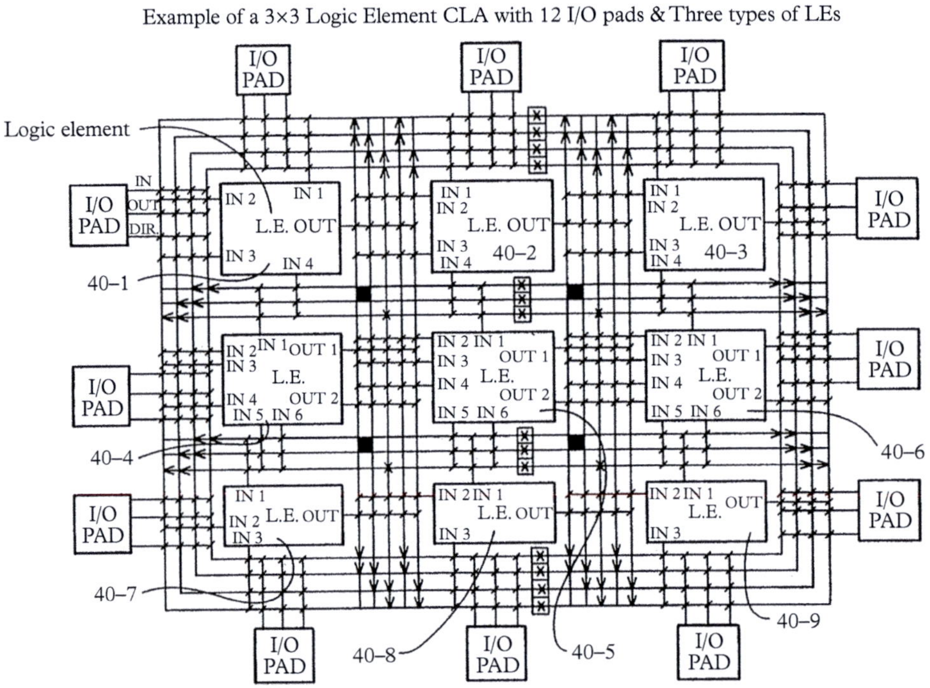

Figure 9.1 *Schematic diagram of a 3 × 3 FPGA as proposed in the original patent application (after [4]).*

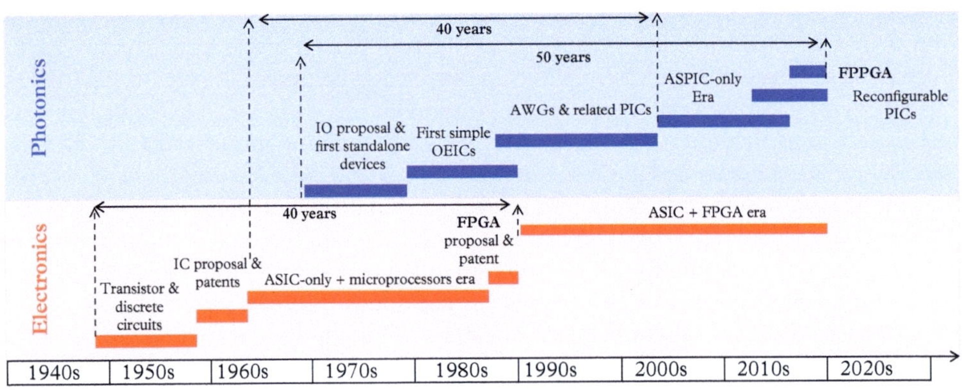

Figure 9.2 *Significant milestones and evolution periods in integrated electronics and photonics.*

This analysis also proves useful in explaining why a similar behaviour might be expected in the context of integrated photonic circuits when comparing application-specific photonic integrated circuits (ASPICs) with field programmable photonic gate arrays (FPPGAs).

In the 1980s ASIC companies enabled the 'build to order', custom integrated circuits featuring low cost, high capacity and speed as winning attributes. When FPGAs first appeared, they compared poorly with ASICS in all these features, yet they succeeded for a variety of reasons. In ASICs the functionality is determined by custom mask tooling and customers must pay for those masks with an upfront non-recurring engineering (NRE) charge. Conversely, FPGAs do not have custom tooling and hence their upfront NRE costs are reduced or even negligible since the custom mask can serve hundreds or thousands of users and are thus amortised over all of them. Up-front cost reductions in NRE guarantees that, up to a certain fabrication volume, FPGAs are more cost-effective than ASICs. To explain this point, we assume that ASICs have an initial *NRE* cost and that each fabricated unit U adds its unit cost $C_{U,ASIC}$ to the total. FPGAs have no NRE costs and each subsequent unit U adds its cost $C_{U,FPGA}$ (higher than that of ASIC units) to the total. Thus

$$\begin{aligned} C_{ASIC} &= C_{U,ASIC}U + NRE \\ C_{FPGA} &= C_{U,FPGA}U \\ C_{U,ASIC} &< C_{U,FPGA} \end{aligned} \tag{9.1}$$

Figure 9.3 illustrates the model given by Eqn. (9.1).

The cross-point that gives the number of maximum fabricated units U_{cr} for which FPGAs are a more cost-effective than ASICS is given by

$$U_{cr} = \frac{NRE}{\left(C_{U,FPGA} - C_{U,ASIC}\right)} \tag{9.2}$$

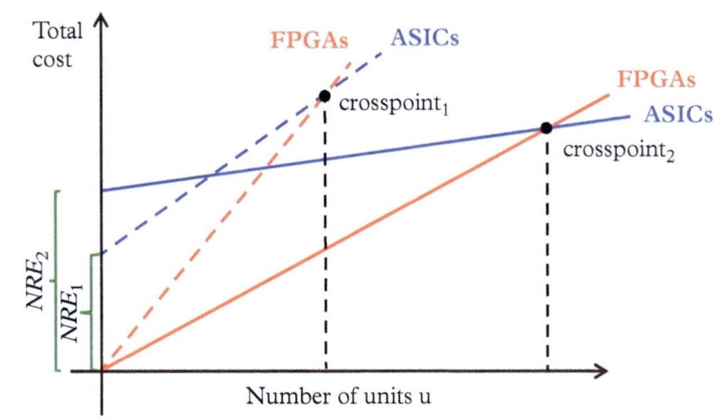

Figure 9.3 *Illustrates the model given by (9.1).*

With time, unit costs (slopes) decreased while NRE increased as circuits became more complex (and mask tooling more expensive), shifting cross-points to the right and thus making FPGAs a more interesting option for increased fabrication volumes (Figure 9.3).

While fabrication cost was a determinant issue during the first years of coexistence between ASICs and FPGAs, this is no longer the case when deciding between these two options, as other indicators like performance, time to market, power consumption and input/output (I/O) capacity (among others) are deemed more priority.

It was not only that FPGAs eliminated up-front costs, but also they decreased design costs by eliminating a wide class of design problems, e.g. transistor-level design, testing, crosstalk, signal integrity, etc. Equally important, FPGAs provide instant availability and reduced visibility of failures. ASICs were rarely correct the first time they were designed and fabricated despite the use of intensive simulation. Since wafer turnaround times are on the order of weeks or months, this impacts negatively the final product delivery and raises costs due to the need of repeated fabrication. Furthermore, ASICs demand extensive chip verification. On the contrary, FPGAs can be reworked in minutes so they are not subject to the same lengthy delay as a consequence of a failure or fabrication error and can offer verification by self-emulation.

Another key point is related to the fact that ASICs were subject to high production risks. An ASIC company only made money when its customer designs went into production and in reality, only around one-third of designs fulfilled this condition (i.e. two-thirds of ASIC designs lost money). Losses were incurred not only by ASIC customers, but also by suppliers as the NRE charges hardly covered actual costs. On the other hand, companies producing programmable circuits (and their customers) could make money on small volumes, while small errors could be corrected without assuming costly mask making.

9.1.2 The Three Ages of FPGA Development

FPGAs have undergone three different development phases [3], each one lasting around seven years. In the first phase between 1984 to 1991, the critical requirement was cost containment. This implied the need to maximise efficiency in cost and functionality and to eliminate waste in logic blocks. It was during this phase that efficient interconnection architectures would require 2D topologies and short connections between logic blocks. FPGAs were small and their design problems were small, so component synthesis, placement and routing were done manually. In fact, the many different FPGA architectures available at the time precluded the existence of universal design tools, which are commonplace in ASIC design. Since FPGAs were smaller than required, multiple-FPGA systems were popular. Manufacturing companies developed their captive software, the cost of which was amortised among multiple users.

The second phase of development or age of expansion was characterised by the fact that Moore's law enabled doubling the size of the largest FPGAs and halving the cost per function; therefore, the available area became a less precious resource. This also enabled smaller and simpler logic units and the inclusion of more complex logic blocks. Furthermore, the stacking of more metal layers allowed reducing the costs of interconnects, and longer connections—not only nearest neighbours—became feasible. With increased complexity and size, design automation (placing, routing synthesis) became an essential requirement and companies that developed and controlled their own software for these tasks were eventually those that survived; those who relied on open software solutions disappeared. It was during this period where, after a hesitant start, FPGAs started to address and absorb part of the ASIC market.

Finally, in the third phase of development between 2000 to 2007, larger-size FPGAs were no longer considered essential, despite Moore's law still enabling growth in capacity and design size; this was mainly because customers had to pay for a premium prize in return for the extra capacity offered. At the low end, customers started to substitute ASICs by FPGAs even in systems with low complexity, and they demanded a lower-cost product. At the high end, customers were interested in aggregating multiple and potentially unrelated components (e.g. dedicated logic block, memories, microprocessors, extra I/O ports, analogue interfaces, etc.) known as high-performance blocks (HPBs) into the FPGA. Hence, the FPGA no longer was a collection of logical functions and millions of gates, but rather a set of predefined high-performance dedicated blocks integrated with programmable logic. FPGA business grew in this period from general ASIC replacement to a full embrace of customised communications transceivers.

All in all, this thirty-year period comprised between their inception in 1984 and 2014 FPGAs have increased their capacity by a factor of 10,000, their speed by a factor of 100 and decreased their cost and power consumption by a factor of 100.

9.2 Beyond a Programmable Integrated Waveguide Mesh

Throughout this book we stress the objective that lies behind programmable integrated photonics (PIP): the design of common integrated optical hardware configurations able to implement a wide variety of functionalities by suitable programming. From Chapter 3 to Chapter 6 we studied configurations and design principles for programmable circuits based on waveguide meshes built upon the cascade of either integrated 3-dB Mach–Zehnder tunable couplers, beamsplitters and directional couplers. While these offer versatile hardware solutions to the implementation of programmable circuits, none of them defines a complete architectural solution of a photonics device that could be programmed for the implementation of arbitrary simple, complex or even simultaneous circuits. Clearly stated, PIPs need more than just a programmable integrated waveguide mesh.

In electronics, this concept is sustained by FPGAs reviewed in Section 9.1 and by field programmable analogue arrays (FPAAs) [6]–[9] and following a similar rationale behind the principles of these devices one could consider the implementation of a similar concept that can be realised by combining a set of programmable photonics analogue blocks (PPABs) and a set of reconfigurable photonic interconnects (RPIs) implemented over a photonic chip. The FPPGA [10] can be able of implementing one or various simultaneous photonics circuits and/or linear multi-port transformations by the appropriate programming of its resources (i.e. PPABs and RPIs) and the selection of its input and output ports. We first provide the description of the high-level FPPGA concept, the minimum basic functionalities that both PPABs and RPIs need to provide as well as the required architecture. We then discuss possible compacts implementations of this concept by means of two-dimensional integrated photonic waveguide meshes and additional high-performance processing blocks.

9.2.1 The High-level Concept of FPPGAs

The high-level concept of the proposed FPPGA core is schematically shown in Figure 9.4[10].

It consists of a set of programmable photonic analogue blocks (PPABs) and a set of RPIs implemented through an array of photonic waveguide elements grown on a photonic chip substrate. The waveguide elements that compose the RPIs have programmable features as well and can propagate light in both directions. Note that the layout in Figure 9.4 does not presuppose any particular waveguide array geometry and that the square layout depicted there is just for illustration purposes.

The FPPGA concept brings a series of advantages inherent to field programmable hardware approaches, some of which are similar to that of FPGA devices in electronics. These advantages include:

1. Shorter times for production and to market.
2. Lower prototype development and NRE costs.

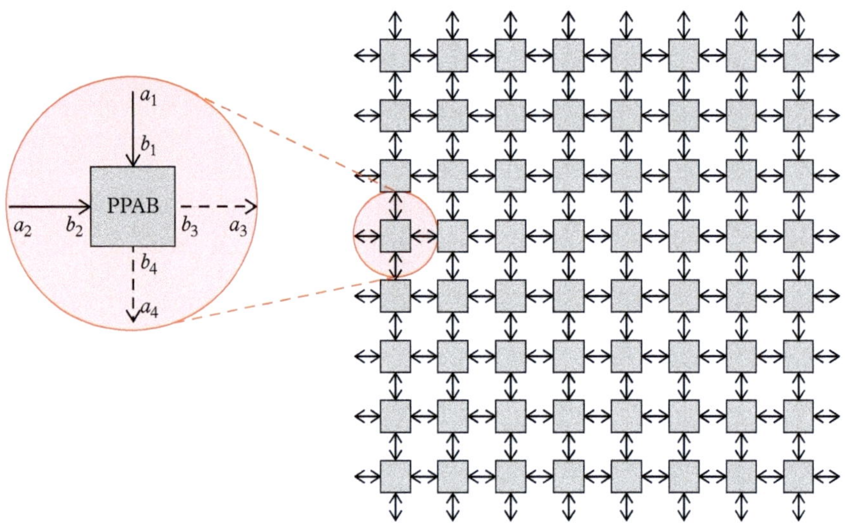

Figure 9.4 *Schematic diagram example of the proposed FPPGA device. The zoom shows a detail of the Programmable Photonic Analog Block as it pertains to the left-up to right-bottom direction of propagation. (After [10]).*

3. Reduced financial risk in developing ideas and translating them into ASPICs.
4. Multi-functional and multi-task operation.

Some of its potential applications are listed below:

1. **Aerospace and defence** (avionics, communications, secure solutions, space).
2. **Automotive** (high-resolution video, image processing vehicle networking and connectivity, automotive infotainment).
3. **Data centres** (servers, routers, switches, gateways).
4. **High-performance computing** (servers, super computers, SIGINT systems, high-end photonic RADARs, high-end optical beamforming systems, quantum computing, high-speed neural networks).
5. **Integrated circuit design** (ASPIC prototyping, integrated photonics hardware emulation).
6. **Wired and wireless communications** (optical transport networks, network processing 5G connectivity interfaces, mobile backhaul).

9.2.2 Reversible Gates and Rotation Matrices

In digital electronic FPGAs it is customary to employ elementary irreversible gates [2], [11] for bit processing using Boolean logic as building blocks for CLEs [4]. These gates are characterised by the fact that the number of input ports (or FAN-IN) is 2 while the number of output ports of FAN-OUT is 1. Figure 9.5(a)–(b) shows some basic examples of these gates, which are characterised by their so-called truth tables. The term irreversibly reflects the fact that the input cannot be deduced from the output unambiguously. By cascading thousands of these gates it is possible to build extremely complex combinatorial and sequential Boolean circuits.

Similarly, we propose PPABs in FPPGAs to be implemented by means of 2×2 *reversible gates* [11], [12], which feature the same number of input and output ports and are characterised as well by truth tables. For example, Figure 9.5(c) shows the example of a SWAP or Pauli X gate. However, in this case, the input state can be deduced from the gate output as the gate operation can also be described by a unitary matrix transformation U. If $\mathbf{I}$ and $\mathbf{O}$ denote respectively the input and output vectors, the $\mathbf{O} = U\mathbf{I}$, hence $\mathbf{I} = U^{-1}\mathbf{O}$, but since U is unitary its inverse is given by the Hermitian conjugate. Reversible gates can be employed to perform digital Boolean operations but this process is inefficient compared to the use of irreversible gates as it entails the use of fixed or ancilla bits and produces as well garbage bits, which are not useful to the rest of the computation [13]. As a consequence, reversible gates are not employed in digital electronics.

Reversible gates have found however an application niche in the field of quantum computation [11]–[14]. The two main reasons are firstly that quantum computation does not rely on Boolean logic, but rather on the use of linear unitary transformations. This is

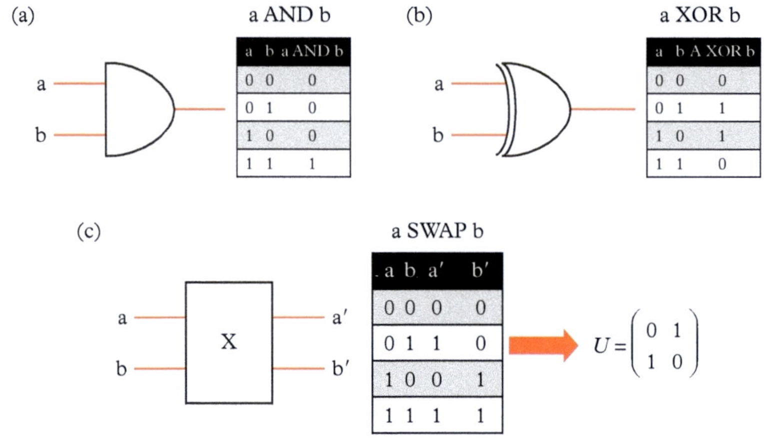

Figure 9.5 *Examples of irreversible AND (a) and XOR (b) gates and of a reversible SWAP gate (after [11], [12]).*

due to the fact that a qubit (as defined in Chapter 7) resembles more an analogue than a digital signal as the multiplying coefficients α and β of the $|0>$ and $|1>$ states can be continuously changed between 0 and 1. Indeed, the qubit processing by the logic gate is really carried altering its wave-like, and not its particle-like, properties. The second (and equally important) reason is that reversible gates can be engineered to exploit the garbage bits as heralding ports [11] to certify the correct operation of the gate.

Integrated photonic circuits for classical signal processing applications handle analogue signals (as seen in Chapter 7) and therefore it makes sense to consider the use of reversible gates as a basic building block (BBB) to implement complex circuit structures. We now develop the formalism for the implementation of complex programmable photonic circuits taking as BBBs a special family of 2×2 reversible gates known as *rotation gates*[11], [12]. We first illustrate how these gates isomorphically correspond to well-known photonic components such as directional and 3-dB tunable couplers and push-pull phase shifters. We then show how these can be concatenated in two compact forms to implement tunable arbitrary 2×2 unitary transformers. These unitaries can be, in turn, assembled into two-dimensional configurations to feature FPGAs.

We consider the basic 2×2 Pauli matrices given by [11], [12]:

$$\begin{aligned} \sigma_0 = I = \begin{pmatrix} 1 & 0 \\ 0 & 1 \end{pmatrix}; \ \sigma_1 = X = \begin{pmatrix} 0 & 1 \\ 1 & 0 \end{pmatrix} \\ \sigma_2 = Y = \begin{pmatrix} 0 & -j \\ j & 0 \end{pmatrix}; \ \sigma_3 = Z = \begin{pmatrix} 1 & 0 \\ 0 & -1 \end{pmatrix} \end{aligned}. \tag{9.3}$$

The rotation matrices (by an angle θ) around axes x, y and z are given by:

$$R_x(\theta) = e^{-j\frac{\theta}{2}X}; \quad R_y(\theta) = e^{-j\frac{\theta}{2}Y}; \quad R_z(\theta) = e^{-j\frac{\theta}{2}Z} \tag{9.4}$$

and it can be readily shown that

$$R_x(\theta) = \begin{pmatrix} \cos(\theta/2) & j\sin(\theta/2) \\ j\sin(\theta/2) & \cos(\theta/2) \end{pmatrix}$$
$$R_y(\theta) = \begin{pmatrix} \cos(\theta/2) & -\sin(\theta/2) \\ \sin(\theta/2) & \cos(\theta/2) \end{pmatrix} \tag{9.5}$$
$$R_z(\theta) = \begin{pmatrix} e^{-j(\theta/2)} & 0 \\ 0 & e^{-j(\theta/2)} \end{pmatrix}$$

Now, if we recall the models developed in Section 2.4 and Section 2.5 we see that these rotations can be implemented in the photonics domain respectively by a directional coupler, a 3-dB Mach–Zehnder tunable coupler (3dB-MZI) and two parallel waveguides including two phase shifters configured in push-pull mode as shown in Figure 9.6(a).

By virtue of the Euler theorem [11], [12] any arbitrary 2×2 unitary matrix U can be expressed in either of the two following sequential rotation matrix concatenation forms:

$$U = e^{j\delta} R_z(\alpha) R_y(\beta) R_z(\gamma)$$
$$U = e^{j\delta} R_z(\alpha) R_x(\beta) R_z(\gamma) \tag{9.6}$$

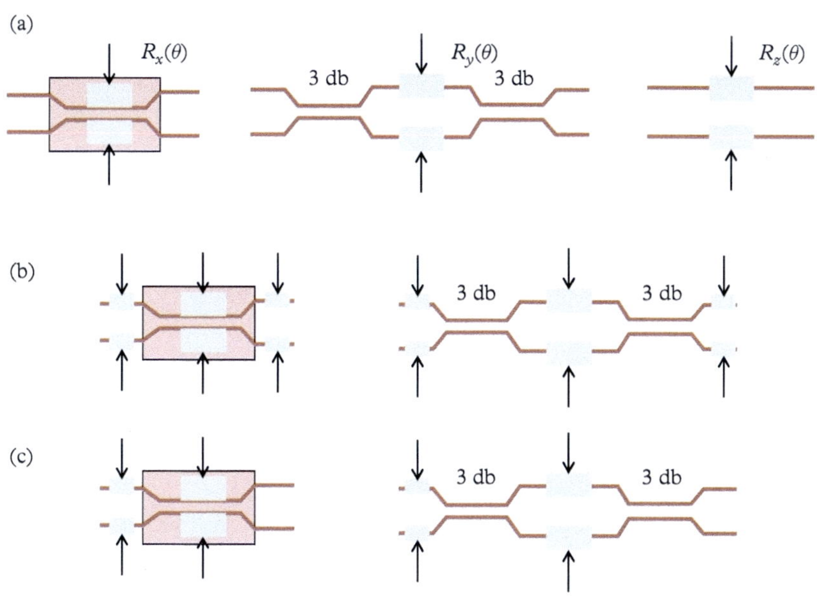

Figure 9.6 *(a) Isomorphism between 2 × 2 rotation operators and integrated waveguide components. (b) two alternatives for the implementation of arbitrary programmable 2 × 2 unitary transformers. (c) a reduced version of the configurations shown in (b) where one phase shifting stage has been supressed.*

where α, β, γ and δ are a set of four real and independent numbers. This means that compact configurations of a preceding and succeeding differential phase shifting units enclosing either a dual-drive directional coupler (DD-DC) or a 3-dB-MZI unit (Figure 9.6(b)) can be employed to implement these arbitrary units.

To construct 2D programmable arrays we consider a reduced version of the above configurations, which we call *gates*

$$\begin{aligned} G &= e^{j\delta} R_y(\beta) R_z(\gamma) \\ G &= e^{j\delta} R_x(\beta) R_z(\gamma) \end{aligned} \tag{9.7}$$

Here, one of the phase-shifting units is not included (in this case, the one succeeding the tunable coupler) as shown in Figure 9.6(c). Programmable arrays can then be constructed by means of 2D assemblies of 2×2 gates given by Eqn. (9.5). This is shown in Figure 9.7(a). Note that as shown in Figure 9.7(b) since for every $2 \times 2G$

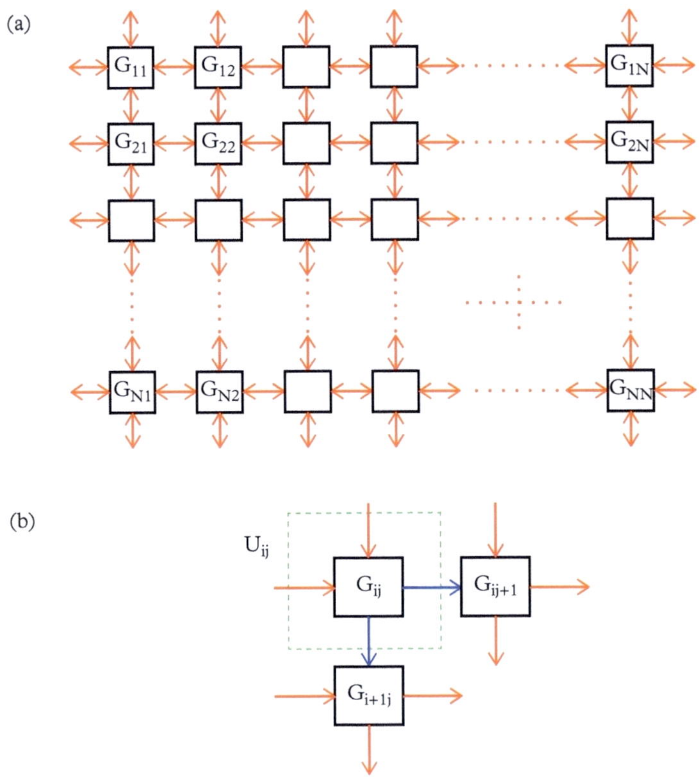

Figure 9.7 *(a) Two-dimensional array arrangement of reversible G gates. (b) Detailed explanation of of how the concatenation of a given Gate G_{ij} actually results in the possibility of implementing a unitary arbitrary transformation by the $_{ij}$ element.*

unit its input phase shifter (shown in blue) can be programmed to incorporate the effect of the required preceding phase shift as well as the fictitious succeeding phase shifter of the preceding 2×2 G unit then this configuration is actually equivalent to that of an array of arbitrary 2×2 unitary transformers. This structure can then be programmed to implement any arbitrary $N \times N$ unitary transformation (see Chapter 4). Furthermore, because of the reversible nature of the gates, both feedforward and feedbackward operations are possible, even simultaneous. This enables the possibility of implementing resonant and cavity structures with this field programmable reversible gate array configuration.

9.3 Programmable Photonic Analogue Blocks

Although several configurations can be considered for the PPABs, here we illustrate the case corresponding to the most basic configuration, which is obtained by employing any of the 2×2 rotation gates introduced in the previous sections. PPABs with increased complexity and number of input/output ports can of course be designed by suitable concatenation of gates. In general, the system's performance depends on the interconnection scheme between PPABs and RPIs. The scheme of the basic PPAB is shown in the inset of Figure 9.4 for a particular axis orientation and with no internal coupling paths. For illustrative purposes, we will consider the following four options obtained from this configuration by rotating $0°, 90°, 45°$ and $-45°$ and denote them as type A, B, C, and D, respectively. Figure 9.8 shows these possible options. The high-level role of the PPAB is to provide tunable independent power coupling ratios and phase shifts.

The standalone operation of the PPAB is illustrated in Figure 9.9 for the Type A case (the description for the other types follows the same line of reasoning). Figure 9.9 shows the layout of the Type A PPBA with indication of the optical fields at the input and output ports (b_1, b_2, b_3, b_4) and also the external fields at the input/output RPI elements enclosing the PPAB (a_1, a_2, a_3, a_4).

The most basic PPAB is a 2×2 photonic component that is capable of implementing a rotation matrix $R_x(2\theta)$ or $R_y(2\theta)$ affected by a common tunable phase shift Δ_{PPAB} and a tunable optical power splitting ratio K among its input optical waveguide input

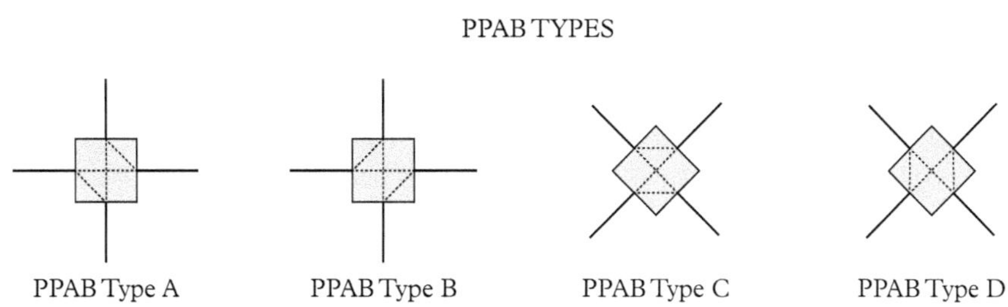

Figure 9.8 *Four types of 2×2 PPAB basic units considered and their internal signal coupling layouts shown in broken lines.*

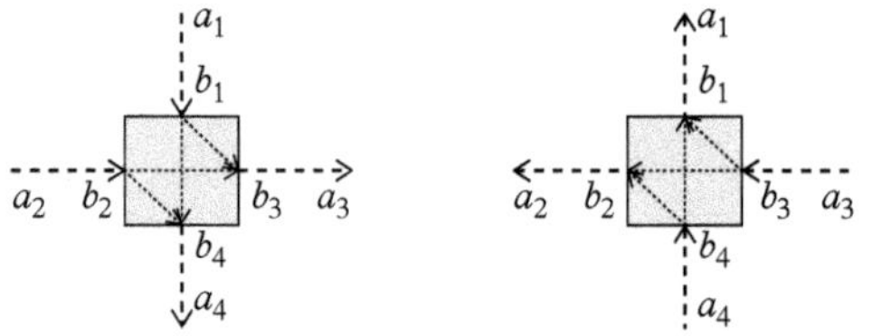

Figure 9.9 *Type A PPAB unit including the internal and external optical fields at its ports for the two directions of signl propagation.*

fields b_1, b_2 and its output optical waveguide output fields b_3, b_4. Two propagation directions are possible as shown in the figure. The first is from the left and up ports to the right and down ports and is characterised in the case of implementing the PPAB with a 3-dB tunable coupler by the following transmission matrix (if a $R_y(2\theta)$ rotation is considered)

$$\begin{pmatrix} b_3 \\ b_4 \end{pmatrix} = -je^{j\Delta_{PPAB}} \begin{pmatrix} \sin\theta & \cos\theta \\ \cos\theta & -\sin\theta \end{pmatrix} \begin{pmatrix} b_1 \\ b_2 \end{pmatrix}, \tag{9.8}$$

Both K and Δ_{PPAB} can be changed by means of two external (electronic, mechanic, acoustic) control signals through a linear relationship. The second is from the *right and down ports to the to left and up ports* and is characterised by the following transmission matrix:

$$\begin{pmatrix} b_1 \\ b_2 \end{pmatrix} = -je^{j\Delta_{PPAB}} \begin{pmatrix} \sin\theta & \cos\theta \\ \cos\theta & -\sin\theta \end{pmatrix} \begin{pmatrix} b_1 \\ b_2 \end{pmatrix}. \tag{9.9}$$

Figure 9.10 shows some examples of simple programming of the Type A PPAB leading to very basic operations required in photonic signal processing.

Similar operation modes and colour codes can be defined for type B, C, and D PPABs. For example, Figure 9.11 shows the layout of the Type D PPBA with indication of the optical fields at the input and output ports (b_1, b_2, b_3, b_4) and also the external fields at the input/output RPI elements enclosing the PPAB (a_1, a_2, a_3, a_4).

Two propagation directions are possible: the first is from the *left-up and right-up ports* to the *to left-down and right-down ports*. The second is from the *to left-down and right-down ports* to the *left-up and right-up ports*. They are characterised by the following transmission matrices, respectively (see Chapter 2):

$$\begin{pmatrix} b_2 \\ b_4 \end{pmatrix} = e^{j\Delta_{PPAB}} \begin{pmatrix} \sin\theta & j\cos\theta \\ j\cos\theta & \sin\theta \end{pmatrix} \begin{pmatrix} b_1 \\ b_3 \end{pmatrix}, \tag{9.10}$$

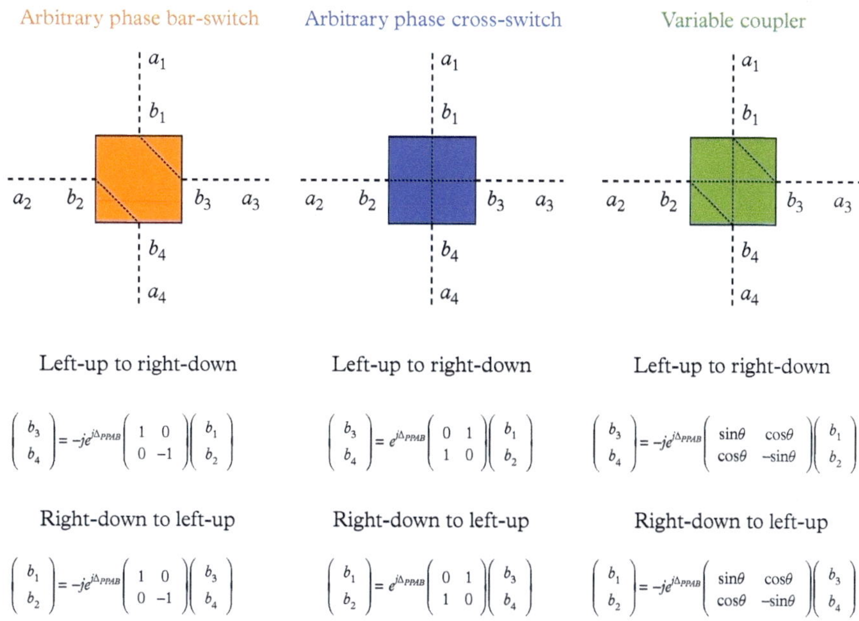

Figure 9.10 *Different functionality programming, resulting transmission matrices and associated code colours for a Type A PPAB.*

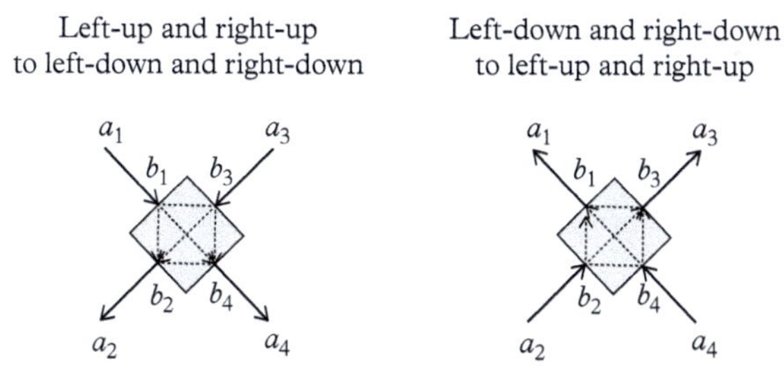

Figure 9.11 *Type D PPAB unit including the internal and external optical fields at its ports for the two directions of signal propagation.*

$$\begin{pmatrix} b_1 \\ b_3 \end{pmatrix} = e^{j\Delta_{PPAB}} \begin{pmatrix} \sin\theta & j\cos\theta \\ j\cos\theta & \sin\theta \end{pmatrix} \begin{pmatrix} b_2 \\ b_4 \end{pmatrix}. \tag{9.11}$$

Figure 9.12 shows some examples of simple programming of the Type D PPAB leading to very basic operations required in photonic signal processing.

Note that in addition to the functionalities displayed, the PPAB can operate in conjunction with an RPI as a phase shifter, as shown in the next section.

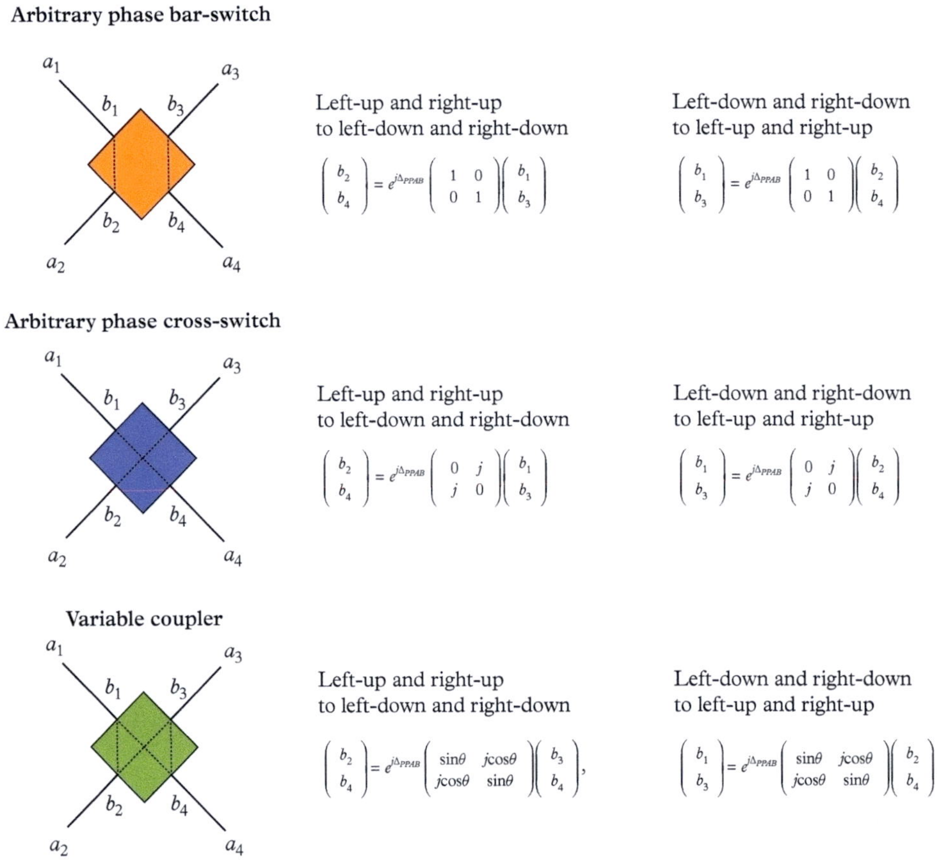

Figure 9.12 *Different functionality programming, resulting transmission matrices and associated code colours for a Type D PPAB.*

9.4 Reconfigurable Photonic Interconnects

The RPI elements are assumed to provide a lossless tunable phase shift and their combination with the PPAB elements provide an extra degree of flexibility in the 2×2 transmission matrix. Figure 9.13 shows this feature for a Type A PPBA element under propagation from left and up ports to right and down ports (a similar procedure can be established for the reverse propagation direction and for PPBA types B, C and D). The optical waveguide RPI elements can provide independent and tunable differential phase shifting ϕ over a common value Δ_{RPI} in the two input and/or output waveguides accessing the PPAB. For instance, and referring to the upper part of Figure 9.13:

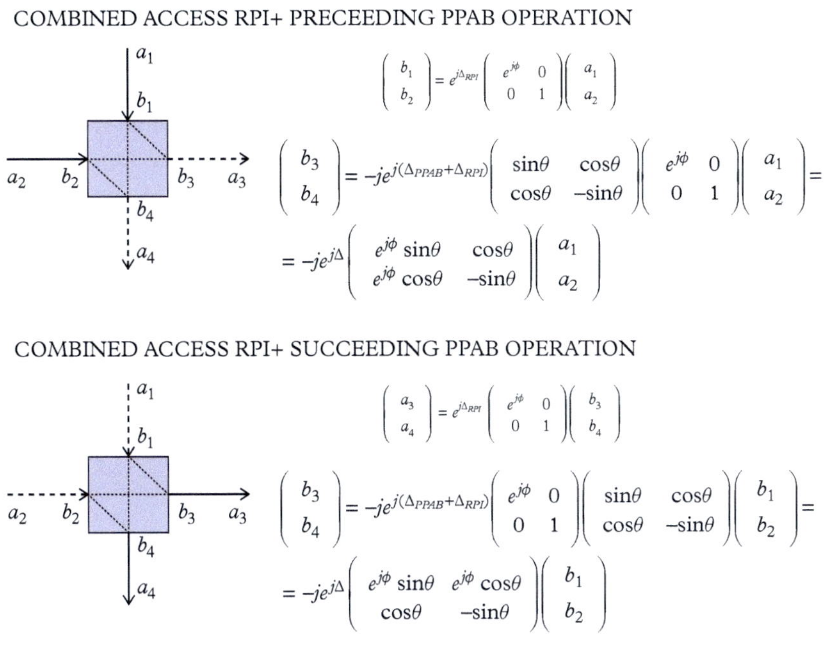

Figure 9.13 *Combined operation of a PPAB and an RPI for operation as a phase shifter.*

$$\begin{pmatrix} b_1 \\ b_2 \end{pmatrix} = \begin{pmatrix} e^{j(\phi+\Delta_{RPI})} & 0 \\ 0 & e^{j\Delta_{RPI}} \end{pmatrix} \begin{pmatrix} a_1 \\ a_2 \end{pmatrix} = e^{j\Delta_{RPI}} \begin{pmatrix} e^{j\phi} & 0 \\ 0 & 1 \end{pmatrix} \begin{pmatrix} a_1 \\ a_2 \end{pmatrix}. \tag{9.12}$$

The combined action of a PPAB element and its preceding RPI element can then be cast as follows:

$$\begin{aligned} \begin{pmatrix} b_3 \\ b_4 \end{pmatrix} &= -je^{j(\Delta_{PPAB}+\Delta_{RPI})} \begin{pmatrix} \sin\theta & \cos\theta \\ \cos\theta & -\sin\theta \end{pmatrix} \begin{pmatrix} e^{j\phi} & 0 \\ 0 & 1 \end{pmatrix} \begin{pmatrix} a_1 \\ a_2 \end{pmatrix} = \\ &= -je^{j\Delta} \begin{pmatrix} e^{j\phi}\sin\theta & \cos\theta \\ e^{j\phi}\cos\theta & -\sin\theta \end{pmatrix} \begin{pmatrix} a_1 \\ a_2 \end{pmatrix}, \end{aligned} \tag{9.13}$$

where the common phase factor is given by $\Delta = \Delta_{RPI} + \Delta_{PPAB}$.

In a similar way, the combined action of a PPAB element and its succeeding RPI element (shown in the right part of Figure 9.13) is given by:

$$\begin{aligned} \begin{pmatrix} a_3 \\ a_4 \end{pmatrix} &= -je^{j(\Delta_{PPAB}+\Delta_{RPI})} \begin{pmatrix} e^{j\phi} & 0 \\ 0 & 1 \end{pmatrix} \begin{pmatrix} \sin\theta & \cos\theta \\ \cos\theta & -\sin\theta \end{pmatrix} \begin{pmatrix} b_1 \\ b_2 \end{pmatrix} = \\ &= -je^{j\Delta} \begin{pmatrix} e^{j\phi}\sin\theta & e^{j\phi}\cos\theta \\ e^{j\phi}\cos\theta & -\sin\theta \end{pmatrix} \begin{pmatrix} b_1 \\ b_2 \end{pmatrix}. \end{aligned} \tag{9.14}$$

9.5 FPPGA Core Architectures, Design Flow and Technology Mapping

It is by adequate concatenation of successive RPI + PPAB and/or PPAB + RPI units into core architectures and subsequent programming that complex standalone and/or parallel photonic circuits and signal processing transformations can be implemented by the FPPGA. This process entails a design flow stage and a technology mapping.

9.5.1 Core Architectures

Figure 9.14 depicts three generic layouts of FPPGA cores built upon assembling RPI + PPAB and/or PPAB + RPI units. For illustrative purposes, Figures 9.14(a)–(b) show the layouts of a square type FPPGA design. Here, type A and type B PPAB elements are interleaved in every column and row of the device. We denote this design by class ABAB.

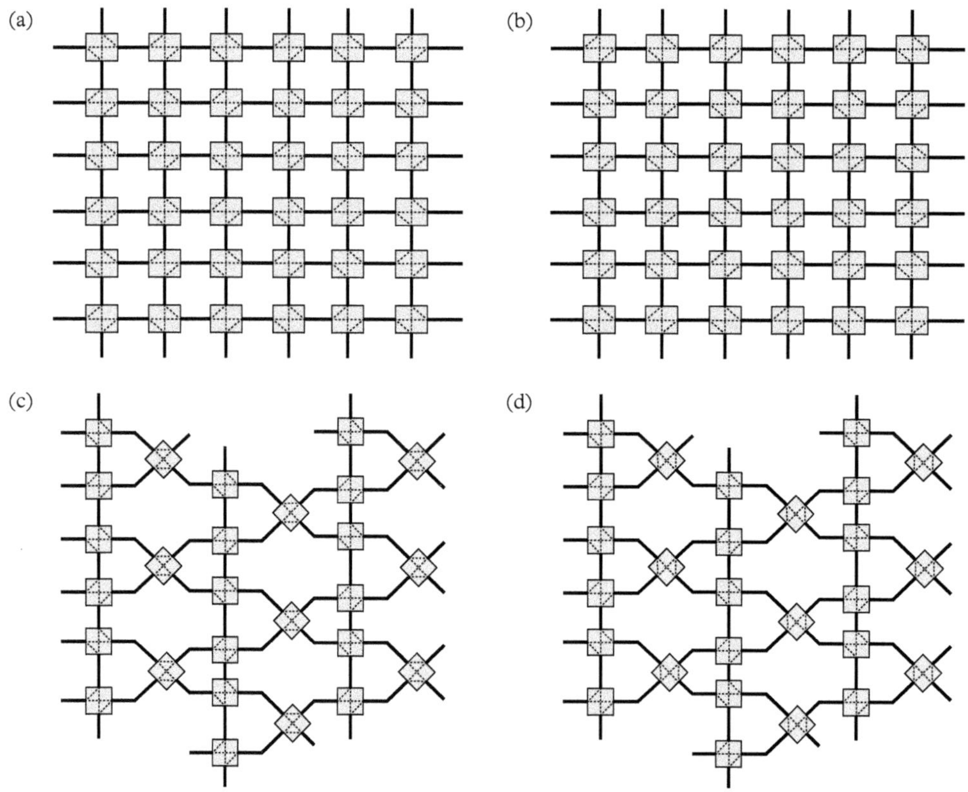

Figure 9.14 *Layouts of class ABAB FPPGAs. (c) Layout of a class ABCC FPPGA. (d) Layout of a class ABDD FPPGA.*

A second class is obtained by combining columns of interleaved AB PPABs and columns of type C PPABs, we denote this class (Figure 9.14(c)) by ABCC. Finally, Figure 9.14(d) shows the layout of class ABDD obtained by combining columns of interleaved AB PPABs and columns of type D PPABs. Other classes, which are not shown here, lead to non-uniform FPPGA design patterns, defined by suitable combinations of different types of PPAB elements. Note again that the main contribution to the system's performance is coming from the selected interconnection pattern between PPABs, rather than their block orientation. Section 9.6 on physical implementation provides further discussion on these.

9.5.2 Design Flow and Technology Mapping

The most general type of programmable devices consists of an array of uncommitted elements that can be interconnected according to a user's specifications and configured for a wide variety of applications. An FPPGA combines the programmability of the most basic reconfigurable photonic integrated circuits in a scalable interconnection structure, allowing programmable circuits with much higher processing density. Thus, processing complexity comes from the interconnectivity.

The left part of Figure 9.15 shows the main steps of the design flow process. The starting point for the design flow is the initial application entry or circuit configuration to be implemented. The specifications are then processed to optimise the area and the performance of the final circuit. Then, specifications are transformed into a compatible circuit of FPPGA processing blocks (technology mapping), optimising attributes such as delay, performance, accumulated loss or number of blocks.

The technology mapping phase transforms the optimised network into a circuit that consists of a restricted set of circuit elements (FPPGA processing blocks). This is done by choosing a set from the available PPABs and specifying how these will be interconnected. This interconnection step implies the setting of several RPI elements physically connecting the selected PPABs. This determines the total number of processing blocks (PPABs and RPIs) to be activated by programming. In a second stage, the processing block configurations (i.e. types of PPABs and RPIs) are chosen and performance calculation and design verification are carried out. This can be done either physically by feeding all the necessary configuration data to the programming units to configure the final chip or, more commonly, by iteratively employing accurate models of the FPPGA in the software plane.

The next step assigns each processing block to a specific location in the FPPGA core including, as well, the choice of the processing units that route the input signals to the core to the input/s of the programmed circuit and the output/s of the programmed circuit to the core outputs. Note that in contrast to FPGAs [2], [7], the proposed structure does not physically differentiate between processing blocks and interconnection resources.

From the aforementioned description, it can be appreciated that the FPPGA device involves considering not only the physical hardware of the photonic and control electronic tier, but also a software layer (see upper right part of Figure 9.15). For the control

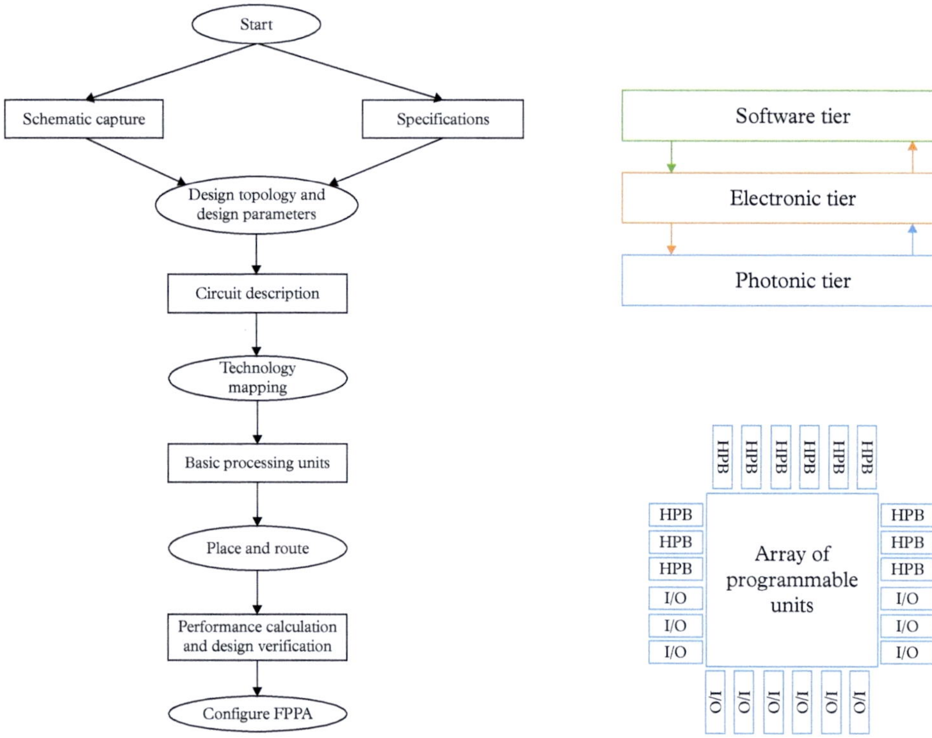

Figure 9.15 *Left: main steps involved in the design Flow of a FPPGA device. Right: FPPGA soft and hard tiers and expanded layout including peripheral high-performance blocks.*

electronics, the same scheme as the one proposed in Section 6.6 applies. For dynamic operation, the steps contained in the generic design flow can be done automatically either by the software layer, the user, or by a mixture of both, depending on the autonomy and the capabilities of the FPPGA. In addition, a failure in any of the steps will require an iterative process till the specifications are accomplished successfully. Additional parallel optimisation process (mainly self-winding), enable robust operation, self-healing attributes and additional processing power to the physical device (Section 6.7).

Similarly to modern FPGA families, FPPGA can include peripheral high-performance blocks (HPB) to expand its capabilities to include higher-level functionality fixed into the chip. This is shown schematically in the lower right part of Figure 9.15. Having these common functions embedded into the chip reduces the area required and gives those functions increased performance compared to building them from primitives. Moreover, some of them are impossible to be obtained by a discretised version of basic processing blocks. Examples of these include high-dispersive elements, generic modulation and photo detection subsystems, optical amplifiers and source subsystems and high-performance filtering structures to cite a few.

9.5.3 Operation and Programming Examples

Figures 9.16–9.18 provide some examples where FPPGAs of different types are programmed to emulate and implement simultaneously different photonic circuits. In each case the figure includes the FFPA layout with coloured PPBAs according to the code defined earlier and the layouts of the implemented circuits.

In all the cases shown, the FPPGA programming is set to implement the same circuits shown in the right parts of the figures. We have chosen a representative example of an infinite impulse response (IIR) filter (i.e. the optical ring cavity), a finite input response (FIR) filter (the Mach–Zehnder interferometer) and a multiple input/multiple output (MIMO) interferometer. Coloured shading is added to help the reader in the identification of which part of the FPPGA is implementing each circuit. Other functionalities are of course possible. For instance, tunable true-time delay lines can be implemented by switching different sets of RPIs using the PPABs in crossbar switch mode [15].

9.6 Physical Implementation, Experimental Results and Discussion

The physical implementation of the FPPA device calls for an integrated optics approach, either based on silicon photonics [16], [17] or a hybrid/heterogeneous III–V and silicon photonics platforms [18]. Figure 9.19 provides some information about the physical options available.

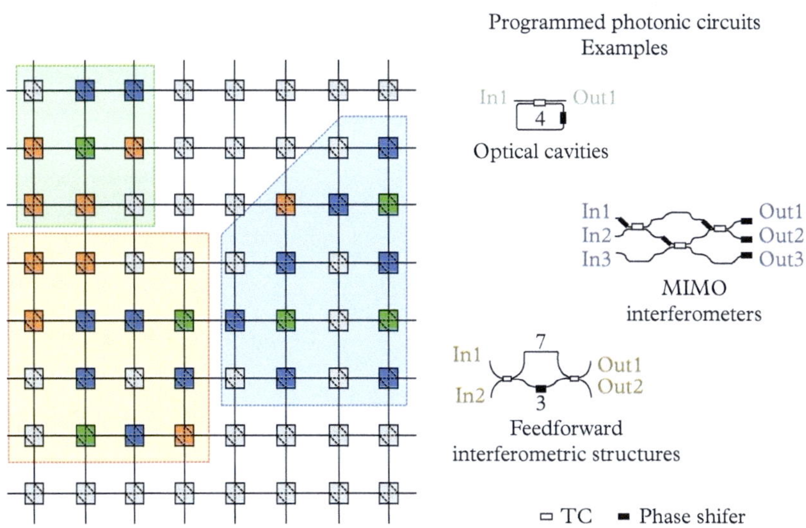

Figure 9.16 *Simultaneous implementation of a Ring cavity (green shaded area) a Mach-Zehnder Interferometer (orange shaded region) and a 3×3 multiple port interferometer (blue shaded region) using an ABAB FPPGA. TC: Tunable Coupler.*

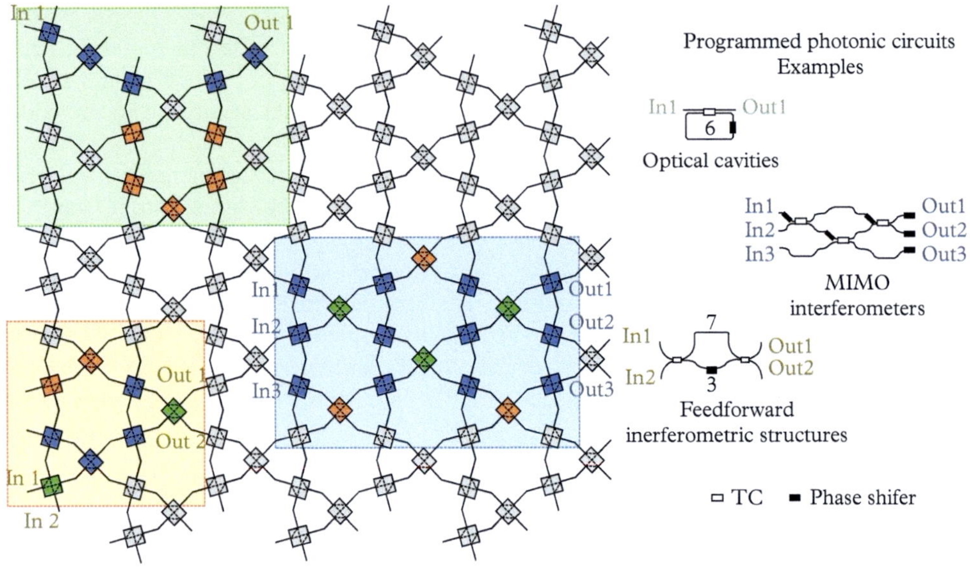

Figure 9.17 *Simultaneous implementation of a Ring cavity (green shaded area), a Mach-Zehnder Interferometer (orange shaded region) and a 3×3 multiple port interferometer (blue shaded region) using an ABCC FPPGA. TC: Tunable Coupler.*

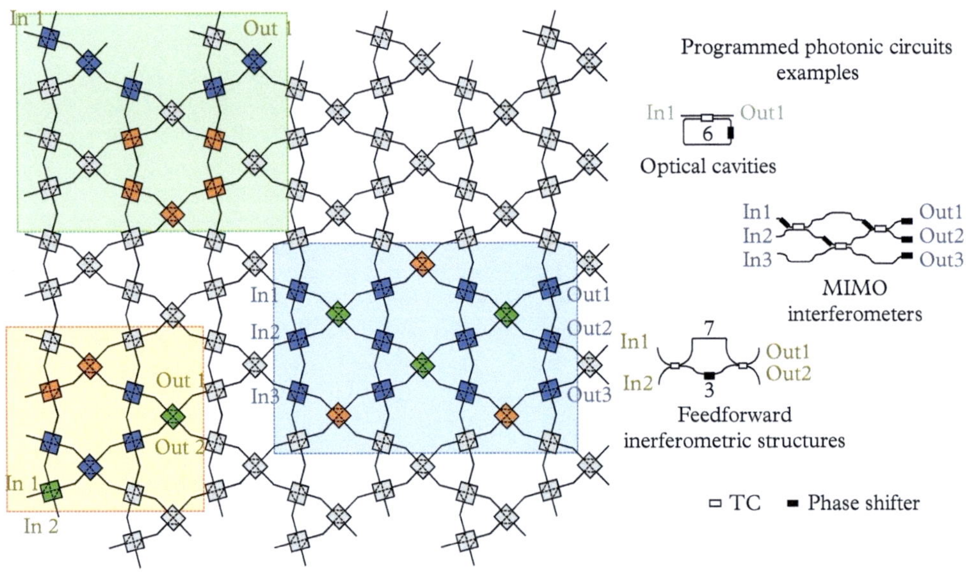

Figure 9.18 *Simultaneous implementation of a Ring cavity (green shaded area), a Mach-Zehnder Interferometer (orange shaded region) and a 3×3 multiple port interferometer (blue shaded region) using an ABDD FPPGA. TC: Tunable Coupler.*

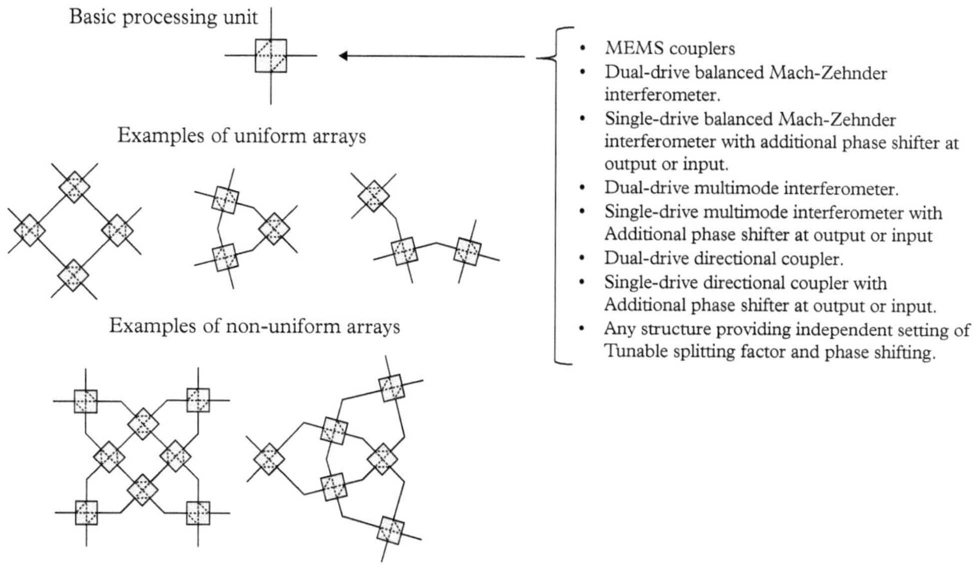

Figure 9.19 *Technology options for the implementation of the PPAB elements (upper), FPPGA layouts (intermediate) and other FFPA possible configurations (lower).*

As for the PPBA elements, the currently available photonics technology options are listed in the upper part of Figure 9.19. Regarding the implementation of ABAB, ABCC and ABDD FPPA layouts, the intermediate part of Figure 9.19 shows the basic replicating blocks. These correspond to the unit blocks of the square, hexagonal and triangular waveguide meshes that were introduced in Chapter 5, as shown in Figure 9.20. Therefore, uniform waveguide meshes provide a natural and compact option for their implementation. However, it should be stressed again that the main system's performance comes from the interconnection scheme between PPBA ports rather than the PPBA orientation. As an example, it is possible to employ the flattened, longitudinally parallel waveguide mesh arrangements presented in Chapter 5 for the FPPGA core.

Finally, as mentioned, more complex FPPA layouts can be designed by interleaving the proposed types of PPBAs. However, in this case, they are implemented by non-uniform waveguide meshes. Some examples are shown in the lower part of Figure 9.19.

Recent reports describe the implementation of an ABCC class FPPA using the silicon photonics hexagonal waveguide mesh composed of seven cells (Chapter 5). Figures 9.21 (a)–(b) show the equivalence between the waveguide mesh and the FPPA design. To facilitate the identification, the different building blocks are shown in different colours. Figures 9.21 (c)–(d) show two implementations of the FPPA core in silicon photonics and silicon nitride, respectively. The silicon chip occupies a surface of 15×20 mm^2, includes 30 3-dB MZIs (featuring a length of 975μm), sixty thermal tuners and 120 pads, and features twenty-four optical I/O ports mounted on a printed circuit board (PCB) that occupies a surface of 60×120 mm^2. We programmed the FPPA operation by acting over

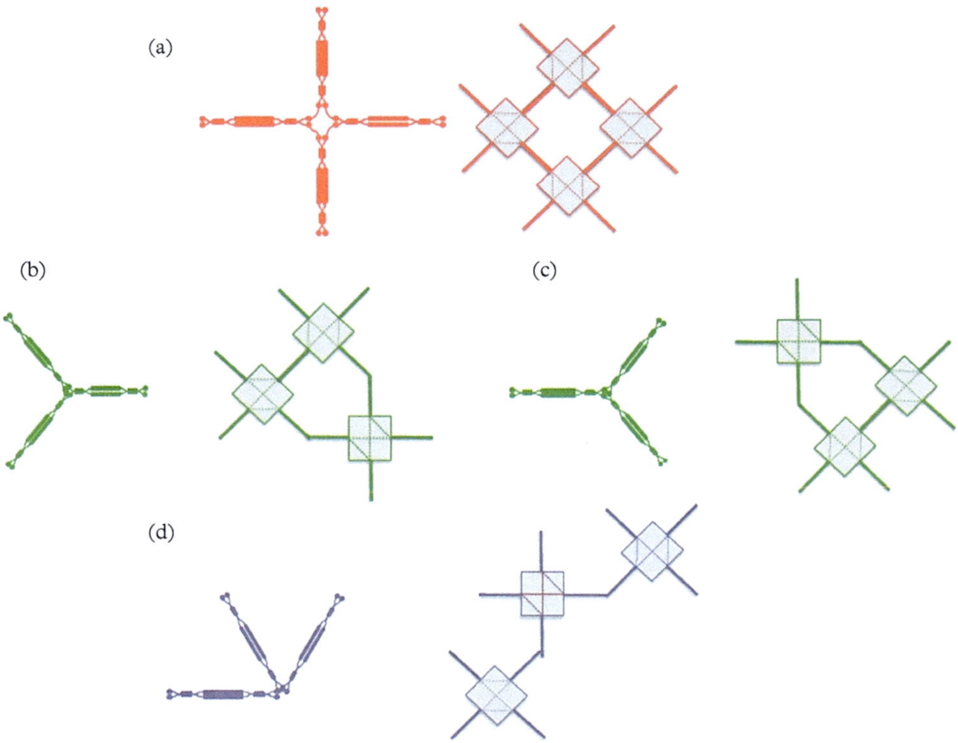

Figure 9.20 *Identification between the fundamental unit blocks employed to construct integrated waveguide meshes and the main RPI+PPAB blocks studied in this chapter. (a) Square waveguide mesh and ABAB block (b), (c) Hexagonal trilattices and the ABCC block. (d) Triangular waveguide mesh and the ABDD block.*

the bias currents of the MZIs. The interested reader is referred to Pérez and colleagues [19] for further details on the chip, and to Capmany, Gasulla and Pérez [20] for details on the measurement and characterisation setup. The silicon nitride chip occupies a surface of 5.5 x 11 mm^2, and includes 40 3-dB MZIs (featuring a length of 1297 μm) with eighty thermal tuners.

Due to the restricted number of cells in the waveguide mesh design, there were not enough PPBAs and RPIs to simultaneously implement the three circuits shown in Figure 9.16. Here we show the results obtained in the programming of each circuit separately.

Figure 9.22 shows the programming (a), equivalent circuit (b) as well as measured modulus (c) and phase (d) of the transfer function for an unbalanced (by 4×975 μm) MZI. Note that two PPABs implement the functionality of tunable couplers K_1 and K_2.

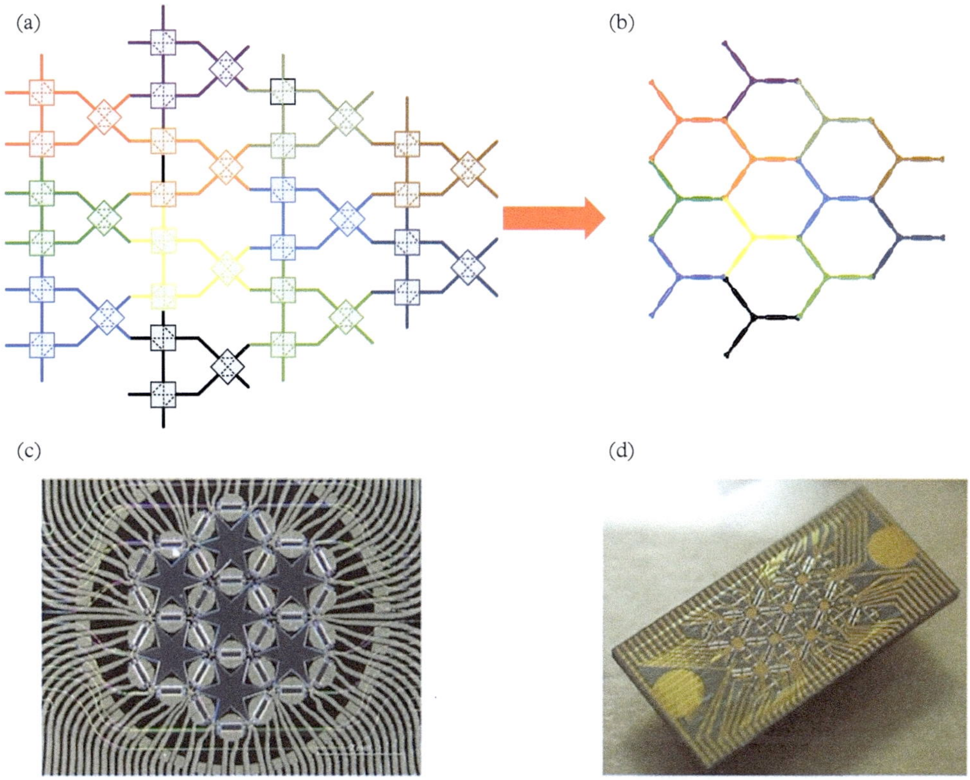

Figure 9.21 *(a) Layout of the ABCC class FPPA and isomorphic implementation by means of a 7-cell hexagonal waveguide mesh displayed in (b). Silicon (c) and Silicon Nitride (d) chips implementing the 7-cell hexagonal waveguide mesh.*

In fact, the experimental results show different spectra corresponding to different values of K_1 and K_2. The free spectral range (FSR) corresponds to the path unbalance.

Figure 9.23 shows the programming (a), equivalent circuit (b) and measured modulus of the reflected (c) and transmitted (d) signals in a double coupler ring cavity (cavity length $= 6 \times 975$ μm) resonator. Again, two PPABs implement the functionality of tunable couplers K_1 and K_2 with the experimental results showing different spectra corresponding to different values of K_1 and K_2. The FSR corresponds to the path unbalance.

Finally, Figure 9.24 shows the results corresponding to a 3×3 MIMO interferometer programmed to implement a 3×3 splitter (tritter) or a discrete Fourier transform. In this case, there is enough room left in the FPPA to accommodate a second circuit (a Hadamard gate), which we also programmed to show simultaneous circuit implementation.

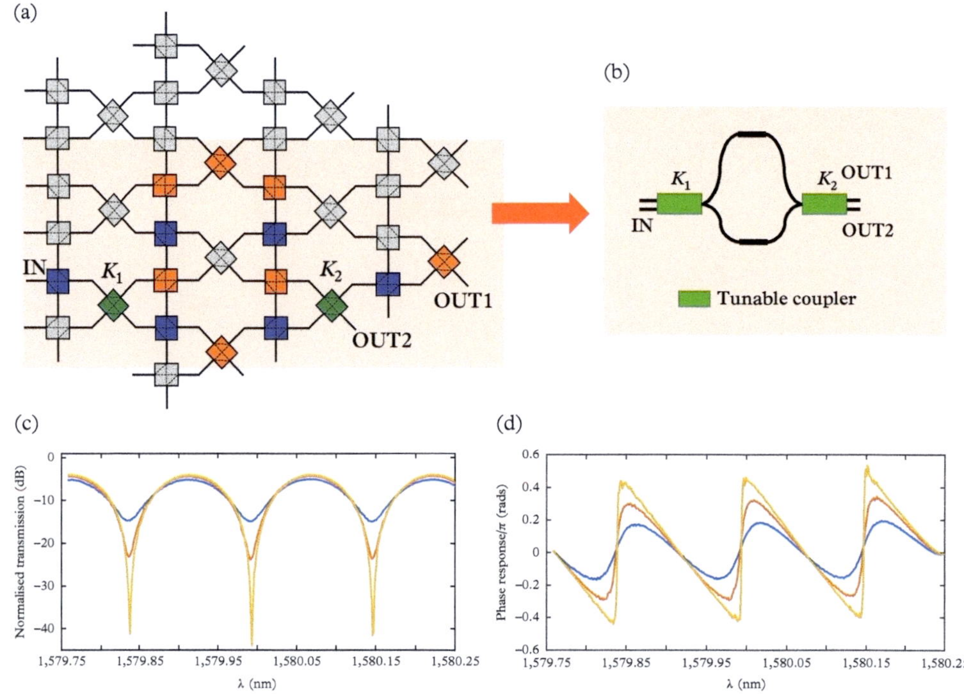

Figure 9.22 *Programming (a), equivalent circuit (b), as well as measured modulus (c) and phase (d) of the transfer function for an unbalanced (by 4×975 μm) Mach-Zehnder interferometer. The different curves in (c) and (d) correspond to different values of K_1 and K_2.*

The specific transformations are given in this case by

$$U_{Hadamard} = \frac{1}{\sqrt{2}}\begin{pmatrix} 1 & 1 \\ 1 & -1 \end{pmatrix},$$
$$U_{Tritter} = \frac{1}{\sqrt{3}}\begin{pmatrix} 1 & 1 & 1 \\ 1 & e^{i\frac{2\pi}{3}} & e^{i\frac{4\pi}{3}} \\ 1 & e^{i\frac{4\pi}{3}} & e^{i\frac{8\pi}{3}} \end{pmatrix}. \tag{9.15}$$

The versatility of the FPPGA is directly proportional to the number of PPABs and RPIs contained in the integrated chip. However, the scalability of these systems is limited by different factors: PPAB and RPI insertion losses, power consumption, optical crosstalk/signal leakage, footprint and the complexity of its control electronics. Of these, the dominant limit is the insertion loss, which is mainly generated by the inner coupling structures and phase-tuning mechanisms. In order to compare them with conventional PICs, we split the total insertion loss per PPAB and RPIs as the sum of the propagation loss and the additional losses (couplers and tuning mechanism). Even using

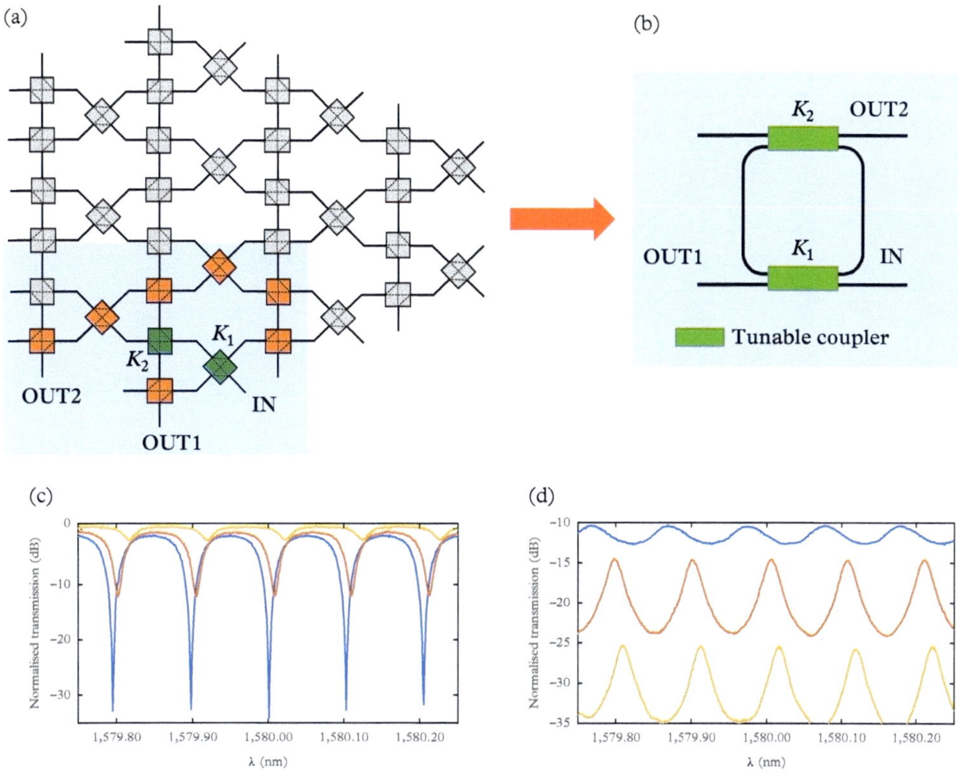

Figure 9.23 *Programming (a), equivalent circuit (b) and measured modulus of the reflected (c) and transmitted (d) transfer functions for a double coupler ring resonator (cavity length = 6×975 µm). The different curves in (c) and (d) correspond to different values of K_1 and K_2.*

state-of-the-art tunable couplers and fabrication procedures to implement the PPABs and RPIs, achieving a value below 0.2-dB additional loss unit is a current challenge.

With these numbers, we can estimate that a programmed light-path crossing fifty PPABs + RPIs will introduce 10-dB additional loss, setting a scalability limit of the size of the programmed circuits and a miniaturisation trade-off [20]. Eventually, these losses may be compensated by the incorporation of semiconductor optical amplifiers (SOAs) as peripheral high-performance blocks (HPB) outside the FPPA core. The FPPGA is implemented by means of a hexagonal waveguide mesh arrangement and five high-performance building blocks (Figure 9.25). Note that the flattened versions of the core reviewed in Chapter 5 would be eligible as well. Regarding the power consumption per PPAB (Pπ, PPAB), and RPI (Pπ, RPI), exploring tuning mechanism approaches will be fundamental to find power-efficient, low-loss, reduced-size, focalised and low-crosstalk phase shifters. In this sense, thermal tuners have been optimised in the last years to open the path for either sub-milliwatts power consumptions [21] or reduced footprint

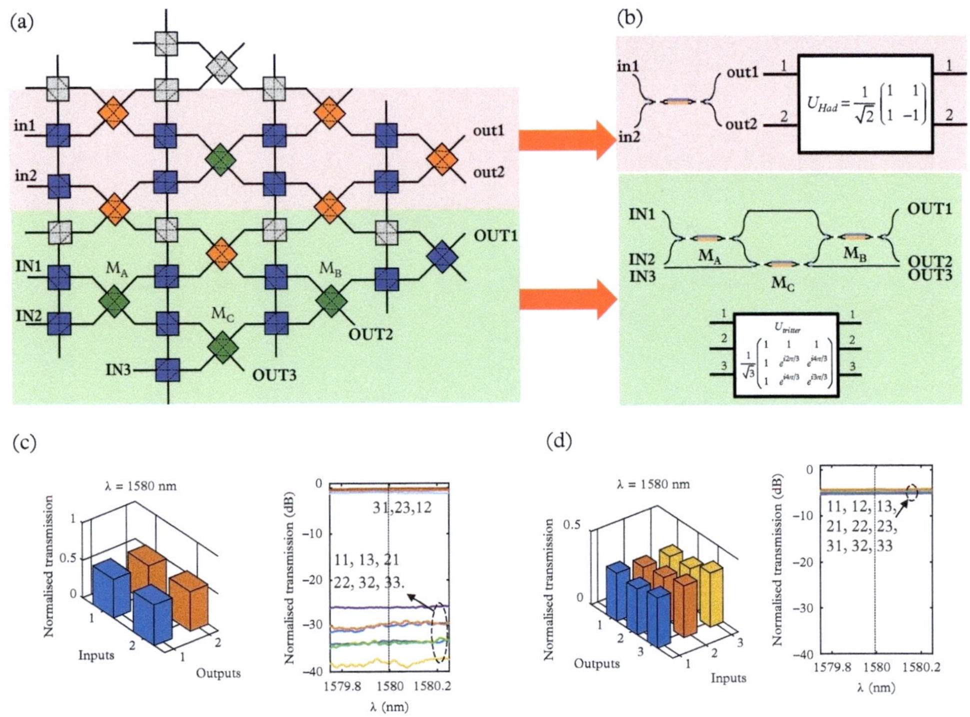

Figure 9.24 *Programming (a), equivalent circuits (b) and measured bar chart (@1580 nm) and modulus of the transfer functions of a (c) 2×2 MIMO interferometer implementing a Hadamard gate and (d) a 3×3 MIMO interferometer implementing a tritter operation.*

structures [22]. For a FPPGA with N active PPABs and M active RPIs, the average power consumption is less than N · Pπ, PPAB + M · Pπ, RPI. Additionally, low-loss alternatives enabling the speed increment of the tuners and non-volatile actuators would open the path to a wider range of application in optical/quantum information processing.

An additional concern is the non-desired side effects related to the use of non-ideal components like the optical crosstalk due to the drift in the configured coupling value and to fabrication or design errors. The optical crosstalk produces signal leaking through the FPPA core that causes reflections inside the circuits, creating ripples in the spectral responses and even lasing phenomena. This has been addressed in other types of complex circuits [20], [23]. Here, the unused PPABs and RPIs can be smartly configured to remove the leaked signal draining optical ports to radically improve the system performance and relax the PPAB specifications to an optical crosstalk <20 dB to assure a good circuit performance. If ultra-low-loss, low-power PPABs and RPIs are obtained, future FPPAs will require an increment of the integration densities to further enlarge their performance in a similar way as the number of transistors per chip rate rises

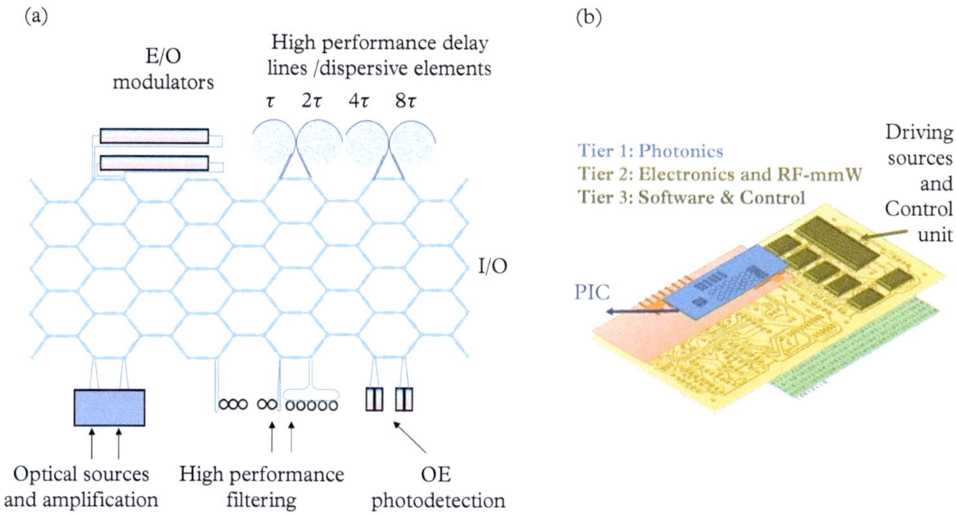

Figure 9.25 *Schematic of the FPPGA architecture implemented with an hexagonal waveguide mesh core (a flattened one can be employed) and different labelled high-performance building blocks, (b) Definition of the different tiers required for the practical driving of the circuit.*

in electronic processors. To overcome the PPAB miniaturisation trade-offs, 3D silicon photonics platforms can be considered [24].

9.7 Chapter Summary

This final chapter addressed the FPPGA, which plays a fundamental role in near- and middle-term research activity in the field of integrated programmable photonics. The FPPGA is inspired by similar principles of those of FPGAs and FPAAs in electronics. It described the main high-level concept of the device, together with a description of its basic constituents or photonic building blocks, which are based on reversible 2×2 gates. These pertain to two different groups, the PPABs, mainly 2×2 tunable photonic components in charge of the basic operations over the amplitude and phase of the optical waveguide modes, and the RPIs, in charge of controlling and routing the signal flow between PPABs. It showed how to assemble the different types of PPABs to form different FPPA classes and discussed general guidelines for design and technology implementation. The physical implementation of the FPPGA, which is a key aspect, has also been considered and the main FPPGA classes can be implemented by means of integrated waveguide meshes. Finally, it provided a simple experimental proof of concept to support the viability of this concept. The future evolution of this concept requires further investigation to address several issues, most importantly scalability, which is required to implement FFPAs with enough PPAB and RPI elements to implement complex and simultaneous operations. The interface with electronic control signals is another important topic, together with the link of the latter to reconfiguration software. Last, but not least, the investigation on different alternatives to provide low-power consumption PPAB elements should be seriously addressed in order to achieve low-power FFPA devices.

REFERENCES

[1] S.D. Brown, R.J. Francis, J. Rose and Z. Vranesic, *Field-Programmable Gate Arrays*, Kluwer, 1992.

[2] S.M. Trimberger (Ed), *Field-programmable Gate Array Technology*, Springer, 1994.

[3] S.M. Trimberger, Three ages of FPGAs: A Retrospective on the First Thirty Years of FPGA Technology. (2015) *Proc IEEE* 103, 318–31.

[4] R.H. Freeman, *Configurable Electrical Circuit Having Configurable Logic Elements and Configurable Interconnects*, US Patent US4870302, (1989).

[5] T.E. Reid, *The Chip: How Two Americans Invented the Microchip and Launched a Revolution*, Random House, 2001.

[6] K. Lee and P. Gulak, Field Programmable Analogue Array Based on MOSFET transconductor. (1992) *Electro Lett* 28, 28–9.

[7] K. Lee and P. Gulak, A Transconductor-based Field-programmable Analog Array, Proceedings of the IEEE International Solid-State Circuits Conference Digest of Technical Papers, 198–9. 15–17 February 1995, San Francisco, CA.

[8] T.S. Hall, C.M. Twigg, P. Hasler, and D.V. Anderson, Developing Large-scale Field-programmable Analog Arrays for Rapid Prototyping. (2004) *Int J Embed Syst* 1(3/4), 179–92.

[9] T.S. Hall, et al., Large-Scale Field-Programmable Analog Arrays for Analog Signal Processing. (2005) *IEEE Trans Circuits Systems and Systems I*, 52, 2298–308.

[10] D. Pérez, I. Gasulla and J. Capmany, Field-programmable Photonic Arrays. (2018) *Opt Express* 26, 27265–78.

[11] E. Desurvire, *Classical and Quantum Information Theory: An Introduction for the Telecom Scientist*, Cambridge University Press, 2009.

[12] M.A. Nielsen and I.L. Chuang, *Quantum Computation and Quantum Information*, Cambridge University Press, 2001.

[13] E. Knill, R. Laflamme and G.J. Milburn, A Scheme for Efficient Quantum Computation with Linear Optics. (2001) *Nature* 409, 46–52.

[14] J. Carolan, et al., Universal Linear Optics. (2015) *Science* 349, 711–16.

[15] D. Pérez-López, E. Sánchez, and J. Capmany, Programmable True Time Delay Lines Using Integrated Waveguide Meshes. (2018) *J Lightwave Technol* 36(19), 4591–601.

[16] D. Thomson, et al., Roadmap on Silicon Photonics. (2016) *J Opt* 18(7), 073003.

[17] W. Bogaerts and L. Chrostowski, Silicon Photonics Circuit Design: Methods, Tools and Challenges. (2018) *Laser Photon Rev* 12(4), 1700237.

[18] L.M. Augustin, et al., InP-based Generic Foundry Platform for Photonic Integrated Circuits. (2018) *IEEE J Sel Topics Quantum Electron* 24(1), 1–10.

[19] D. Pérez, et al., Multipurpose Silicon Photonics Signal Processor Core. (2017) *Nat Commun* 8(636), doi:10.1038/s41467-017-00714-1.

[20] J. Capmany, I. Gasulla and D. Pérez, Microwave Photonics: The Programmable Processor. (2016) *Nat Photonics* 10, 6–8.
[21] Q. Fang, et al., Ultralow Power Silicon Photonics Thermo-Optic Switch with Suspended Phase Arms. (2011) *IEEE Photonics Tech L* 23(8), 525–7.
[22] N. Harris, et al., Large-scale Quantum Photonic Circuits in Silicon. (2016) *Nanophotonics* 5, 456–68.
[23] D.A.B. Miller, Self-configuring Universal Linear Optical Component. (2013) *Photon Res* 1, 1–15.
[24] J.K.S. Poon and W.D. Sacher, Multilayer Silicon Nitride-on-silicon Photonic Platforms for Three-dimensional Integrated Photonic Devices and Circuits, *Proceedings of 75th Annual Device Research Conference (DRC)*, June 25–28 2017, South Bend, IN.

Index

Introductory Note: References such as '178–9' indicate (not necessarily continuous) discussion of a topic across a range of pages. Wherever possible in the case of topics with many references, these have either been divided into sub-topics or only the most significant discussions of the topic are listed. Because the entire work is about 'programmable integrated photonics', the use of this term (and certain others which occur constantly throughout the book) as an entry point has been minimised. Information will be found under the corresponding detailed topics.